Study Guide and Solutions Manual

William Siebler and Jane Siebler

Mathematics
in Life, Society, & the World
Second Edition

Harold Parks

Gary Musser

Robert Burton

William Siebler

PRENTICE HALL, Upper Saddle River, New Jersey 07458

Executive Editor: Sally Yagan
Supplement Editor: Meisha Welch
Special Projects Manager: Barbara A. Murray
Production Editor: Wendy Rivers
Supplement Cover Manager: Paul Gourhan
Supplement Cover Designer: PM Workshop Inc.
Manufacturing Buyer: Alan Fischer

Printed in the United States of America

10 9 8 7 6 5 4 3 2 1

ISBN 0-13-014928-4

Prentice-Hall International (UK) Limited, London
Prentice-Hall of Australia Pty. Limited, Sydney
Prentice-Hall Canada, Inc., Toronto
Prentice-Hall Hispanoamericana, S.A., Mexico
Prentice-Hall of India Private Limited, New Delhi
Pearson Education Asia Pte. Ltd., Singapore
Prentice-Hall of Japan, Inc., Tokyo
Editora Prentice-Hall do Brazil, Ltda., Rio de Janeiro

Contents

Preface

This Study Guide and Solutions Manual is a supplement to *Mathematics in Life, Society, and the World, Second Edition* by Hal Parks, Gary Musser, Bob Burton, and Bill Siebler. As a study guide, this book has been written to provide a comprehensive review of the material in the text, and also give added depth and insights for important topics and concepts. As a solutions manual, it is designed to help you improve your problem-solving ability by providing complete solutions to all odd numbered problems. Hints to get you started on many of the problems are also included.

The material in this supplement has been organized in the following manner:

Summaries of the content of each section in the text

The summaries include:
        Chapter Goals that are related to the section
        Key Ideas and Questions
        Vocabulary and Notation
        An overview of all topics and key concepts
        Additional comments or examples expanding on the presentation in the text

Suggestions and Comments for solving problems in the problem sets

Hints and suggestions are provided for many specific problems, as well as general comments for certain types of problems. This section will often include some review of skills and concepts related to the problems.

Written solutions to odd-numbered problems

All problem sets in the text include exercises for each of the examples in the related section. The solutions to these problems will generally include the key elements in the problem and any essential calculations. As the level of difficulty in the problems increases, more detail is provided. In addition to the extra detail, comments that relate the problem to other parts of the section or previous material in the text may be included. When it is appropriate, alternate problem solving strategies may be included, as well as alternate solutions.

To make the best use of these solutions, it is important that you make your best effort to solve a problem before consulting the solutions provided here. If a problem "stumps" you, review the examples and read any hints that are provided. Then read part of the solution and try to complete the problem yourself. Even if you have already solved the problem correctly on your own, you may find the solutions helpful since you may see a different way to analyze a problem or an alternate approach for solving it.

The best way to learn the mathematical concepts and skills presented in the text is by applying them to the solution of problems. The best way to learn to solve problems is by actually working as many as you can, not by simply seeing how others have solved them.

Acknowledgments

I wish to thank the following people for their assistance in the development of the *Study Guide and Solutions Manual*: Gary Musser for his leadership and tireless efforts throughout the entire textbook project and his encouragement and suggestions for this book; Ann Heath, our editor at Prentice-Hall for her patience and assistance; and , most importantly, my wife Jane for her general support and help. In addition to writing initial drafts for the summaries in most of the sections and spending numerous hours proof-reading and editing, Jane (who has an undergraduate degree in literature and a graduate degree in management) played the role of a student in reading the comments and solutions for clarity and completeness.

We hope that our collective efforts in the text and this book help make mathematics more understandable to you and a more relevant part of your academic and personal life.

Any errors in the preparation of this book or in the details of the solutions are my sole responsibility. All questions, comments or suggestions, and possible corrections can be sent to me via e-mail: billsiebler@netscape.net

Bill Siebler
September, 1999

1 Mathematical Structures and Methods

Section 1.1 Sets

Goals
1. Learn the language of sets as a means of describing objects that belong together.
2. Learn to use set concepts and notation to analyze relationships among objects in a set.

Key Ideas and Questions
1. What terminology is used to describe sets, membership in sets, and relationships among sets?
2. How are Venn diagrams used to represent and work with sets?

Vocabulary

Set	One-to-one Correspondence	Complement
Element (or Member)	Equivalent Sets	Union
Well Defined	Universal Sets	Intersection
Roster	Subset	Set Difference
Set-Builder Notation	Proper Subset	De Morgan's Laws
Empty Set (or Null Set)	Venn Diagram	Number of Elements in the
Equal Sets	Disjoint Sets	Union of Two Sets

. . . .

Overview

The most elementary mathematics is counting the number of objects that are together on some basis. Any collection of objects is called a **set**, and the objects are called **elements** or **members** of the set.

Sets and Elements

In order to be useful, a set must be **well-defined**; that is, it must be clear whether any object belongs to the set or not. Sets can be defined in three basic ways: (1) a verbal description, (2) a **roster**, or listing, of the members, or (3) a description of the characteristics of the elements using **set-builder notation**.

In the case of a verbal description or the use of set-builder notation, there may be several possibilities for adequately describing a set. When using a roster, the order in which the elements are listed doesn't matter, so there is essentially only one roster.

Often it is not convenient, or even possible, to list all the elements in a set using the roster method or set-builder notation.

If the set is too large to conveniently list all the elements, but we can establish the general nature of the elements, we can use an **ellipsis** (". . .") to indicate missing elements in the list. Thus the set of the first hundred even whole numbers can be given as {2, 4, 6, . . . , 200}. This approach can also be used for certain infinite sets; we can indicate the set of *all* even whole numbers as {2, 4, 6, . . . }. In these cases, set-builder notation could have been used. Sets are usually denoted by capital letters such as A, B, W. The symbols $x \in A$ and $y \notin A$ are used to indicate that "x is an element of set A" and "y is not an element of A", respectively. The set without any elements is called the **empty set** or **null set** and is denoted by { } or $\varnothing$.

Equality and Equivalence of Sets

Two sets A and B are **equal**, written $A = B$, if and only if they have precisely the same elements. If the two sets A and B are **not equal**, written $A \neq B$, we must be able to show that there is an element in one set that is not in the other. That is why there is only *one* empty set; there are no elements to check.

Sets can also be compared by matching up elements from the two sets. A **one-to-one (or 1-1) correspondence** between two sets A and B is a pairing of the elements of A with those of B so that each element of A is paired with exactly one element of B, and vice versa. If we can establish a one-to-one correspondence between sets A and B, then we say the sets are **equivalent**, written $A \sim B$. Equal sets are always equivalent (each element can be matched with itself), but equivalent sets are not necessarily equal. *Equal* sets have the *same* elements, *equivalent* sets have the *same number* of elements.

Universal Set and Subsets

The set that contains all the elements under consideration in a given discussion is called the **universal set**, and is usually denoted by U, although other letters may be used to suggest the nature of the set. For every problem, a universal set must be specified or implied, and it must remain fixed for that problem.

If one set A is "part" of another set B (in the sense that every element in A is also in B), we say that A is a **subset** of B, written $A \subseteq B$. If there is at least one element of A that is not an element of B, then A is not a subset of B. If $A \subseteq B$, and B has an element that is not in A, then A is a **proper subset** of B, written $A \subset B$. Every set is a subset of itself ($A \subseteq A$), and the empty set is a proper subset of every non-empty set ($\varnothing \subset A$).

Venn Diagrams Relationships among sets can be studied using pictures called Venn diagrams.

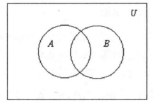
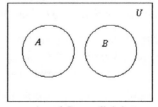
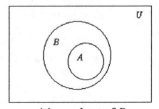

The general Venn diagram A and B are disjoint A is a subset of B
for two sets, A and B

Operations on Sets

When the elements of sets are well-defined, we can determine what elements belong to other related sets. Given a set A, we can determine which elements are not in A. Given two sets A and B, we can determine elements that are common to both, the elements that are in at least one of the sets, and the elements that are in one set but not the other.

Complement
The complement of a set A consists of all elements from the universal set that are *not* elements of A.
$$A' = \{x \mid x \notin A)$$

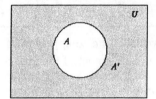

Intersection
The intersection of two sets A and B consists of all elements the two sets have in common.
$$A \cap B = \{x \mid x \in A \text{ and } x \in B\}$$

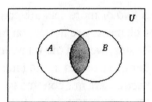

Union
The union of two sets A and B consists of all elements that are in at least one of the two sets.
$$A \cup B = \{x \mid x \in A \text{ or } x \in B\}$$

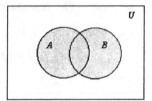

Difference
The set difference (or relative complement) of B with respect to A consists of all the elements of A that are not elements of B
$$A - B = \{x \mid x \in A \text{ and } x \notin B\}$$

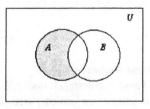

Combined Operations on Two Sets

Two or more operations on sets can be combined. In some cases, the results are quite similar to those of arithmetic, although some only look similar. For example:

$(A')' = A$ is similar to $-(-a) = a$, and

$A \cap (B \cup C) = (A \cap B) \cup (A \cap C)$ is similar to $a \times (b + c) = (a \times b) + (a \times c)$

However, we see that

$A \cup (B \cap C) = (A \cup B) \cap (A \cup C)$, while $a + (b \times c) \neq (a + b) \times (a + c)$.

De Morgan's Laws

The combination of negation with either set union or intersection has results that are useful in logic and probability.

De Morgan's Laws

For any sets A and B, $(A \cup B)' = A' \cap B'$ and $(A \cap B)' = A' \cup B'$

In words, "The complement of the union is the intersection of the complements" and "The complement of the intersection is the union of the complements".

Sets and Counting

When two or more sets are combined in a union of sets, a basic question is "How many elements are in the combined sets?" To find the number of elements in the union of two sets, we add the number of elements in each of the sets and then *subtract* the number of elements in the intersection of the sets so that the elements in the intersection are not counted twice.

The Number of Elements in the Union of Two Sets

For any two sets A and B, $n(A \cup B) = n(A) + n(B) - n(A \cap B)$

$n(A \cup B) = n(A) + n(B)$ if and only if $A \cap B = \varnothing$

. . . .

Suggestions and Comments for Odd-numbered problems

General Comments

These problems all deal with the language (the terminology, symbols, and notation) that is used when working with sets. Most of the language deals with whether or not an object belongs with other objects (person, place, or thing; animal, vegetable, or mineral; or on whatever basis). It is used to describe whether one set is part of another; whether or not sets have common members, and how sets can be combined. An important part of mathematics is the language, and you should fully understand the language being used so that you can understand the concepts you are trying to learn. This is, of course, true of all your classes.

11. and 13.

If sets are *equivalent*, they have the *same number* of members. If they are *equal*, they have exactly *the same* members.

17. and 19.

There are 2^n subsets that can be formed for a set having n elements. There are $2^n - 1$ <u>proper</u> subsets.

25. through 29.

Venn diagrams are very useful tools although they become difficult to work with when there are four or more sets involved and you want to show all possible relationships. Venn diagrams will be used again in several sections.

29. through 33.

De Morgan's laws have a counterpart in mathematical logic (focusing on truth and validity), which is covered in Chapter 10.

In the context of sets, "The element is not in A or B" is equivalent to "The element is not in A, and the element is not in B". In the context of logic "Neither statement p nor q is true" is equivalent to "p is not true, and q is not true."

35. through 41.

Always start with a general sketch, using the Venn diagrams in the text as models. Try to identify a "region" that is separate or not composed of any sub-regions. Practice on the easier problems will make work on the harder ones go more smoothly.

Solutions to Odd-numbered Problems

1. **(a)** Not well-defined   **(b)** Well-defined
   (c) Well-defined   **(d)** Well-defined

3. **(a)** {0, 1, 4, 9, 16, 25}
   (b) {Alaska, California, Hawaii, Oregon, Washington}
   (c) {3, 5, 6, 9, 10, 12, 15, 18, 20, 21, 24, 25, 27}
   (d) {0, 1, 8, 27, 64, 125, ...}

5. **(a)** { $x \mid x$ is a vowel}
   (b) { $y \mid y = x^3$; x is a whole number, $1 \le x \le 5$}
   (c) { $x \mid x$ is a Republican President, 1950 - 1992}
   (d) { $x \mid x$ is a letter in "mathematics"}

7. **(a)** True   **(b)** False   **(c)** False   **(d)** False

9. **(a)** True   **(b)** True   **(c)** True   **(d)** True

11. **(a)** The sets are equivalent.   **(b)** The sets are not equal.

13. **(a)** The sets are not equivalent   Note: The Big-10 conference has 11 members
   (b) The sets are not equal.

15. **(a)** A = {2, 6, 10, 12, 16}   **(b)** B = {3, 6, 9, 12, 15}

17. **(a)** $\varnothing$, {a}, {2}, {#}, {a, 2}, {a, #}, {2, #}, {a, 2, #}   **(b)** $\varnothing$, {0}

19. **(a)** $\varnothing$, {0}, {1}   **(b)** $\varnothing$, {a}, {b}, {c}, {a, b}, {a, c}, {b, c}

21. (a) True   **(b)** False; it is not a "proper" subset
(c) False (if you distinguish between numerals and words; True if both sets indicate numbers)
(d) True   **(e)** True

23. (a) ∉   **(b)** ⊂   **(c)** ⊄   **(d)** ⊆   **(e)** ∈

25. (a)   **(b)**

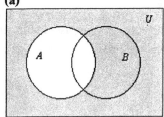   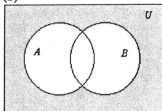

(c)   **(d)**

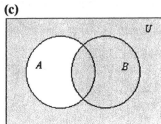   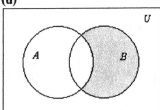

27. (a)   **(b)**

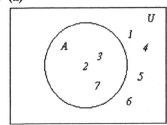   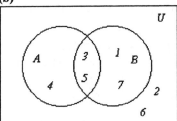

(c)

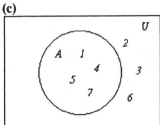

29. (a)

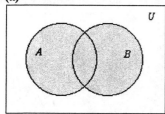

(b)

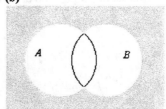

(c)

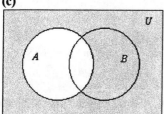

(d)

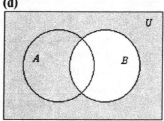

(e) Look for the common regions in A' and B'

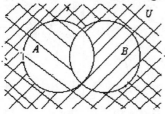
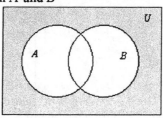

31. **(a)** $A \cup B = \{a, b, c, d f, h\}$ **(b)** $(A \cup B)' = \{e, g\}$ **(c)** $A' = \{c, d, e, g\}$
    (d) $B' = \{a, e, f, g, h\}$ **(e)** $A' \cap B' = \{e, g\}$

33. **(a)** $A \cap B = \{b\}$ **(b)** $(A \cap B)' = \{a, c, d, e, f, g, h\}$
    (c) $A' = \{c, d, e, g\}$ **(d)** $B' = \{a, e, f, g, h\}$ **(e)** $A' \cup B' = \{a, c, d, e, f, g, h\}$

35. 35 of the freshmen participated in both activities in high school. Call the sets A and M.
We know that $n(A) + n(M) = 120 + 105 = 225$, while $n(A \cup M) = 190$

37. First, use the given information to separate those in geometry into two subsets with
respect to biology. Then separate those in biology into subsets with respect to geometry.

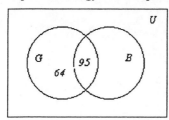
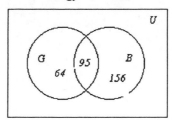

(a) 95   Geometry and Biology: 159 - 64 = 95
(b) 315   Geometry or Biology = 251 + 159 - 95 = 315
(c) 33   348 - 315 = 33

39. Let B = the set of people who enjoy biking, S = the set of people who enjoy swimming.

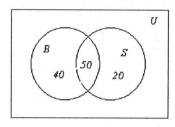

(a) 40	90 - 50 = 40	$n(B) - n(B \cap S)$
(b) 20	70 - 50 = 20	$n(S) - n(B \cap S)$
(c) 110	90 + 70 - 50 = 110	$n(B \cup S) = n(B) + n(S) - n(B \cap S)$
(d) 40	150 - 110 = 40	$n(U) - n(B \cup S)$ or $n((B \cup S)')$

41. Let L = set of students who read the local; M = set of students who read the metropolitan; and N = the set of students who read the national.

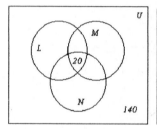 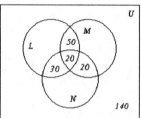 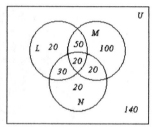

(a) 70   120 - 50 = 70   n(L) - n(L∩N)

(b) 150   190 - 40 = 150   n(M) - n(M∩N)

(c) 50   90 - 40 = 50   n(N) - n(M∩N)

(d) 140 $n(N \cup M) = n(N) + n(M) - n(N \cap M) = 90 + 190 - 40 = 240$

There are 240 who read the national or the metropolitan paper.
We want to exclude from this set those who also read the local;
They belong to the set $(L \cap N) \cup (L \cap M)$

$$n((L \cap N) \cup (L \cap M)) = n(L \cap N) + n(L \cap M) - n((L \cap N) \cap (L \cap M))$$
$$= n(L \cap N) + n(L \cap N) - n((L \cap N \cap M))$$
$$= 50 + 70 - 20 = 100$$

240 - 100 = 140

Section 1.2   Integers and Rational Numbers

Goals
1. Learn to visualize numbers and numerical relationships with a number line.
2. Understand what integers and rational numbers are, and learn the properties of integer and rational number operations.

Key Ideas and Questions
1. What properties do whole number addition and multiplication share?
2. Explain how 1 plays the same role for multiplication that 0 plays for addition.
3. Explain why rational numbers are needed to have multiplicative inverses for integers.

Vocabulary

Number	Integers	Subtraction and Division
Numeral	Integer Number Line	Positive Rational Numbers
Cardinal Number	Additive Identity	Negative Rational Numbers
Ordinal Number	Additive Inverse	Fractions
Identification Number	Properties of Integer	Simplest Form
Whole Numbers	Operations	Rational Number
Addition	Rational Numbers	Simplification
Properties of Whole	Numerator	Rational Number Equality
Number Addition	Denominator	Properties of Rational
Multiplication	Distributivity: Multiplication	Number Operations
Properties of Whole	over Addition	Multiplicative Inverse
Number Multiplication		

· ·

Overview

Essentially, all number systems began with the need for counting and expanded as other computational needs arose, We begin with the natural numbers, and consider the integers and rational numbers. The properties of all numbers are based on the properties of the natural numbers.

Integers

The set of natural numbers (also known as the counting numbers) is the set of numbers {1, 2, 3, . . .}. This set together with zero, namely {0, 1, 2, 3, . . .) , is called the set of **whole numbers**. The set of **integers**, {. . . , -3, -2, -1, 0, 1, 2, 3, . . . }, can be used to represent both positive and negative numbers.

The integers can be represented using the **integer number line.**

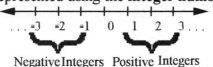

Negative Integers Positive Integers

On the integer number line, two numbers are said to be **opposites** if they are the same distance from zero, but on the opposite sides of zero. For example, 3 and -3 are opposites. The symbol -a is used to represent the opposite of a. Notice that -(-3) = 3; that is, the opposite of the opposite of 3 is 3. In general, we have: -(-a) = a.

Integers: Addition and Subtraction

Integer arithmetic is a extension of whole number arithmetic. Operations such as addition and multiplication of whole numbers *as integers* are exactly as they are in the set of whole numbers. A number plus its opposite is zero. Symbolically, $a + (-a) = 0$. The number -a is called also called the **additive inverse** of a since it adds to a to produce 0. The number zero is called the **additive identity** since for any integer a, $a + 0 = a$. Subtracting an integer is the same as adding its opposite (additive inverse).

$$\text{Subtraction of Integers} \quad a - b = a + (-b)$$

Integers: Multiplication and Division

Integer multiplication and division are extensions of whole number multiplication and division. In general, *the product of a positive integer and a negative integer is negative*. Also, *a negative integer times a negative integer is positive*. Finally, **zero times any integer is zero.**

Integer division is derived from integer multiplication by extending the definition of division of whole numbers. That is, $8 \div 2 = n$ if and only if $8 = n \times 2$. In general, *the quotient of two positives (or two negatives) is positive* and *the quotient of a positive and a negative is negative*.

The following properties hold for integer addition and multiplication.
Note: The symbol '±', read "plus or minus", is a convenient notation used to represent two statements at once, one with '+' and one with '-'.

Properties of Integer Operations

$a + b = b + a$ Commutative Property of Addition
$a \times b = b \times a$ Commutative Property of Multiplication
$a + (b + c) = (a + b) + c$ Associative Property of Addition
$a \times (b \times c) = (a \times b) \times c$ Associative Property of Multiplication
$a \times (b \pm c) = (a \times b) \pm (a \times c)$ Distributivity of Multiplication over Addition
$\qquad\qquad\qquad$ (and Subtraction)
$(a \pm b) \div c = (a \div c) \pm (b \div c)$ Right Distributivity of Division over Addition
$\qquad\qquad\qquad$ (and Subtraction)
$a + 0 = 0 + a = a$ Identity for Addition Property (Zero)
$a \times 1 = 1 \times a = a$ Identity for Multiplication Property (One)
$a + (-a) = (-a) + a = 0$ Additive Inverse Property

The additive inverse property is a property of integers, but not of whole numbers since the set of whole numbers does not include negative numbers.

Rational Numbers

Fractions are used to represent parts of a whole. The set of rational numbers, which we will study next, extends the integers *and* the fractions.

The Rational Numbers

The set of **rational numbers** is the set of all numbers that can be represented in the form $\dfrac{a}{b}$, where a and b are integers and $b \neq 0$.

The a in $\dfrac{a}{b}$ is called its **numerator** and b is called its **denominator**.

The denominator of a rational number can never be zero, since $a \div 0 = c$ would require $a = c \times 0$. Rational numbers whose numerators and denominators are both positive or both negative are called the **positive rational numbers**; the rational numbers where one of the numerator or denominator is positive and the other is negative are the **negative rational numbers**. The integers are a part of the set of rational numbers.

Rational Numbers: Simplest Form

A rational number is said to be written in **simplest form** if the numerator and denominator have only 1 or -1 as common factors *and* the denominator is positive. The following is a generalization of this idea.

Rational Number Simplification

$\dfrac{na}{nb} = \dfrac{a}{b}$ for integers a, b, and n, where b and n are nonzero.

The above property can be used to determine if two rational numbers are equal. Another way to check for equality is to find common denominators. You can also find equality by "cross-multiplying" the fraction. To do this, find the product of the numerator of one rational number times the denominator of the other. This leads to the following generalization.

Rational Number Equality

$\dfrac{a}{b} = \dfrac{c}{d}$ if and only if $a \times d = b \times c$.

Rational Numbers: Addition and Subtraction

Rational Number Addition and Subtraction

$$\frac{a}{c} + \frac{b}{c} = \frac{a+b}{c} \quad \text{and} \quad \frac{a}{c} - \frac{b}{c} = \frac{a-b}{c}$$

There are three alternate forms for negative rational numbers:

$$-\frac{a}{b} = \frac{-a}{b} = \frac{a}{-b}.$$

The form that is used depends upon the situation, but the first form is generally preferred for expressing the final results in calculations or when the number stands alone. Since rational number subtraction extends integer subtraction, an alternate way to define subtraction is

$$\frac{a}{c} - \frac{b}{c} = \frac{a}{c} + \left(-\frac{b}{c}\right).$$

If the denominators of two rational numbers are different, find a common denominator before adding or subtracting.

$$\frac{a}{b} + \frac{c}{d} = \frac{ad}{bd} + \frac{bc}{bd} = \frac{ad + bc}{bd} \quad \text{and} \quad \frac{a}{b} - \frac{c}{d} = \frac{ad}{bd} - \frac{bc}{bd} = \frac{ad - bc}{bd}$$

Rational Numbers: Multiplication and Division

Multiplication in the rational numbers has both inverse and identity elements. That is, if $b \neq 0$, then $\frac{a}{b} \times \frac{b}{a} = 1$. The number $\frac{b}{a}$ is called the **reciprocal** or **multiplicative inverse** of $\frac{a}{b}$. In words, a rational number times its multiplicative inverse is 1. Also, since $\frac{a}{b} \times 1 = \frac{a}{b}$, 1 is called the **multiplicative identity**.

Division of rational numbers extends integer division. Recall that with integers and whole numbers, $a \div b = c$ if and only if $a = c \times b$. For example, $12 \div 4 = 3$ since $12 = 3 \times 4$. Next, we extend this idea to determine how to divide rational numbers.

Rational Number Multiplication and Division

$$\frac{a}{b} \times \frac{c}{d} = \frac{ac}{bd} \quad \text{and} \quad \frac{a}{b} \div \frac{c}{d} = \frac{a}{b} \times \frac{d}{c} = \frac{ad}{bc}, \text{ when } c \neq 0.$$

The product of two rational numbers is the rational number whose numerator is the product of their numerators and whose denominator is the product of their denominators. The quotient of one rational number divided by a second is the product of the first number and the reciprocal of the second number.

The properties listed for integer addition and multiplication (including subtraction and division) also hold for rational number addition and multiplication. However, while division could not always be performed solely with integers, all operations can be completed in rational numbers, provided that division by zero is never possible.

The multiplicative inverse property is a property of rational numbers, but not of integers. The set of integers only includes reciprocals for 1 and -1.

. . . .

Suggestions and Comments for Odd-numbered problems

19. (a) When the *identity element* for an operation is combined with another number, the result of the operation has the value of the second number. The result is "identical" to the second number.

(b) The *opposite* of a number refers to addition. The opposite of a number is on the opposite side of zero from the number when they are placed on the number line. It has the "opposite" sign.

23. Write out the sequence of transactions so that you can tell if the balance becomes negative.

25. First, calculate the recycling rate for the two states.
$$\text{Rate} = \frac{\text{Recycled waste}}{\text{Total waste}}$$
Compare the two fractions to see which is bigger.

31. All sums have to be identical, so they are either all even or they are all odd. What are the consequences of putting an even number in the center square?
If an odd number has to be in the center square, can any odd numbers be in the corners? What are the consequences?

Solutions to Odd-numbered Problems

1. (a) -14   (b) 5   (c) -21   (d) -23   (e) 21   (f) 7

3. (a) -147   (b) -84   (c) 90   (d) 9   (e) -9   (f) -9

5. (a) Distributivity of multiplication over addition.

(b) Associativity of addition.

(c) Commutativity of multiplication.

(d) Commutativity of addition.

7. (a) $\frac{4}{5}$ (b) $\frac{5}{6}$ (c) $\frac{1}{3}$ (d) $\frac{3}{2}$ (e) $\frac{7}{3}$ (f) $\frac{9}{11}$

9. (a) $\frac{6}{9}$ and $\frac{25}{45}$ are not equivalent. $6 \times 45 = 270$; $25 \times 9 = 225$

(b) $\frac{8}{10}$ and $\frac{12}{15}$ are equivalent. $8 \times 15 = 120$; $10 \times 12 = 120$

(c) $\frac{33}{121}$ and $\frac{45}{143}$ are not equivalent. $33 \times 143 = 4719$; $121 \times 45 = 5445$

11. (a) $\frac{9}{12} = \frac{3}{4} = \frac{6}{8} = \frac{15}{20} = \frac{18}{24} = \frac{21}{28} = \cdots$ (b) $\frac{1}{5} = \frac{2}{10} = \frac{3}{15} = \frac{4}{20} = \frac{5}{25} = \cdots$

13. To combine the fractions, we first get a common denominator. Always try to reduce the answer to simplest terms.

(a) $\dfrac{24}{30} + \dfrac{9}{30} = \dfrac{33}{30} = \dfrac{11}{10}$ **(b)** $\dfrac{5}{18} + \dfrac{2}{18} = \dfrac{7}{18}$ **(c)** $\dfrac{28}{84} - \dfrac{12}{84} = \dfrac{16}{84} = \dfrac{4}{21}$

(d) $\dfrac{4}{5} + \dfrac{2}{15} = \dfrac{12}{15} + \dfrac{2}{15} = \dfrac{14}{15}$ **(e)** $\dfrac{5}{6} - \dfrac{4}{21} = \dfrac{35}{42} - \dfrac{8}{42} = \dfrac{27}{42} = \dfrac{9}{14}$

(f) $\dfrac{13}{24} - \dfrac{13}{28} = \dfrac{91}{168} - \dfrac{78}{168} = \dfrac{13}{168}$ $24 = 4 \times 6;\ 28 = 4 \times 7;\ \text{LCD} = 4 \times 6 \times 7 = 168$

or $= \dfrac{364}{672} - \dfrac{312}{672} = \dfrac{52}{672} = \dfrac{13}{168}$ $24 \times 28 = 672$

15. (a) We can cancel sequentially, or multiply and then reduce the answer

$\dfrac{-4}{9} \times \dfrac{15}{8} = \dfrac{-1}{9} \times \dfrac{15}{2} = \dfrac{-1}{3} \times \dfrac{5}{2} = \dfrac{-5}{6}$ or $\dfrac{-4}{9} \times \dfrac{15}{8} = \dfrac{-60}{72} = \dfrac{-5}{6}$

(b) $\dfrac{-6}{15} \times \dfrac{10}{-21} = \dfrac{2}{15} \times \dfrac{10}{7} = \dfrac{2}{3} \times \dfrac{2}{7} = \dfrac{4}{21}$

(c) $\dfrac{4}{21} \div \dfrac{-2}{3} = \dfrac{4}{21} \times \dfrac{-3}{2} = \dfrac{2}{21} \times \dfrac{-3}{1} = \dfrac{2}{7} \times \dfrac{-1}{1} = \dfrac{-2}{7}$

(d) $\dfrac{-2}{3} \div \dfrac{-5}{12} = \dfrac{-2}{3} \times \dfrac{-12}{5} = \dfrac{-2}{1} \times \dfrac{4}{5} = \dfrac{8}{5}$

(e) $\dfrac{8}{15} \times \dfrac{-25}{16} = \dfrac{1}{15} \times \dfrac{-25}{2} = \dfrac{1}{3} \times \dfrac{-5}{2} = \dfrac{-5}{6}$

(f) $\dfrac{-8}{25} \div \dfrac{-12}{35} = \dfrac{-8}{25} \times \dfrac{-35}{12} = \dfrac{-2}{25} \times \dfrac{-35}{3} = \dfrac{-2}{5} \times \dfrac{-7}{3} = \dfrac{14}{15}$

17. (a) Multiplicative commutativity, multiplicative identity.

(b) Right distributivity of division over addition.

(c) Multiplicative commutativity.

(d) Multiplicative inverse.

19. (a) 0 is the additive identity. $\dfrac{m}{n} + 0 = \dfrac{m}{n} + \dfrac{0}{n} = \dfrac{m+0}{n} = \dfrac{m}{n}$

(b) $\dfrac{-m}{n}$ is the opposite of $\dfrac{m}{n}$. $\dfrac{m}{n} + \dfrac{-m}{n} = \dfrac{m + -m}{n} = \dfrac{0}{n} = 0$

21. The air temperature would drop about 60 degrees (3×20) to a temperature of $-10°$ F.

23.

$$115 + 384 = \$499$$
$$499 - 153 = \ \ 346$$
$$346 - 86 = \ \ 260$$
$$260 - 196 = \ \ \ \ 64$$
$$64 - 34 = \ \ \ \ 30$$
$$30 - 79 = \ -49 \quad \text{(overdrawn)}$$
$$-49 - 10 = \ -59 \quad \text{(service charge assessed)}$$
$$-59 + 123 = \ \$64 \quad \text{Final balance}$$

25. Texas has the higher recycling rate.

Texas: $\dfrac{1440000}{18000000} = \dfrac{144}{1800} = \dfrac{8}{100}$

Illinois: $\dfrac{786000}{13100000} = \dfrac{786}{13100} = \dfrac{6}{100}$

27. (a) $2 \times 1\frac{1}{4} = (2 \times 1) + (2 \times \frac{1}{4}) = 2 + \frac{1}{2}$; $2\frac{1}{2}$ cups   **(b)** $\frac{1}{2} \times 1\frac{1}{4} = \frac{1}{2} \times \frac{5}{4} = \frac{5}{8}$; $\frac{5}{8}$ cups

(c) We want $\dfrac{5}{3}$ of a standard recipe: $\frac{5}{3} \times 1\frac{1}{4} = \frac{5}{3} \times \frac{5}{4} = \frac{25}{12}$; $2\frac{1}{12}$ cups

29. First, we begin in the bottom row and fill in the squares we can according to the directions.

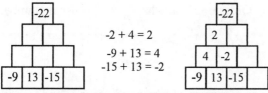

$-2 + 4 = 2$

$-9 + 13 = 4$

$-15 + 13 = -2$

Next we label the unknown squares and identify the relationships to other squares based on the directions.

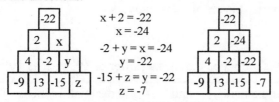

$x + 2 = -22$

$x = -24$

$-2 + y = x = -24$

$y = -22$

$-15 + z = y = -22$

$z = -7$

31. (a) and **(c)**   There are eight possible ways, and three are shown. Each of the eight may be obtained from the first by reflections and/or rotations.

8	3	4
1	5	9
6	7	2

8	1	6
3	5	7
4	9	2

4	3	8
9	5	1
2	7	6

(b) Since there are 8 identical sums in a magic square (3 columns, 3 rows, 2 diagonals) the sum of all the rows, columns, and diagonals must be even. If an even number is in the middle square, some of the sums will be forced to be odd, while others are even, so they can't be identical. An odd number must be in the middle square. Since all sums must be odd or all sums must be even, we are forced to put the odd numbers in the squares that are not on the diagonal. The only way we can arrange the five odd numbers so that the middle row and the middle column have the same total (15) is to put the 5 in the middle. The other numbers are then placed through trial and error.

37. All of the properties for integer operations hold for this system.
0 is the identity element for the + operation, and 1 is the identity element for the * operation. 1 is the inverse element for 1 with respect to the + operation.

Section 1.3   Decimals and Real Numbers

Goals
1. Learn about real numbers and decimals and how to change from one form to the other.

Key Ideas and Questions
1. Explain why decimals are needed to represent all numbers.
2. Why is every rational number either a rational decimal or a repeating decimal?

Vocabulary

Decimal	Terminating Decimal	Irrational Number
Decimal Point	Repeating Decimal	Real Number
Decimal Fraction	Ordering Decimals	Real Number Line

- -

Overview

The set of decimals is an extension of the set of integers, and is a convenient numeration system for fractions. In this section, we also discuss rational and irrational numbers which comprise the set of real numbers.

Decimals

A **decimal fraction** is a fraction that has a power of 10, such as 10, 100, or 1000, for its denominator; for example $\frac{37}{100}$ and $\frac{291}{10000}$ are decimal fractions. Decimal fractions can be expressed in decimal notation which is an extension of the Hindu-Arabic system of place values. A **decimal point** is a period that separates the whole number on the left from the decimal fraction on the right. The first place to the right of the decimal point has the place value of $\frac{1}{10}$. The place value of each subsequent place to the right is one tenth that of the place to its immediate left. The term **decimal** is often used to refer to any number written using a decimal point and the preceding place value notation; thus we consider 13.0 and 0.0 to be decimals. A fraction is converted to a decimal by performing the indicated division, either by hand or with a calculator.

A decimal such as 0.125 is called a **terminating decimal** because there are only a finite number of non-zero digits to the right of the decimal point, in this case 3. The decimal 0.3333 . . . has infinitely many non-zero digits, but they occur with a simple pattern--three is repeated over and over. A decimal that repeats indefinitely is called a **repeating decimal**. Every rational number can be represented by either a terminating decimal or a repeating decimal. For a repeating decimal, a bar is put over the set of digits that repeat. Thus we write $\frac{1}{3} = 0.\overline{3}$

Ordering Decimals

When you want to determine which decimals are larger than others in a group, first be sure that a number that can be written in a terminating form is written in that way. Then compare digits place by place, starting at the highest place value. When a place is reached where the digits disagree, the number with the larger digit is the larger number overall.

Multiplying and Dividing Decimals by 10, 100, 1000, etc.

Multiplication and Division by Powers of Ten

To multiply a decimal by the number represented by 1 followed by n zeros, move the decimal point n digits to the right.

To divide a decimal by the number represented by 1 followed by n zeros, move the decimal point n digits to the left.

Converting a Repeating Decimal to a Fraction

Every repeating decimal can be changed to a rational number by an algebraic procedure using multiples of powers of ten chosen to move the decimal point so that subtraction of the two multiples eliminates the decimal portion of the result. For example, if $x = 0.27272727\ldots$ we multiply by 100.

$$100x = 27.272727\ldots$$
$$-x = 0.272727\ldots$$
$$99x = 27 \qquad \text{or} \quad x = \frac{27}{99} = \frac{3}{11}$$

Real Numbers

Any nonterminating, non-repeating decimal represents a number, but a number which is not rational. Such numbers are called **irrational numbers**. The number $\sqrt{2}$ is an irrational number. The set of numbers consisting of both the rational numbers and the irrational numbers is called the **real numbers**. A common way to visualize the real numbers is with the aid of the **number line**. Assume the line extends infinitely both to the left and to the right. Any real number r corresponds to a point on the line at a distance $|r|$ from the point labeled 0 - with the positive numbers corresponding to points on the right of 0, and the negative numbers corresponding to points on the left.

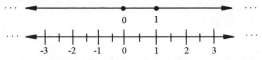

The arithmetic of the real numbers is an extension of the rational numbers. Since irrational numbers cannot be expressed in a fraction form, decimal approximations are usually used to represent real numbers. $\sqrt{2} \approx 1.414$ is usually used.

Suggestions and Comments for Odd-numbered problems

21. Call the number x. If the repeating decimal does not begin immediately after the decimal point, multiply by the smallest power of 10 that will move the decimal point in front of the repeating block of digits. Then get another multiple of x in which the decimal is moved immediately behind the first repeated block in the decimal expression. The numbers that result should have decimal fractions that are identical, and the decimal will be eliminated when the multiples are subtracted.

23. Call the initial amount of money x, then make the calculations step by step. After the last transaction, equate the expression to $21.45 and solve for x.

25. What is the smallest number (in ten-thousandths) that would round to 28.67? What is the largest?

33. In a magic square, the sum in each row, each column, and each diagonal must all be the same. Compare the three rows and then the three columns to identify the position with the incorrect entry.

Solutions to Odd-numbered Problems

1. (a) Three and forty-seven thousandths.

   (b) Three hundred and forty-seven hundredths.

   (c) Three hundred forty-seven thousandths.

3. (a) Seventy-five thousandths.   **(b)** One hundred fifty and twenty-five hundredths.

   (c) Five and one hundred seventy-five ten thousandths.

5. (a) 3.0105   **(b)** 327.27   **(c)** 0.000085

7. (a) 0.0305   **(b)** 0.004038   **(c)** 25.05287

9. (a) 413.46   **(b)** 413   (With a whole number, there is no decimal point or part)

11. (a) 28.875   **(b)** 87.70

13. (a) 0.68   **(b)** $\frac{3}{8}$ $0.375 = \frac{375}{1000} = \frac{3 \times 125}{8 \times 125}$

   (c) 0.8333

15. (a) $2\frac{3}{8}$ $2.375 = 2 + 0.375$ (See 13 b.)   **(b)** 0.5714

   (c) $\frac{21}{25}$ $0.84 = \frac{84}{100} = \frac{4 \times 21}{4 \times 25}$

17. (a) 23.346517   **(b)** 6.34277

   (c) $412.3\overline{45}$ 412.3454545... vs. 412.34454545...

19. (a) 37,500   **(b)** 5652.9   **(c)** 0.00000045

21. (a) Let x = $0.3\overline{27}$ = 0.3272727. . .

   First, we move the decimal to the front of the repeating block:
       10x = 3.272727. . .
   Then we move the decimal behind the first repeating pair:
       1000x = 327.272727. . .
   We subtract the first number from the second
       1000x = 327.272727. . .
       - (10x =   3.272727. . .)
         990x = 324 $x = \dfrac{324}{990} = \dfrac{18}{55}$

(b) Let x = $32.\overline{7}$ = 32.77777. . .

   Since the decimal is already in front of the repeating block, we have only to find a multiple with the decimal behind the first repeated block.
       10x = 327.77777. . .
   Next, we subtract:
       10x = 327.77777. . .
       - (x =   32.77777. . .)
         9x = 295 $x = \dfrac{295}{9}$

23. Call the initial amount of money x, then make the calculations step by step:
       x + $29.35 = x + 29.35
         - 2(1.95) = x + 25.45
            - 5.95 = x + 19.50
            - 5.98 = x + 13.52
       x + 13.52 = 21.45   or   x = 7.93   He had $7.93 to begin with.

25. (a) The minimum it could weigh is 28.6650 grams, which rounds up to 28.67.
   (28.6649 would round down to 28.66)

(b) The maximum it could weigh is 28.6749 grams.

27. (a) 344,800 Belgian francs   (34.48 per $1, times 10,000)

   (b) 15,400 Canadian dollars   (1.54 per $1, times 10,000)

   (c) 55,200 French francs   (5.52 per $1, times 10,000)

29. 6.21 miles   (0.62137 mi per km, times 10 = 6.2137 and rounded)

31. (a) 5.3 miles   (0.62137 × 8.488 = 5.28946696)

   (b) 6.9 miles   (0.62137 × 11.034 = 6.87605778)

33. 0.647 should be changed to 0.657.

Section 1.4   Solving Percentage Problems

Goals
1. Understand percents and learn how to solve the three types of percentage problems.

Key Ideas and Questions
1. Explain how to convert between decimals, percents, and fractions.
2. Describe how markdowns and markups are percents of change.

Vocabulary

Percent	Percent Change	Percent Markup
Percentage	Percent Increase	Percent Markdown
Fraction Equivalents	Percent Decrease	

. . . .

Overview

Percents provide another common way to represent fractions. Percents are well-suited for easy comparisons and for use in business applications.

Percent

The term "percent" comes from the Latin phrase *per centum* which means "per hundred" or "for each hundred." The definition of *N* **percent** is $\frac{N}{100}$, written as N%.

Because the denominator is 100, it is very convenient to use decimals to represent a percent. To convert a fraction to a percent, first convert the fraction to a decimal (usually with a calculator) and then move the decimal point two places to the right. Any number can be expressed as a percent.

Decimal-Percent-Fraction Conversions

1. To convert between a decimal and a percent, move the decimal point two two places (to the right when converting from decimal to percent and to the left when converting from percent to decimal).

2. To change a fraction to a decimal, divide the numerator by the denominator. To change a fraction to a percent, change it to a decimal and use step 1.

3. To change a percent to a fraction, express the percent in its decimal equivalent, then change the decimal to its fraction form.

Percent of a Number

A **percentage** is a rate or proportion per hundred, as in the percentage of a group falling into a category. The most common use of percentage is finding the percent one number is of another number. A second type of computation is finding what percentage of a group fall into some category. A third type of percent problem gives the "part" and the percent it represents, and asks for the original value.

Three Types of Percent Problems

1. To find $p\%$ of n, multiply n by $\frac{p}{100}$.

   (Find n [×] p [÷] 100 [=] on a calculator.)

2. To find what percent m is of n, express $\frac{m}{n}$ as a percent.

   (Find m [÷] n [×] 100 [=] on a calculator

3. To find $p\%$ of what number is n, calculate $n \div p\%$.
   (Find n [÷] p [×] 100 [=]on a calculator.)

Percent Change

We often hear of the **percent change** of some quantity. Symbolically, the percent of change is found as follows. If the present value is greater than the original value, the percent change is often called the **percent increase**. If the original value is the greater than the present value, then the percent change is called the **percent decrease**.

Percent Change

$$\frac{\text{Final Value - Original Value}}{\text{Original Value}} \times 100\%$$

Almost everyone worries about changes in the prices they pay for all kinds of things, since prices tend to increase rather than decrease. The most widely used measure to keep track of price level changes is the Consumer Price Index published monthly by the U.S. Department of Labor's Bureau of Labor Statistics. The Consumer Price Index (CPI) is a monthly index based on the composite cost of selected goods and services such as housing, food, transportation, and utilities used by working class households. The CPI is often referred to as the "cost-of-living" index.

The **selling price** of an item is made up of two amounts: the **cost** and the **markup**. Most merchants define their markup to be the selling price minus the dealer's cost, divided by the selling price. This is then converted to a percent If an item is marked up 50%, however, it's not an increase of 50%. A markup of 50% represents 50% of the *selling price*. This means it equals 100% of the *cost* of the item to the merchant. Note: There are exceptions; many merchants base their markup on their costs.

A **markdown** is computed on the original selling price or on the manufacturer's suggested retail price (often abbreviated MSRP).

Percent Markup and Markdown

$$\text{Percent Markup} = \frac{\text{Selling Price - Cost}}{\text{Selling Price}} \times 100\%$$

$$\text{Percent Markdown} = \frac{\text{Selling Price - Sale Price}}{\text{Selling Price}} \times 100\%$$

Calculating Markup and Markdown

Markup and markdown represent an increase and decrease, but both are based on the established selling price. As a result, there are two types of confusion as to what happens when two or more are applied, compared to the way in which a usual sequence of percentage changes affects the result.

Example 1a: If a merchant with a 40% markup buys an item for $30, what is the selling price?

If the markup is 40%, then the cost represents 60% of the selling price, which is $30, This is the third type of percentage problem: $30 is 60% of what amount? We find that the selling price is $50 ($30 ÷ 0.6).

Example 1b: If the $30 technology fee a student pays for computer resources is increased 40%, what is the new technology fee?

In this case, the increase is 40% of the current amount, which is $30. This is the second type of percentage problem: What is 40% of $30? We see that the increase is $12 (0.4 × $30), and the new technology fee is $42.

Example 2a: A distributor gets an item from a manufacturer for $30, and sells it to a merchant after adding a 40% markup. The merchant also takes a 50% markup when pricing the item. What's the selling price of the item?

As we saw in Example 1a, the cost to the merchant would be $50. A 50% markup represents a doubling of the cost of the item, which means that the merchant buys the item for $50 and then prices it at $100.

Example 2b: Suppose the price of a (very successful) high-tech stock was $30 at the end of 1997, and then increased by 40% in 1998 and 50% in 1999. What was the value of the stock at the end of 1999?

If the stock increased by 40% in 1998, then, as we saw in example 1b, the price at the end of 1998 would have been $42. The 50% increase for 1999 would be 50% of $42, or an increase of $21, making the final price $63.

. . . .

Suggestions and Comments for Odd-numbered problems

11. First determine the amount of increase or decrease. Then compare that to the base (the beginning, or original, value).

13. and 15.

Remember that percentage change $= \dfrac{\text{final value - original value}}{\text{original value}}$.

Use a variable if one of the amounts is not known, and solve the equation.

17. If purchasing power is to be maintained, then the ratio of the salaries must be the same as the ratio of the related CPI values.

21. In situations like this, you cannot add the percentages since they refer to different bases (beginning values). For example, a 10% decrease followed by a 10% increase (over the new value) is not a break-even situation. Do the problem in stages.

23. Remember that percentage markup is based on the eventual selling price, not the cost of the item to the retailer (Note: Not ALL retailers define the markup the same way).

25. First, you need to find the selling price of the stereo. If we call the stereo price 100%, then the stereo plus tax is 106% of the stereo price.

27. Cost + Markup = Selling Price or "% cost" + "% markup" = 100%
The cost is what percent of the selling price?

29. Work backward from the information you have. If 6% tax is added, then the final amount is 106% of the sale price. With a discount of 20%, how does the sale price compare to the listed price?

Solutions to Odd-numbered Problems

1. (a) $25\% = \dfrac{25}{100} = \dfrac{1}{4}$ or 0.25　　　　**(b)** $4.25 = 425\%$ or $4\dfrac{25}{100} = 4\dfrac{1}{4}$

　(c) $\dfrac{2}{5} = 0.4 = 0.40 = \dfrac{40}{100}$ or 40%

3. (a) $\dfrac{3}{8} = 0.375$ or 37.5%　　　　**(b)** $112\% = 1.12$ or $1\dfrac{12}{100} = 1\dfrac{3}{25}$

　(c) $0.875 = 87.5\%$ or $\dfrac{875}{1000} = \dfrac{25 \times 35}{25 \times 40} = \dfrac{35}{40} = \dfrac{7}{8}$

5. Convert the fraction to a decimal and move the decimal two places to the right (or words to that effect).

7. (a) 120　　40% of 300: $0.40 \times 300 = 120$

　(b) $266\dfrac{2}{3}$　　30% of A = 80: $.30 \times A = 80$ or $A = \dfrac{80}{.3} = 266.6666\ldots$

　(c) 64　　200% of A = 128: $2.00 \times A = 128$ or $A = \dfrac{128}{2} = 64$

9. (a) 100   Estimate: $50\% \left(\frac{1}{2}\right)$ of 200   **(b)** 150   Estimate: $33\frac{1}{3}\% \left(\frac{1}{3}\right)$ of x is 50

(c) 25%   Estimate: 20 is about what percent of 80; 20 is $\frac{1}{4}$ of 80.

11. (a) 40% increase   The increase is 20; 20 is what percent of 50?

(b) $33\frac{1}{3}\%$ decrease   The decrease is 25; 25 is what percent of 75?

(c) 200% increase   The increase is 60; 60 is what percent of 30?

13. (a) 104   Amount of decrease is 20% of 130.
    20% of 130 = .2 × 130 = 26;   130 - 26 = 104

(b) 25%   Percentage increase $= \dfrac{300 - 240}{240} = \dfrac{60}{240} = 0.25$

(c) 300   Percentage increase $= \dfrac{390 - A}{A} = 30\% = 0.3$

$\dfrac{390 - A}{A} = 0.3$ or $390 - A = 0.3A$ or $390 = 1.3A$

$A = \dfrac{390}{1.3} = 300$ (With a 30% increase 390 is 130% of A = 300)

15. (a) 156   Amount of increase is 30% of 120; .3 × 120 = 36
    Alternatively: What number is 130% of 120?

(b) 14% decrease   Percent of change $= \dfrac{180 - 210}{210} = \dfrac{-30}{210} = -0.1429$

(c) 56   See 13(c): 70 is 125% of what number?

125% of A = 70 or 1.25 × A = 70; $A = \dfrac{70}{1.25}$

17. If we use the CPI as our gauge for maintaining purchasing power, then
$\dfrac{1990\ salary}{1985\ salary} = \dfrac{1990\ CPI}{1985\ CPI}$ or $\dfrac{1990\ salary}{1850} = \dfrac{131}{108}$

1990 salary $= \$1850 \times \dfrac{131}{108} \approx \$2244$

19. Percentage change $= \dfrac{1990\ CPI - 1970\ CPI}{1970\ CPI} = \dfrac{131 - 39}{39} = 2.3589...$
236% increase.

21. (a) Original salary of $2000; decrease of 10% ($200) to a salary of $1800.
    A 20% increase from $1800 is $360 (.2 × 1800)
    The final salary is $2160

(b) Percentage change $= \dfrac{2160 - 2000}{2000} = \dfrac{160}{2000} = 0.08$ or 8%

23. Percentage Markup $= \dfrac{60 - 36}{60} = \dfrac{24}{60} = 0.4 = 40\%$

25. The stereo (100%) plus the sales tax is 106% of the stereo price alone.

$418.70 is 106% of what number? $\frac{418.70}{1.06} = 395.00$

The amount of the tax is $418.70 - $395.00 = $23.70

27. If the art dealer has a 50% markup, then the markup is half of the selling price (the cost to the dealer is the other 50%). If the dealer pays $1275 for an item, then she will price the item to sell at 2 × $1275 = $2550.

With a 10% discount (.10 × 2550 = 255) the price to the patron is $2295.

29. Working backwards, what was the sale price of the table? Or, $296.80 is 106% of what number? Sale price $= \frac{296.80}{1.06} = 280.00$.

Still working backwards, if the listed price is discounted 20%, then the sale price is 80% of the listed price. 280 is 80% of what number?

Listed price $= \frac{280}{0.8} = 350.00$. If the listed price is $350, and the markup is 40% of the

listed price, then the amount of markup is 0.4 × $350 = $140.00.

The cost of the table to the dealer is $210. Alternatively, if the markup is 40%, then the cost is 60% of the listed price. 60% of $350 = $210.

Section 1.5 Exponents

Goals

1. Learn about integer exponents and the rules for their applications.
2. Understand the concept of a root and its expression with a radical or a rational exponent.
3. Understand rational and real exponents and their properties.
4. Solve problems using exponents and roots.

Key Ideas and Questions

1. What are the main properties of exponents?
2. Discuss the similarities and differences between the properties of exponents and properties of multiplication.

Vocabulary

Exponentiation	Natural Logarithms	Radicand
Positive Integer Exponents	Negative and Zero	Index
Power	Exponents	nth Root
Base	Scientific Notation	Radical
Five general Rules	Squaring	Unit Fraction Exponents
for Exponents	Square Root	Rational Exponent
Common Logarithm	Principal Square Root	Properties of Real
Base of Natural Logarithms	Cube Root	Exponents

Overview

Exponents are used in many important applications. The key to understanding all exponents and their use begins with positive integer (counting number) exponents, such as in the expression a^4, which stands for $a \times a \times a \times a$. In this case, we are counting the number of common factors, an operation called **exponentiation.** The development to more general exponents proceeds much like the development of the real number system - negative exponents, rational number exponents, and real number exponents. As with the development of the real number system, not all extensions are possible. With some restrictions, the following properties apply to all exponents.

General Properties for Exponents

The number m, in a^m, is called the **power** applied to a, or the **exponent** of a, and a is called the **base.** The number a^m is read "a to the power m" or "a to the *mth* power," or simply as "a to the *mth*." When the exponent has the particular values 2 and 3, the terms "squared" and "cubed" are used. The same is true for certain rational numbers.

Properties of Real Exponents

Let a, b represent positive real numbers, and m, n any real exponents. Then

$$a^m \times a^n = a^{(m+n)} \qquad a^m \div a^n = \frac{a^m}{a^n} = a^{m-n}$$

$$a^m \times b^m = (a \times b)^m$$

$$(a^m)^n = a^{m \times n}$$

$$\frac{a^n}{b^n} = \left(\frac{a}{b}\right)^n$$

Negative and Zero Exponents

Negative and Zero Exponents

Let a be a nonzero real number and let m be any nonzero whole number.

Then $\qquad\qquad a^0 = 1$ and $a^{-m} = \dfrac{1}{a^m}$; 0^0 is not defined.

Scientific Notation

Scientific notation uses integer exponents to represent very large and very small numbers. A number is written in **scientific notation** if it is written in the form $a \times 10^n$ where $1 \le a < 10$ and n is an integer. If $n > 0$, then $a \times 10^n \ge 1$; if $n < 0$, then $0 < a \times 10^n < 1$. Numbers written in their scientific notation are easy to compare. For example, 8.73×10^7 is less than 4.31×10^9 since $7 < 9$. Properties of exponents are useful when calculating products and quotients of large or small numbers using scientific notation.

Roots and Radicals

In this section, we consider the conditions under which we can reverse the relationship between a number and a power of that number. This leads to the idea of a *root*, which we express first with a radical and then with a rational numbers exponent. Finally, we see that with a suitable interpretation any real number can be used as an exponent.

Remember that we refer to b^2 as "*b* squared." Often it is convenient to think of **squaring** as an operation which when applied to the number *b* produces the number b^2. The **square root** is the operation which undoes squaring. We write the square root of a number by using the radical sign '$\sqrt{}$'. For example, we have $\sqrt{9} = \sqrt{3^2} = 3$. Notice that since both $(-3)^2$ and 3^2 equal 9, there are actually two choices for the square root of 9, either -3 or 3. The symbol $\sqrt{9}$ represents the *nonnegative* choice 3. (Note: $-\sqrt{9} = -3$.) The number $\sqrt{a}$, called the **principal square root of *a***, is the nonnegative number whose square is *a*, *if such a real number exists*. Since b^2 is nonnegative for any real number *b*, a negative number cannot have a real number square root.

Square Root

Let a be a nonnegative real number.

Then the **principal square root** of *a*, written $\sqrt{a}$, is defined by
$$\sqrt{a} = b \text{ where } b^2 = a \text{ and } b \geq 0.$$

Next we generalize the definition of square root to more general types of roots.

---

*n*th Root

Let *a* be a real number and let *n* be a positive integer.
1. If $a \geq 0$, then $\sqrt[n]{a} = b$ if and only if $b^n = a$ and $b \geq 0$.
2. If $a < 0$, then $\sqrt[n]{a} = b$ if and only if $b^n = a$.

The number *a* in $\sqrt[n]{a}$ is called the **radicand** and *n* is called the **index**. The symbol $\sqrt[n]{a}$ is read **the *n*th root of *a*** and is called a **radical**. The expression $\sqrt[n]{a}$ has not been defined for the case when *n* is even and *a* is negative. This is because $b^n \geq 0$ for any real number *b* when *n* is an even positive integer. For example, there is no real number *b* such that $b = \sqrt{-1}$ (or $b^2 = -1$).

Rational and Real Exponents

Using the concept of radicals, we can now proceed to define rational exponents.

Unit Fraction Exponent

Let a be a real number and let n be any positive integer.

Then $a^{1/n} = \sqrt[n]{a}$ where

1. n may be any positive integer when $a \geq 0$, and
2. n must be an odd positive integer when $a < 0$.

The combination of the last definition with the definitions for integer exponents leads us to the definition for any rational exponent. We restrict our definition to rational exponents of nonnegative real numbers.

Rational Exponent

Let a be a nonnegative real number, and let $\dfrac{m}{n}$ be a rational number. Then

$$a^{m/n} = (a^{1/n})^m$$

Note: $a^{m/n}$ also equals $(a^m)^{1/n}$, provided that $(a^m)^{1/n}$ is defined.

Real-number exponents, such as π in 2^π, are defined using more advanced mathematics. Exponentials involving real-number bases and/or real-number exponents can be approximated well by approximating the real numbers by rational numbers. The properties for real exponents (which follow) apply equally well for all rational exponents. However, there are special considerations for the meaning of b^x if b is a negative number. For example, some calculators may return an answer for $(-2)^{1.6}$, while others will correctly inform you that there is an error involved.

· ·

Suggestions and Comments for Odd-numbered problems

13. and 15.
Remember the definitions of zero as an exponent and negative integers as exponents.

15. When negative numbers are used as exponents, they have a special meaning related to numerators and denominators of fractions. However, the numerators or denominators could be negative numbers. Pay close attention to how the number is being used.

37. Use the alternate form: $x^{m/n} = (x^{1/n})^m$

41. Apply the properties one step at a time. Remember that the denominator in the exponent means that a root is involved.

43. Use the properties of exponents to get a single exponential expression on the left-hand side of the equation. Express the right-hand side in the same base, and equate the exponents when the bases are equal.

Solutions to Odd-numbered Problems

1. **(a)** 2^{10} **(b)** 5^6 **(c)** b^9

3. **(a)** 8^4 **(b)** 8^3 **(c)** 30^2

5. **(a)** 5^6 **(b)** 3^8 **(c)** b^{18}

7. **(a)** 3^5 **(b)** 5^4 **(c)** b^4

9. **(a)** $(\frac{5}{8})^3$ **(b)** $(\frac{4}{7})^2$ **(c)** $(\frac{x}{y})^5$

11. **(a)** 3802.04032   **(b)** 20107.5858   **(c)** 9.0658

13. **(a)** $2^{-3} = \dfrac{1}{2^3} = \dfrac{1}{8}$ **(b)** $3 \times 5^{-2} = 3 \times \dfrac{1}{5^2} = \dfrac{3}{25}$

    (c) $\dfrac{4}{2^{-2}} = 4 \times \dfrac{1}{2^{-2}} = 4 \times 2^2 = 4 \times 4 = 16$

15. $3^0 \times 2^{-2} \times \dfrac{1}{5^{-1}} = 1 \times \dfrac{1}{2^2} \times 5 = \dfrac{5}{4}$

17. **(a)** 8.25×10^8 **(b)** 237,000   **(c)** 2.53×10^{-2} **(d)** 8.05×10^5

19. $4^x \times 4^2 = 4^{x+2} = 4^{12}$; $x + 2 = 12$ or $x = 10$

21. $(b^x)^2 = b^{2x} = b^{18}$; $2x = 18$ or $x = 9$

23. 484,000,000 (8 places) $= 4.84 \times 10^8$

25. 0.00000028 (7 places) $= 2.8 \times 10^{-7}$

27. **(a)** Approximately 200,000 $10^{5.3} = 199,526.23...$
    Note: A reading on the Richter scale would be an approximation, as are most
    measurements. An *exact* value is not justified.

    (b) Approximately 600,000 $10^{5.8} = 630,957.34...$

    (c) Approximately 60,000,000 $10^{7.8} = 63,095,734$
    Note: $10^{7.7} = 50,118,723.36...$ A reading of 7.8 on the scale represents an
    increase of 26% over the intensity of an earthquake that registers 7.7.

29. **(a)** 6   **(b)** 8   **(c)** -5

31. (a) $\sqrt[4]{16} = 2$ since $2^4 = 16$.   **(b)** $\sqrt[3]{-8} = -2$ since $(-2)^3 = -8$.

     (c) $\sqrt[4]{-81}$ is undefined since $x^4 \geq 0$ for all x.

33. (a) $27^{1/2}$ **(b)** $15^{1/3}$ **(c)** $x^{3/4}$

35. (a) $\sqrt[3]{25}$ **(b)** $\sqrt[3]{100}$ **(c)** $\sqrt[3]{8^2} = \sqrt[3]{64} = 4$

37. (a) $27^{2/3} = (27^{1/3})^2 = (\sqrt[3]{27})^2 = 3^2 = 9$ **(b)** $16^{3/2} = (16^{1/2})^3 = (\sqrt{16})^3 = 4^3 = 64$

     (c) $16^{-5/4} = (16^{1/4})^{-5} = (\sqrt[4]{16})^{-5} = 2^{-5} = \dfrac{1}{2^5} = \dfrac{1}{32}$

39. (a) 1.442 **(b)** 8.550 **(c)** 181.019

41. (a) $b^{1/2} \times b^{1/6} = b^{1/2+1/6} = b^{4/6} = b^{2/3}$

     (b) $\left(\dfrac{4x}{9}\right)^{1/2} = \sqrt{\dfrac{4x}{9}} = \dfrac{\sqrt{4x}}{\sqrt{9}} = \dfrac{2\sqrt{x}}{3} = \dfrac{2}{3}\sqrt{x}$

     (c) $12x^{25} \div 4x^{15} = \dfrac{12x^{25}}{4x^{15}} = 3x^{25-15} = 3x^{10}$

     (d) $(9y^2)^{3/2} = 9^{3/2} \times (y^2)^{3/2} = (9^{1/2})^3 \times (y^2)^{3/2} = 3^3 \times y^3 = 27y^3$

43. (a) $4^{2/3} \times 4^k = 4^{2/3+k} = 4^2;$ $\dfrac{2}{3} + k = 2$ or $k = \dfrac{4}{3}$

     (b) $5^{3/2} \div 5^k = 5^{3/2-k} = 25 = 5^2;$

         $5^{3/2-k} = 5^2; \dfrac{3}{2} - k = 2$ or $k = \dfrac{-1}{2}$

     (c) $2^k \times 3^k = (2 \times 3)^k = 6^k = 36 = 6^2; 6^k = 6^2$ or $k = 2$

45.

Planet	k	Planet	k
Mercury	1.0021		
Planet	k	Planet	k
Venus	1.0008	Earth	1.0000
Mars	1.0016	Jupiter	0.9988
Saturn	0.9991	Uranus	0.9987
Neptune	0.9971	Pluto	1.0002

$k = (0.241^2) \div (0.387^3) = 1.002077221$

Yes, k is approximately 1.0 in each case.

47. We make the assumption that $T^2 \div R^3 = 1$ for this planet; that is $T^2 = R^3$.
If the mean distance, R, is 48.125 AU, then $T^2 = (48,125)^3$
$T^2 \approx 111,485$ or $T \approx 333.85$ years.

Section 1.6   The Concept of Function

Goals
1. Learn the language and notation of sequences and functions.
2. Use the concept and terminology of functions to express and work with relationships between variable quantities.

Key Ideas and Questions
1. Explain what it means to say that one variable quantity is a function of another.
2. Explain how functions can be represented using tables, arrow diagrams, and graphs.
3. How can a function be viewed as a quantitative machine?

Vocabulary

Terms	Function	Functions as Tables
Initial Term	Function Notation	Functions as Machines
Counting Numbers	The Value of a Function	Functions as Ordered Pairs
Arithmetic Sequence	Domain	Functions as Graphs
Common Difference	Codomain	The Graph of a Function
Geometric Sequence	Range	Functions as Formulas
Common Ratio	Arrow Diagrams	

• •

Overview

The concept of a function is one of the most powerful and useful in all mathematics. The study of different types of functions, their properties and applications, is the focus of most mathematics study by college students. A function is a special case of what we call a relation between sets. A *relation* is a means of associating, or matching up, the elements from two or more sets.

Suppose a demographer (a social scientist who studies populations) is studying the birth records from 1990 through 1999 in a given county. There are several sets of data (information) that might be collected, including the set of all children, the set of all mothers (all fathers, all doctors, all hospitals or places of birth, etc.). We will consider only the sets of children and mothers.

Suppose we pick a child and ask "Who is the mother of this child?" The records will provide us with a single name. If, however, we pick a mother and ask "Who is the child of this mother?", we may get more than one name since the records cover ten years, and many women will give birth to several children over this period. Although this is useful information, it is not as useful or as easy to work with as the information in the first situation in which there was always one, and only one, response.

Whenever two sets are related so that for each element in the first set there is one, and only one, element in the second set, we call this a function from the first set to the second set.

Sequences

The simplest type of function is a **sequence**, in which the first set is the set of natural numbers or whole numbers (natural numbers plus zero). Any set that is arranged in order can be considered as a sequence. If we looked at the order of birth in the previous section, there would be a *first* child, *second* child, and so on. In the notation of sequences, these children would be labeled (named) a_1, a_2, etc.

A mathematical sequence is defined as a list of numbers, or **terms**, arranged in order, where the first term is called the initial term. Some sequences are defined in such a way that after the first term has been given, the succeeding terms can be found systematically. If the succeeding terms can be found by adding a **common difference**, the sequence is called **arithmetic**; if succeeding terms are found by multiplying by a **common ratio**, the sequence is called **geometric**.

Arithmetic Sequence: $a, a + d, a + 2d, a + 3d, \ldots$; in general, $a_0 = a$, and $a_n = a + nd$

Geometric Sequence: $a, ar, ar^2, ar^3, \ldots$; in general $a_0 = a$, and $a_n = ar^n$.

Function

Function

A **function** is a rule that assigns each element of a first set to an element of a second set in such a way that no element of the first set is assigned to two different elements from the second set.

Note: In most discussions of functions (the one in the text, for example) the definition will say "... assigns **to** each element of a first set an element of a second set... ". This is a matter of perspective. What matters is how you *think* about the function and the nature of the relationship. In our opening example with mothers and children, each element in one set (children, C) is matched with one element in the second set (mothers, M), and each element of the second set, M, is matched with at least one element from the first set, C. Now comes the question: How do we represent these relationships mathematically - with symbols and/or diagrams?

Representations of Functions

First, there is the symbolism for functions, themselves: We will use f to represent the function in question, and A and B to represent the first and second sets, respectively. We show this as $f: A \rightarrow B$ Note the implication that we start in A and end in B. The first set is called the **domain**; the second set is called the **co-domain**. The second element is also called the **image** of the first element, and the set of all images is called the **range**. The element from the first set is call the **independent variable**, since it's selected first; the element from the second set is called the **dependent variable**.

Consider the following situation: A group of students (*S*) arrives at a camp for an extended field trip. As the thirty students step off the bus, the facility manager tells each of them which of the three dormitories (*D*) they will room in. A student is likely to say, "I was assigned to dorm *2*." The manager, however, might say, "I assigned dorms to each of the students." It's a matter of perspective. What is important is that each student ends up in a dorm. For each student, there is one, and only one, dorm; that's a function. For the manager looking at a given dorm, however, there may be several students (although dorms with no students or one student may be possible); that's not a function. However, if each student had been told which bed to sleep in, then the relationship would be a function both ways.

What's important about a function is that when we want to know "what's the second element that belongs with the first element," the answer is definite - no ambiguity.

That's the reason we say $\sqrt{9} = \sqrt{3^2} = 3$, although $(-3)^2 = 9$, as well.

The "tools" that have been developed to deal with functions are so powerful that we tend work in terms of functions whenever possible. In cases where a function isn't present, we'll work with a portion of the relation (that *is* a function) at a time.

Function Notation

Continuing with the field trip example, suppose Ari (*a*) is in dorm *1*, Barry (*b*) is in dorm *1*, and Carrie (*c*) is in dorm *2*. Using the standard function notation, with *d*, for *dorm*, representing the function, we have $d(a) = 1$, $d(b) = 1$, and $d(c) = 2$.

We can represent the function in several other common ways.

Functions as Arrow Diagrams   **Functions as Tables**

$d: S \rightarrow D$

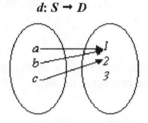

S	D
a	1
b	1
c	2

Functions as Machines

Functions are often thought of as a means or a machinery for taking one set of numbers and producing another set of numbers. Elements from the domain are the input, and elements in the range are the output. Some functions have the property that elements produced as outputs become inputs for the next term in a sequence, such as an arithmetic or geometric sequence.

Functions as Ordered Pairs

A function can be represented by a set of ordered pairs: $d = \{(a,1), (b, 1), (c,2),....\}$

Functions as Graphs and Functions as Formulas

As the size of the sets increases, the difficulty working with arrow diagrams, tables, and sets of ordered pairs increases significantly. If the domain contains any interval of real numbers, then only selected values are useful if they can represent or indicate a pattern. The arrow diagram and table above really aren't useful if you want to know the dorm where you can find Dave (d) or Erin (e).

Most work with functions in mathematics and the sciences involves working with functions that are represented by graphs and/or formulas. The **graph** of a function consists of the points in two-dimensional coordinate system that correspond to the set of ordered pairs that belong to the function. When using a graph, there will be a visual characteristic we look for.

When using a formula or equation, we have to be sure that there is no chance of getting two or more values from a single input from the domain. Using equations has the advantage that you can often manipulate the equation algebraically to answer important questions about the function.

. . . .

Suggestions and Comments for Odd-numbered problems

General Comments
Although the language of functions distinguished between dependent and independent variables, the second variable doesn't need to depend on the first variable in the more usual cause-and-effect type of dependence. The key question to ask is "If a specific value is named from the domain, is there a specific (one and only one) value that can (or could) be associated with it from the co-domain. You don't actually even have to be able to state the value, only be assured that one and only one could exist.

5. and 7.
Knowing only a very few terms in a sequence is not enough to determine the other terms unless you know there is a common difference or a common ratio.

9. through 15.
Write an equation in the form $y = f(x)$ using suitable letters for the variables.

25. and 27.
If there are m elements in the domain, and each of these could be associated with any *one* of the n elements in the domain, why are there n^m possible ways to show a function? Draw some simple arrow diagrams.

33. through 37.
Study the example carefully. Instead of substituting a single value into an equation and carrying out the calculations, you'll be substituting an expression into an equation and simplifying the resulting algebraic expression.

Solutions to Odd-numbered Problems

1. (a) Function   Every circle has a unique value for its area.
   (b) Function   The cost is $C = n \times 79$ cents; a unique value for each value of n.
   (c) Function   For each specific place, there is a unique time assigned. An exact interpretation that this is a function may be debatable.

3. (a) Not a function   The area of a triangle depends upon both the *base* and the *height*.
   (b) Function   The cost is $C = 79 \times n$ cents; a unique value for each value of n.
   (c) Not a function   There are generally *two* low tides each day, and two high tides.

5. (a) 2, 6, 10, 14, 18, 22   For the arithmetic sequence, we want to find a common difference. The difference between the first two terms (2, 6, ...) is 4. We add 4 to get each remaining term in succession.
   (b) 2, 6, 18, 54, 162, 486   For the geometric sequence, we want to find a common ratio. The ratio between the first two terms (2, 6, ...) is 3. We multiply by 3 to get each remaining term in succession.

7. (a) 8, 4, 0, -4, -8, -12   For the arithmetic sequence, we want to find a common difference. The difference between the first two terms (8, 4, ...) is -4. We add -4 (subtract 4) to get each remaining term in succession.
   (b) 8, 4, 2, 1, 1/2, 1/4   For the geometric sequence, we want to find a common ratio. The ratio between the first two terms (8, 4, ...) is $\frac{1}{2}$.

   We multiply by $\frac{1}{2}$ to get each remaining term in succession.

9. Let C = cost, x = weight of item in pounds. The cost is a function of the weight
   $C(x) = \$15.00 + \$0.25x$

11. Let A = area of circle, r = radius of circle. Every circle with a given radius has the same area, which is proportional to the square of the radius; $r > 0$ is required, although r = 0 raises a philosophical point (pun intended).
   $A(r) = \pi r^2$

13. Let C = cost of call, t = time in minutes.   All calls of a given length have the same cost. In contrast to problem 11, $t > 0$ is strictly required.
$C(t) = \$0.99 + \$.10t$

15. C = length in centimeters, x = length in inches
   $C(x) = 2.54x$

17. (a) Function   **(b)** Function

19. (a) Not a function; a appears twice; once associated with 2, and then with 1
   (b) Function

21. (a) Function   **(b)** Not a function; b is associated with both 2 and 3.

23. (a) Function   **(b)** Not a function; c is associated with both 2 and 0

25. 9 possible functions   For each of the three ways a single arrow could be drawn from the first element in the domain to an element in the codomain, there are three ways to draw a single arrow from the second element in the domain to an element in the codomain. The number of acceptable arrow diagrams is $3 \times 3 = 3^2 = 9$

27. 81 possible functions   For each of the three ways a single arrow could be drawn from the first element in the domain to an element in the codomain, there are three ways to draw a single arrow from each of the remaining elements in the domain to an element in the codomain. The number of acceptable arrow diagrams is $3 \times 3 \times 3 \times 3 = 3^4 = 81$

29. (a) Function   **(b)** Function
     (c) Function   **(d)** Not a function <u>from</u> A to B; $f(5)$ not defined

31. (a) $f(a) = 2$ **(b)** $f(b) = 4$

33. Let $G(x) = 5x$ and $F(u) = u - 3.$ Then, $F(G(x)) = 5x - 3$

35. $F(G(x)) = (G(x))^2 + 2 = (3x + 1)^2 + 2 = (9x^2 + 6x + 1) + 2 = 9x^2 + 6x + 3$

37. $F(F(x)) = (F(x))^2 + 2 = (x^2 + 2)^2 + 2 = (x^4 + 4x^2 + 4) + 2 = x^4 + 4x^2 + 6$

Section 1.7   Functions and Their Graphs

Goals
1. Develop skills in graphing linear, quadratic, and exponential functions.
2. Gain a better understanding about the relationship between the equation of a function and the shape of its graph.

Key Ideas and Questions
1. How can you tell if a graph is that of a function?
2. What are the basic properties and equations for linear, quadratic, and exponential functions?
3. How is the form of a function related to the shape of its graph?

Vocabulary

Cartesian Coordinate System	x-coordinate	Increasing/Decreasing
Origin	y-coordinate	Rate of Change
x-axis	Vertical Line Test	Quadratic Function
y-axis	Independent Variable	Parabola
Quadrants	Dependent Variable	Exponential Functions
Coordinates	Linear Functions	Exponential Growth
	Slope	Exponential Decay
	Slope Intercept Equation	

Overview

The **graph** of a function, f, consists of all the points in the plane corresponding to ordered pairs $(x, f(x))$ as x takes on all values in the domain of f. Throughout the text, this will mean we will use the Cartesian coordinate system to represent the ordered pairs. However, in many situations, other systems are possible. For radar, where direction and distance are the measures, a polar coordinate system might be more useful. **The Cartesian coordinate system** is based on two sets of parallel lines that fill the plane, one set perpendicular to the other. Not only is the location of points relatively simple, but any determination of the distance between two points can use "built-in" right triangles and the Pythagorean Theorem to find the needed value. The invention of the Cartesian coordinate system allowed geometry and algebra to be used together, and gave mathematicians powerful new tools to analyze functions and solve related problems.

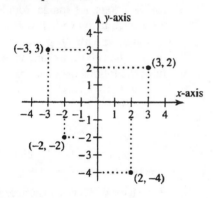

The x-**axis** and y-**axis** combine to divide the plane into four regions called **quadrants**, which are numbered I through IV, going counter-clockwise. The point $(3, 2)$ is in quadrant I.

The x-**coordinate** of the point in the quadrant II is -3; the y-**coordinate** is 3.

Each axis is a *real number* line. The point where the axes intersect, with coordinates of $(0, 0)$, called the **origin.**

Graphs of Functions

The shape of the graph of a function can be quite different with different coordinate systems. Even when only using the Cartesian coordinate system, we can make a graph look different by the scale we use for the axes. It's not always practical (or even wise) to have the scale - the distance between units - be the same for both variables. This will be true for many of the functions we will consider in this section. In the Cartesian coordinate system, however, there is one thing that is true of all graphs of functions: There is no vertical line that intersects the graph in two or more places: for each value of x, there is one, and only one, value for y.

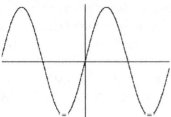
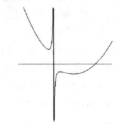
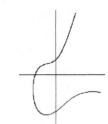

A function in the x-y system   A function in the x-y system   Not a function in the x-y system

Functions and Their Graphs

Certain functions have more practical applications than others and show up in many different settings. In particular, there are three types of functions that we will use and study at several points in the text. The first is the linear function, which is the most common; then we will examine the quadratic and exponential functions. Each of these functions has both a recognizable equation and a graph with a recognizable general shape. There is also a strong relationship between the "particulars" of the equation and the "specific" shape of the graph.

Linear Functions

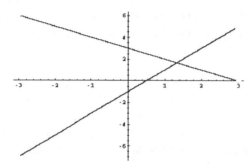

The equation of a linear function can be written in the general form $y = mx + b$, where m is the slope and b is the intercept for the y-axis. As you go from left to right, a linear function with a *positive* slope goes *up*, and one with a *negative* slope goes *down*. The two linear functions shown have the following equations:

$$y = 2x - 1 \quad \text{and} \quad y = -x + 3$$

Quadratic Functions

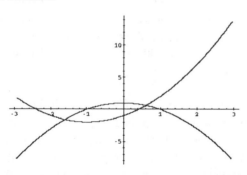

The general equation for a quadratic function is $y = ax^2 + bx + c$

The shape of every equation of this form is called a parabola. Except for scaling and orientation, all parabolas "look alike".

The shape and orientation ("opens up" or "opens down") is completely determined by the value of a in the equation. The two quadratic equations have equations:

$$y = x^2 + 2x - 1 \quad \text{and} \quad y = -x^2 + 1$$

Exponential Functions

An equation of an *exponential* function (the independent variable is used as an exponent) has the general form $y = Ca^x$, where $a > 0$ and $a \neq 1$. Since $a^0 = 1$ for all allowable values of a, $y = C$ when $x = 0$. Depending upon the value of a, the value of an exponential function always increases as x increases (exponential growth) or decreases as x increases (exponential decay). The exponential functions have the feature that over intervals of the same length, the ratio of the beginning and ending terms is constant. Exponential functions are used extensively in applications where values increase or decrease at a constant ratio over time, and t is used as the variable.

Exponential growth

When $a > 1$, the value of a^x increases.

That is, if $x_1 < x_2$, then $a^{x_1} < a^{x_2}$

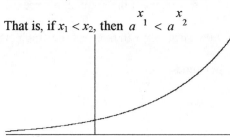

Exponential decay

When $0 < a < 1$, the value of a^x decreases.

That is, if $x_1 < x_2$, then $a^{x_1} > a^{x_2}$

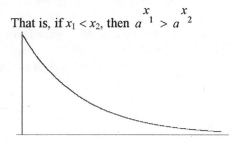

Suggestions and Comments for Odd-numbered problems

General Comments

Most types of equations have a specific type of graph associated with them, although some might be very complicated. Not only do the linear, quadratic, and exponential functions have many applications in real (normal) life, they have graphs that are very consistent. All lines, all parabolas, all exponential curves look the same as the others of their type. They are smooth curves with no corners.

When sketching the graph of a quadratic or exponential function, it is only necessary to plot a few points since you should know how the graph will look except for its actual placement on the coordinate grid.

11. and 13.

You can use the coordinates of any two points on a line to find the slope of the line. If the y-intercept, b, is given, you can use $(0, b)$ as a point as well as using b directly in writing the equation.

15. and 17.

Be careful in changing the form of the equation to $y = f(x)$. One way to check your work is to pick a value of x ($x \neq 0$), and then find $f(x)$. The pair should satisfy the original form of the equation.

19. and 21.

First, make sure the equation is in the form $y = f(x) = ax^2 + bx + c$. The formulas for finding the roots and vertex require it to be in this form.

23. through 27.

When working with an exponential function, always try to identify whether it is a growth function or a decay function by the value of a in $f(x) = Ca^x$.

29. and 31.

How is x used as a variable? Are there multiples of x, powers of x, or exponents with x?

Solutions to Odd-numbered Problems

1. Function   Every *vertical line* will intersect the graph only once..

3. Not a function   Some *vertical lines*, such as $x = 0$, intersect the graph more then once.

5. Let C = cost, x = weight of item in pounds
$C(x) = \$15.00 + \$0.25x$

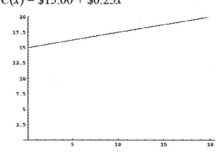

7. C = length in centimeters, x = length in inches
$C(x) = 2.54x$

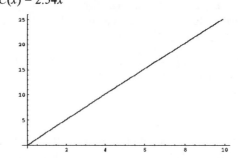

9. **(a)** $C = \frac{5}{9}(F - 32)$ or $C = \frac{5}{9}F - \frac{160}{9}$

   (b)

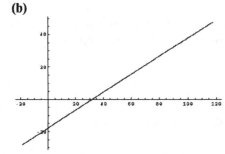

   (c) Slope $= \frac{5}{9}$ **(d)** C-intercept $= -\frac{160}{9} \approx -17.8$

11. $m = \frac{6-0}{4-0} = \frac{3}{2}$; $y = \frac{3}{2}x$

13. The known points, (R, F), are $(0, 32)$ and $(80, 212)$

$m = \frac{212-32}{80-0} = \frac{180}{80} = \frac{9}{4}$ $F = \frac{9}{4}R + 32$

15. (a) $y = -\frac{4}{3}x + 4$ **(b)** Slope $= -\frac{4}{3}$ **(c)** y-intercept $= 4$

(d)

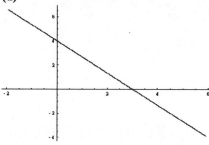

17. (a) $y = \frac{5}{2}x - 9$ **(b)** Slope $= \frac{5}{2}$ **(c)** y-intercept $= -9$

(d)

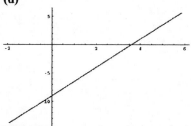

19. (a) Roots are $x = -3$, $x = 1$ **(b)** Vertex is $(-1, -4)$

(c)

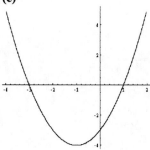

21. (a) Roots are $x = -1$, $x = 4$ **(b)** Vertex is $(\frac{3}{2}, \frac{25}{4})$

(c)

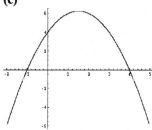

23. (a)

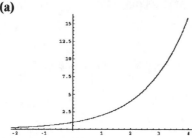

(b) Exponential growth

25. (a)

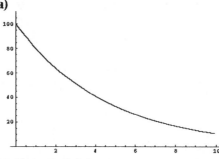

(b) Exponential decay

27. (a)

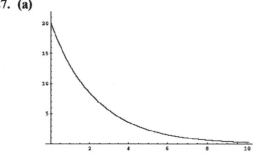

(b) Exponential decay

29. (a) Exponential decay   **(b)** Line
　　(c) Parabola with minimum value **(d)** Line

31. (a) Parabola with minimum value **(b)** Exponential growth
　　(c) Line   **(d)** Parabola with maximum value

33. Let x = miles driven; $C(x) = \$14.95 + \$0.20x$

35. Let x = years from present. $P(x) = 2(1.05)^x$ (in millions)

37. Let t = time in years;
　　(a) $S(t) = 12000(1.08)^t$
　　(b) $S(6) = 19{,}042$

39. $y = \frac{1}{3}x + \frac{4}{3}$

41. $y = -\frac{1}{3}x + \frac{2}{3}$

43.

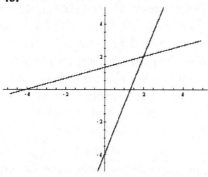

Section 1.8   Solving Equations and Systems of Equations

Goals
1. Learn techniques for solving linear equations and systems of linear equations.

Key Ideas and Questions
1. What are the two standard techniques for solving systems of linear equations?
2. Under what conditions does a system of linear equations have one, no, or infinitely many solutions?

Vocabulary

Linear Equation	Substitution	Inconsistent
System of Linear Equations	Elimination	General Solution
Solution	Back Substitution	Particular Solution
Solution Set		

· ·

Overview

The most common functions in the "real" world are the linear functions; they occur in daily life as well as business and technical applications. Working with these equations, either singly or with two or more as part of a system, is a basic skill that all educated people, not just mathematics and science students, need to master.

Linear Equations

A **linear equation** is any equation (with any number of variables) that can be expressed as the sum (or difference) of multiples of the variable and constants.

Linear equations are classified as first degree equations. No variable occurs to other than the first power, or in a radical or denominator. The term linear is used because when there are two variable (the most usual situation) the graph of the equation will be a line in the plane. When there is only one variable, the graph is a point on a real number line; when there are two variables, the graph is a line on a two-dimensional plane; when there are three variables, the graph is a plane in a three-dimensional coordinate system that is an extension of the Cartesian coordinate system. The general forms for these equations are $ax + b = c$, $ax + by = c$, and $ax + by + cz = d$

Solution Sets for Linear Equations

We will refer to the set of all values of the variable (or combination of variables) that satisfy the equation as the solution set for the equation. As indicated in the preceding paragraph, this will be a single value (corresponding to a point on a real number line) when there is a single variable, and a set of pairs of values (corresponding to the coordinates of points on a line) when there are two variables. In general terms, the solution set and graph are basically synonymous. In order to graph an equation, we find enough members of the solution set to establish a pattern.

Solving First Degree Equations in One Variable

Given a first degree equation in the form $ax + b = c$, we solve by isolating x as
follows:

$$ax + b = c \qquad 4x + 3 = 15 \qquad 3x + 9 = 4$$
$$ax = c - b \qquad 4x = 12 \qquad 3x = \text{-}5$$
$$x = \frac{c - b}{a} \qquad x = 3 \qquad x = \frac{-5}{3}$$

Solution Sets for Linear Equations in Two Variables

For a linear equation in two variables, $ax + by = c$, the solution set is not a single value, but an infinite set of ordered pairs that satisfy the equation; the pairs correspond to the coordinates of the points on the graph. For any value of x, the corresponding value of y is found as follows:

$$ax + by = c \qquad 2x + 3y = 12$$
$$by = \text{-ax} + c \qquad 3y = \text{-}2x + 12$$
$$y = \frac{-a}{b}x + \frac{c}{b} \qquad y = \frac{-2}{3}x + 4$$

The previous solution set can be defined as $\{(x, y) \mid y = \frac{-2}{3}x + 4\}$, the set of coordinates of points on the graph. The last equation is the standard equation of a line that has a slope of $\frac{-2}{3}$ and a y-intercept of 4.

In order to graph the equation of a line, we need only find two points on the graph, since any two points in the plane determine one, and only one, line. There are three methods generally used; we'll illustrate these with $2x + 3y = 12$ and the alternate form.

Method 1: Use the y-intercept (when $x = 0$) and the x-intercept (when $y = 0$).

Given: $2x + 3y = 12$

For $x = 0$ $2(0) + 3y = 12$
$$3y = 12$$
$$y = 4$$

For $y = 0$ $2x + 3(0) = 12$
$$2x = 12$$
$$x = 6$$

The points are $(0, 4)$ and $(6, 0)$

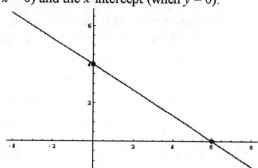

Method 2: Use the equation in the form $y = \dfrac{-2}{3}x + 4$. Substitute two values for x, and find the corresponding values for y. If a fraction is involved, use multiples of the denominator if it will aid computation.

Given: $y = \dfrac{-2}{3}x + 4$

For $x = 3$ $y = \dfrac{-2}{3}(3) + 4$
$$y = -2 + 4 = 2$$

For $x = -3$ $y = \dfrac{-2}{3}(-3) + 4$
$$y = 2 + 4 = 6$$

The points are $(3,2)$ and $(-3,6)$

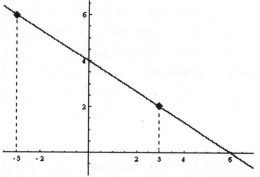

Method 3: In the form $y = \dfrac{-2}{3}x + 4$, the 4 corresponds to the point $(0,4)$. We can use the slope to find a second point. The slope is the "rise" over the "run", so we can use a "run" of 3 and a "rise" (actually a "fall") of -2. From $(0, 4)$ we go three units to the right and down two units to the point $(3, 2)$.

Given: $y = \dfrac{-2}{3}x + 4$

For $x = 0$ $y = \dfrac{-2}{3}(0) + 4$
$$y = 0 + 4 = 4$$

slope $= m = \dfrac{-2}{3} = \dfrac{"rise"}{"run"}$

Let "run" = 3 (3 units to the right)
and "rise" = -2 (2 units down)

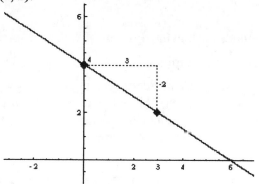

Solution Sets for Systems of Linear Equations

In many applications with two variables, there may be more than one condition the variables have to satisfy, and each can be "modeled" with an equation. The resulting set of equations is called a system of equations. In Chapter 8, Management Mathematics, we'll see how resource allocation problems (and similar problems) are solved using a system of linear inequalities, which in turn depend on solving a related system of linear equations.

Suppose you know that there are 12 items having a total value of $400, and that some cost $30 each, while the rest cost $40. Can we determine how many of each there are?

First, there are two variables - the number of each type of item; call them x and y. The fact that there are 12 means that $x + y = 12$. This is our first equation. The first type of item costs a total of $30x$, while the second type costs a total of $40y$. This means that $30x + 40y = 400$. This is our second equation.

We have the following system of equations for which we want to find a solution,

$$x + y = 12$$
$$30x + 40y = 400$$

A **solution** to a system of linear equations in two variable is any ordered pair (x, y) that satisfies all the equations in the system. In terms of the graphs, a solution is any point whose coordinates satisfy all the equations.

When two (straight) lines are drawn, only one of three things can happen: they intersect (in a single point), they are parallel (and don't intersect at all), or they are coincident (the equations produce the same graph).

$x + y = 12$	$x + 2y = 5$	$x - 2y = 8$
$30x + 40y = 400$	$y = -0.5x + 5$	$y = 0.5x - 4$
The graphs intersect; there is one solution.	The graphs are parallel; there is no solution.	The graphs are coincident; all points are solutions.

Methods for Solving Systems of Linear Equations

Graphing

As a matter of general practice, you should draw a quick, but neat, sketch of the graphs. This will indicate a possible solution, which can be checked by substituting the values into the equations. In the first system above, it appears that $x = 8, y = 4$ is a solution. We confirm that it is: $8 + 4 = 12$, and $30(8) + 40(4) = 400$.

Most systems don't work out so nicely; many solutions include fractions. We can still find approximate values, but must use algebraic methods for accurate values.

Substitution

Solving by substitution usually means solving for one variable in terms of the other in one of the equations and then substituting that expression into the other equation.

$x + y = 12$

$30x + 40y = 400$

First, solve for either x or y in the first equation: $y = 12 - x$,

Then, substitute this expression in place of y in the second equation: $30x + 40(12 - x) = 400$ or $30x + 480 - 40x = 400$ This simplifies to $-10x = -80$, or $x = 8$

To find the corresponding value for y, back-substitute $x = 8$ in the equation $y = 12 - x$. When $x = 8$, $y = 4$.

$x = 8, y = 4$ The solution should be verified by a final check.

Elimination

The elimination method is based on two facts or generalizations. The first is simply that if an ordered pair (x, y) satisfies an equation, then it satisfies any non-zero multiple of the equation. The second says that if "equals are added to equals, the results are equal". In terms of equations, that means if an ordered pair satisfies two equations, it satisfies the result when the two equations are added (or subtracted). The elimination method uses multiples of the equations and adds or subtracts them so that one of the variables is "eliminated". The other is solved for directly, and then the one that was eliminated is found through back-substitution.

$x + y = 12$

$30x + 40y = 400$

We multiply one (or both) equations to produce a "common" coefficient for one of the variables; ideally the coefficients will have the same absolute value but different signs. To achieve this, we can multiply the first equation by -30.

$-30x - 30y = -360$

$\underline{30x + 40y = 400}$ The second equation is unchanged

$10y = 40$ or $y = 4$

$x = 8, y = 4$ Using back substitution, we find (again) that $x = 8$

The solution should be verified by a final check.

The method of elimination is based on the assumption that there is a solution. If the method leads to a **contradiction** or absurdity (such as $0 = 4$), this means there is *no* solution. If it leads to a statement devoid of x and y that is always true (such as $4 = 4$), this means that the two equations are equivalent and there are infinitely many solutions. In this case, the set of all solutions is called the **general solution**, whereas individual solutions are called **particular solutions**.

· ·

Suggestions and Comments for Odd-numbered problems

General Comments

As a rule, you should graph the lines in the system of linear equations and estimate the point of intersection. If the lines intersect (which they usually will) and the solution appears to be a pair of integers, you can substitute the values in the equations to see if they work. However, the exercises are designed to help you learn new techniques and you should substitution or elimination whenever they are called for. If the solution to a system contains fractions, these can only be found accurately by using one of the algebraic techniques. The graphing and algebraic approaches should be used together; you might make a mistake graphing, but you're more likely to make an algebraic error.

When using elimination, first look to see if the coefficients for one of the variables have different signs in the two equations. If they do, multiply the equations to produce a common multiple for the two. Notice how the following system is changed.

$$3x - 8y = 10 \quad \text{multiply by 3} \quad 9x - 24y = 30$$
$$5x + 6y = 12 \quad \text{multiply by 4} \quad 20x + 24y = 48 \quad \text{This means that } 29x = 78$$

Solutions to Odd-numbered Problems

1. **(a)** $x = 3$ **(b)** $x = 2$ **(c)** $x = 7/3$ **(d)** $x = 11/2$

3. **(a)** $x = 0.6$ **(b)** $x = 15$

5. $(3, -1)$ is a solution $3x - 2y = 11$ $3(3) - 2(-1) = 9 + 2 = 11$
$ \quad x + y = 2 \qquad (3) + (-1) = 2$

7. $(2, -1)$ is not a solution $4x - 3y = 11$ $4(2) - 3(-1) = 8 + 3 = 11$
$ \quad 2x + y = 5 \qquad 2(2) + (-1) = 4 - 1 = 3 \neq 5$

9. $x = 23, \ y = -43$

Given $y = -2x + 3$ and $3x + 2y = -17$
Substitute in the second equation
$3x + 2(-2x + 3) = -17$
$3x - 4x + 6 = -17 \Leftrightarrow -x = -23 \text{ or } x = 23$

$y = -2x + 3 \Rightarrow y = -43 \text{ when } x = 23$

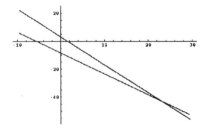

11. $x = 7, \ y = 3$

Given $3x - 2y = 15$ and $x + y = 10$
$ \ y = 10 - x$
$3x - 2(10 - x) = 15$
$3x - 20 + 2x = 15 \Leftrightarrow 5x = 35 \text{ or } x = 7$

$y = 10 - x \Rightarrow y = 3 \text{ when } x = 7$

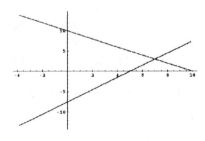

13. $x = 1, \; y = 1$

Given $\quad 6x + y = 7$ and $5x - 2y = 3$
$$y = 7 - 6x$$
$$5x - 2(7 - 6x) = 3$$
$$5x - 14 + 12x = 3 \Leftrightarrow 17x = 17 \text{ or } x = 1$$
$$y = 7 - 6x \Rightarrow y = 1 \text{ when } x = 1$$

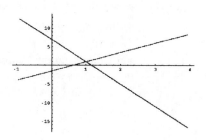

15. $x = 1, \; y = -3$

Given $\quad 4x + y = 1$
$$\underline{3x - y = 6}$$
$$7x \quad\quad = 7 \Leftrightarrow x = 1$$
$$4x + y = 1 \Leftrightarrow y = 1 - 4x$$
$$y = 1 - 4x \Rightarrow y = -3 \text{ when } x = 1$$

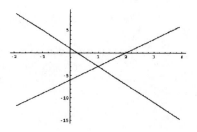

17. $x = 3, \; y = 1$

Given $\quad 2x + y = 7 \;\; (\times 2) \;\; 4x + 2y = 14$
$$\quad\quad\quad x - 2y = 1 \quad\quad\quad \underline{x - 2y = \;\; 1}$$
$$\quad\quad\quad\quad\quad\quad\quad\quad\quad\quad 5x \quad\quad = 15$$
$$5x = 15 \Leftrightarrow x = 3$$
$$2x + y = 7 \Leftrightarrow y = 7 - 2x$$
$$y = 7 - 2x \Rightarrow y = 1 \text{ when } x = 3$$

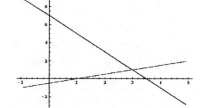

19. $x = 4, \; y = 4$

Given $\quad x + 2y = 12 \quad\quad\quad x + 2y = 12$
$$\quad\quad\quad 3x - y = 8 \;\; (\times 2) \;\; \underline{6x - 2y = 16}$$
$$\quad\quad\quad\quad\quad\quad\quad\quad\quad\quad 7x \quad\quad = 28$$
$$7x = 28 \Leftrightarrow x = 4$$
$$3x - y = 8 \Leftrightarrow y = 3x - 8$$
$$y = 3x - 8 \Rightarrow y = 4 \text{ when } x = 4$$

21. System
$$\quad\quad\quad \text{E1:} \quad x + y = 29$$
$$\quad\quad\quad \text{E2:} \quad x - y = 7$$
E1 + E2: $\quad 2x \quad\quad = 36 \text{ or } x = 18$
Back-substitution in E1: $18 + y = 29$ or $y = 11$
Solution: $x = 11, y = 18$
Check: E1: $18 + 11 = 29 \quad$ E2: $18 - 11 = 7$

23. System: $x + y = 10000 \quad\quad\quad x + \quad y = 10000 \;\; (\times 10) \quad 10x + 10y = 100000$
$$\quad\quad\quad .06x + .1y = 840 \;\; (\times 100) \;\; 6x + 10y = 84000 \quad\quad\quad \underline{6x + 10y = \;\; 84000}$$
$$\quad\quad\quad\quad\quad\quad\quad\quad\quad\quad\quad\quad\quad\quad\quad\quad\quad\quad\quad\quad\quad \text{subtract:} \quad 4x \quad\quad\quad = \;\; 16000$$

Solution: $x = 4000, y = 6000$

25. The system has infinitely many solutions.

The lines are coincident.
   General solution: $y = -2x + 4$
   Particular solutions: (answers will vary)
      $x = -2 \Rightarrow y = 8$
      $x = 0 \Rightarrow y = 4$
      $x = 3 \Rightarrow y = -2$

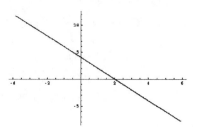

27. The system has no solution.

The two lines are parallel.

The equations can be re-written as
   $y = 2x + 6$ and $y = 2x + 8$

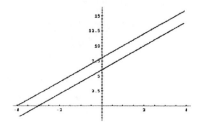

29. The system has a single solution: $x = 9$, $y = 8$

Given $2x - y = 10$ and $2y - x = 7$
   $2x - y = 10 \Leftrightarrow y = 2x - 10$
Substituting in the second equation:
$2(2x - 10) - x = 7 \Leftrightarrow 4x - 20 - x = 7$
$\Leftrightarrow 3x = 27$ or $x = 9$

$y = 2x - 10 \Rightarrow y = 8$ when $x = 9$

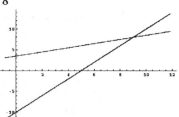

31. The system has two solutions: $x = -2, y = 2$; and $x = -4, y = 0$

Given $y = x + 4$ and $y = x^2 + 7x + 12$
Since $y = y$, $x + 4 = x^2 + 7x + 12$
$x + 4 = x^2 + 7x + 12 \Leftrightarrow x^2 + 6x + 8 = 0$

   This factors as $(x + 2)(x + 4) = 0$
   Either $x = -2$ or $x = -4$
$y = x + 4$
$x = -2 \Rightarrow y = 2$ and $x = -4 \Rightarrow y = 0$

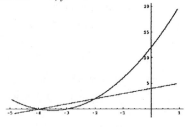

33. The system has no solutions in the real number system.

Given $y = x - 2$ and $y = x^2 - 1$
Since $y = y$, $x - 2 = x^2 - 1$
$x - 2 = x^2 - 1 \Leftrightarrow x^2 - x + 1 = 0$

Since this doesn't factor, we try the
quadratic formula to find x.

$$x = \frac{-(-1) \pm \sqrt{(-1)^2 - 4(1)(1)}}{2(1)} = \frac{1 \pm \sqrt{-3}}{2}$$

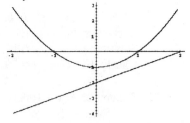

Chapter 1 Review Problems

Solutions to Problems   Note: Generally, we will provide answers to the odd-numbered problems only. However, since this chapter covers many topics, all answers are given to provide adequate coverage.

1. **(a)** False   **(b)** False   **(c)** True

   (d) False   **(e)** True   **(e)** True

2. **(a)** {4, 6, 8, 9, 10, 12, 14}   **(b)** {2, 3, 5, 6,7, 9, 11}   **(c)** {3}

3. **(a)** {a, b, c, e, g}   **(b)** {b}   **(c)** { } or $\varnothing$ **(d)** {a, c}

4. **(a)** {4, 5, 6, 7}   **(b)** {1, 3}   **(c)** {6}   **(d)** {6}

5. The order can't be filled. There are only 10 cars with air conditioning and no custom interior.

6. **(a)** Additive Inverse Property
   (b) Identity for Addition Property (Zero)
   (c) Identity for Multiplication Property (One)
   (d) Commutative Property of Addition
   (e) Distributivity of Multiplication over Addition
   (e) Associative Property of Multiplication

7. **(a)** $\frac{1}{2}$ **(b)** $-\frac{5}{36}$ **(c)** $\frac{1}{6}$ **(d)** $\frac{1}{3}$

8. **(a)** Multiplicative Inverse Property
   (b) Commutative Property of Multiplication
   (c) Distributivity of Multiplication over Addition

9. $\frac{3}{7} = 0.\overline{428571}$

10. $1.12121212\ldots = \frac{37}{33}$ $x = 1.121212\ldots, 100x = 112.121212\ldots; 99x = 111$

11. **(a)** $40\% = 0.4 = \frac{2}{5}$ **(b)** $2.95 = 295\% = 2\frac{19}{20}$ or $\frac{59}{20}$ **(c)** $\frac{3}{8} = 0.375 = 37.5\%$

12. **(a)** $\frac{10}{11}$ **(b)** 3.91540006

13. **(a)** 53.12   **(b)** 171.4%   **(c)** 523.81

14. $11,432.94; $9146.35

15. (a) $30^3 = 27000$ **(b)** $5^6 = 15625$ **(c)** $2^{15} = 32768$ **(d)** $5^5 = 3125$

16. (a) 2.6744×10^{-5} **(b)** 1.8636×10^8

17. (a) $2^2 = 4$ **(b)** $81^{3/4} = (81^{1/4})^3 = 27$

18. (a) 10.926 **(b)** 2.665

19. (a) $4, 6, 8, 10, 12, 14$ **(b)** $4, 6, 9, \frac{27}{2}, \frac{81}{4}, \frac{243}{8}$

20. (a) $P = 4s$ Independent variable is s, dependent variable is P

   (b) This is not a function. The perimeter also depends on the height.

   (c) $V = \frac{4}{3}\pi r^3$ Independent variable is r, dependent variable is V

   (d) This is not a function. The weight also depends on the density

21. (a) Let D = distance traveled, t = time in hours; $D = 55t$

   (b) Let A = area of rectangle, L = length; $A = Lw = L(L - 3)$ or $A = L^2 - 3L$

22. (a) This table represents a function; for each element in the first set, there is one and only one element associated with it from the second set.

   (b) This table does not represent a function. 3 is listed twice, with a different element from the codomain in each case.

23. The Vertical Line Test is used to determine if a graph represents a function. If NO vertical line intersects the graph at two points, then the graph is that of a function.

24. (a) $C = 1200 + 2.25x$, where x is the number of square feet and C is in dollars..

   (b) When $x = 2000$, $C = \$5700$ **(c)**

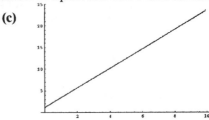

25. (a) Domain: All real numbers   **(b)** Domain: All real numbers
      Range: All real numbers   Range: $\{y \mid y \geq -4 , y \text{ is a real number}\}$

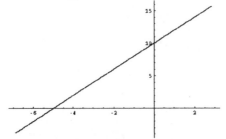   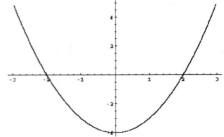

26. (a) Domain: All real numbers
Range: All positive real numbers

(b) Domain: All real numbers
Range: All positive real numbers

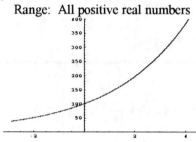

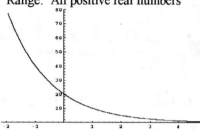

27. (a) Maximum value is $\frac{25}{4}$ **(b)** $f(x) = 0$ when $x = -1$, or $x = 4$

28. There is one solution: $x = \frac{5}{2}$, $y = \frac{7}{2}$

Given $3x + y = 11$ **and** $5x - 3y = 2$
  $3x + y = 11 \Leftrightarrow y = 11 - 3x$
Substituting in the second equation:
$5x - 3(11 - 3x) = 2 \Leftrightarrow 5x - 33 + 9x = 2$
$\Leftrightarrow 14x = 35$ or $x = 5/2$

$y = 11 - 2x \Rightarrow y = 7/2$ when $x = 5/2$

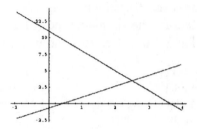

29. (a) Note: We'll go through all the steps, beginning with the graph.

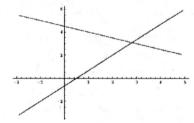

From the graph, we expect the values for x and y to both be approximately 3.
$x = 3$, $y = 3$ is not a solution, since it doesn't satisfy the second equation; $4(3) - 3(3) \neq 2$

Since we can solve for x in terms of y in the first equation, we'll solve by substitution:
From the first equation, we have $x + 2y = 9 \Leftrightarrow x = 9 - 2y$
Substituting in the second equation we have:
  $4x - 3y = 2 \Leftrightarrow 4(9 - 2y) - 3y = 2 \Leftrightarrow 36 - 8y - 3y = 2 \Leftrightarrow 34 = 11y$ or $y = \frac{34}{11}$

Now, we use back-substitution in the first equation: $x = 9 - 2(\frac{34}{11}) = \frac{99}{11} - \frac{68}{11} = \frac{31}{11}$

(b) $x = \frac{31}{11}$, $y = \frac{34}{11}$ You should verify that these values satisfy both equations.

30. (a) $(\frac{5}{2}, 0)$ **(b)** $(1, 2)$ **(c)** $\{(x, 3x - 5) \mid x \text{ is any real}\}$

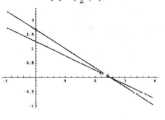

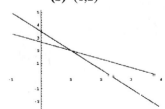

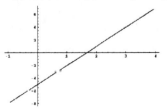

Infinitely many solutions

2 Descriptive Statistics Data and Patterns

Section 2.1   Organizing and Picturing Data

Goals
1. Graph data using various types of graphs to display and compare data.
2. Learn to use the type of graph best suited for certain kinds of data.

Key Ideas and Questions
1. What are the six main types of graphs in common use?
2. What are the strengths and weaknesses of each of these types?

Vocabulary

Data Set	Outliers	Histogram
Dot Plot	Measurement Class	Bar Graph (Bar Chart)
Stem and Leaf Plot	Frequency	Line Graph
Cluster	Frequency Table	Pie Chart (Circle Graph)
Gap	Relative Frequency	

• •

Overview

How can you organize a **data set** (a set of numbers) into a visual pattern or display so that the numbers make sense or satisfy a purpose? This chapter shows ways to do that, and also reveals how the human eye can be fooled by such charts.

Dot Plots

Dot plots can be used to represent a data set simply. It shows the numbers that occur most often in the tallest column of dots. Also, gaps and numbers that are widely separated from each other are shown. You should start a dot plot by putting the numbers in order (usually from lowest to highest), then draw a graph to show the numbers and their frequency (or how often they occur).

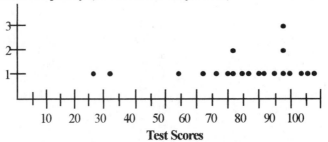

Stem and Leaf Plots

Stem and Leaf Plots use digits to show relative sizes of categories of the numbers. An advantage of the stem and leaf plot is that most of the data is still contained in the graph. If the data consists of two digit numbers, the first digit (tens place) is usually the stem and the units are the leaves. If the data has three digits, we generally round to the nearest 10 and use the hundreds digit for the stem and the tens digits for the leaves. These graphs reveal groupings or **clusters** of data, as well as large **gaps** between data. Scores that are separated from the others by gaps are called **outliers.**

Histograms

Histograms display grouped data. In this type of graph, data are grouped into intervals called **measurement classes** (or *bins*), and the number of data points in each measurement class is called the **frequency** of the interval. This information is collected into **a frequency table**, from which the histogram is plotted. Histograms show the general picture and suppress specifics of the data. A dot plot can be viewed as a histogram where the measurement classes are intervals of length one. It can be tricky to choose the best intervals to use. If the intervals are too small, it can be hard to tell the general pattern easily, and there can be a large number of measurement classes. When the interval is too large, much of the information is obscured.

Bar Graphs

Bar graphs display trends and amounts as lengths. Histograms are a special case of these types of graphs. In bar graphs, the length of the bars is used to represent frequencies or quantities. The lengths of the bars (either vertical or horizontal) provide a visual summary of the data. Such charts can also reveal trends over time.

Line Graphs

Another type of graph that is often used to plot data over time is called a **line graph**. A line graph displays trends and variation. The horizontal axis of the line graph should be in equal units (or periods) of time. If data is missing from a specific period of time, an assumption is usually made that the missing value is the average of the ones before and after it, and the line is draw between the points representing those values. To show variability rather just long term trends, shorter time periods are best.

Pie Charts

Pie charts display relative proportions for related quantities. The quantities should represent all the relevant data. A pie chart is constructed by finding the portion of a circle each quantity should represent. The combined proportions need to add up to 100% (with some allowance for rounding error). Pie charts provide a way to show proportions nearly free of distortion.

· ·

Suggestions and Comments for Odd-numbered problems

3. Since the leaves are to be in hundreds, the stems should be in thousands.

13. If you have trouble reading the information in graphs when horizontal or vertical level lines are not provide in the graph, use a ruler or other straight edge (a folded piece of paper) that is placed parallel to an axis of the graph as an aid in reading correctly from the scale.

15. A graph is a representation of the data, but it cannot be expected to show exact values such as 7.82; especially when the top of the scale will be beyond 32 and the vertical scale will probably be in increments of 2 or 5. Histograms are an exception in that the vertical scale represents frequencies (counts) and there are no fractions. The heights of the bars should be relatively close to the correct values, but the most important considerations are that they show the relative comparative sizes. The height of the bar for 7.82 should be about half way between 5 and 10 and a little less than one-fourth the height of the bar representing 32.44. Similar considerations apply to other types of graphs.

15. and beyond
When constructing graphs with vertical scales, the top of the scale should go beyond the largest value in the data, but the scale need not include the full range of *possible* values for the data. In problem 19, for example, the largest *possible* value is 100%, but the largest *actual* value is 54%. The top of the scale could reasonably be 60%. However, having a vertical scale that goes to 100% is not wrong, but it may give a different impression of the data. More on this is covered in Section 2.3. The point is: You have choices.

25. Making calculations (such as percentage of change) based on information in a graph can be rather tricky. First, you have to read the values carefully and you have to assume that the graph has been prepared correctly and the data is presented fairly. Published materials in books, magazines, television, and newspapers are filled with graphs and charts; many of which are of poor quality with respect to representing the data even though they are "slick" graphics. A reminder:

$$\text{Percentage change from A to B} = \left(\frac{B - A}{A} - 1\right) \times 100\%$$

If B is bigger than A, there is an increase (+); otherwise, a decrease (-).

33. through 37.
In making a pie chart (a circle graph), the area of a sector of the circle is proportional to the value or percentage it represents. This is equivalent to having the central angle of the sector being proportional to the amount or percentage. For example, if the amount involved 16 out of a total of 80, then the central angle should be

$$\frac{16}{80} \times 360° = 72°$$

Because there are limitations to drawing the pie chart accurately (even with a computer) all values can be rounded somewhat before being used. The total of the percentages might not add up to exactly 100% or the degrees might not add to 360, but they should be close, differing by 1 or 2 at the most. When it comes to actually drawing the sectors, however, there is another problem: how can you get the right sized pieces in the picture?

Making pie charts (continued)

First, make a rough sketch to see that the arrangement of sectors has a good look. If they are small sectors, they might be drawn in early so they don't have to be squeezed in at the end. If the larger sectors are off a little, the effect won't be as noticeable. Use the following chart as a rough guide for the degrees.

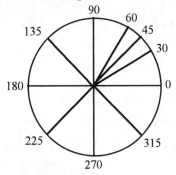

Solutions to Odd-numbered Problems

1. (a)

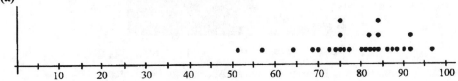

(b)

```
5 | 1 7
6 | 4 8
7 | 0 3 4 5 5 5 6 7
8 | 0 1 2 2 3 4 4 4 6 7 8
9 | 0 2 2 7
```

3. (a)

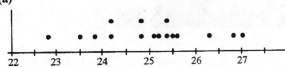

(b) Salaries (in hundreds)

```
22 | 8
23 | 5 8
24 | 2 2 8 8
25 | 1 2 4 4 5 6
26 | 3 8
27 | 0
```

5. Batting Champion Averages
   1975-1998

```
.31
.32 8
.33 2 3 3 3 6 9 9
.34 1 3 3 7
.35 6 7 8 9 9
.36 1 3 3 6 8
.37
.38 8
.39 0
```

Most of the averages are between .332 and .368.

7. Midterm Chemistry Score

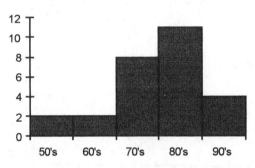

9. Starting Salaries for Recent Accounting Graduates

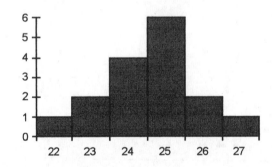

11. (a) Health Scores for the States   **(b)**

```
 8 | 4 5 7 7 8 8
 9 | 0 1 3 3 3 3 4 6 6 6 6 7 8 9 9
10 | 0 0 0 0 1 1 2 3 4 4 5 6 7 8 9
11 | 0 0 0 3 3 4 4 4 6 6 8 9
12 | 0 0
```

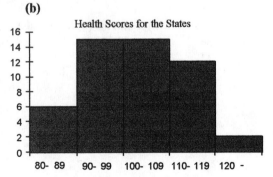

Health Scores for the States

13. (a)

Score	Frequency
3	2
4	5
5	6
6	4
7	4
8	5
9	3
10	1

(b) 30 students took the quiz.

15. Weekly Allowances for Teenagers - 1989

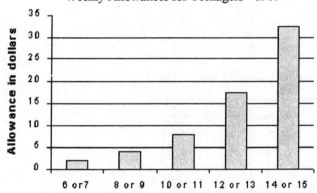

17. Attendance at Major Musical and Sporting Events
Attendance in millions

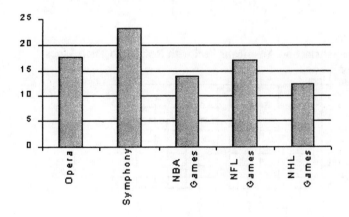

19.   Problems Related to Drugs in the Workplace

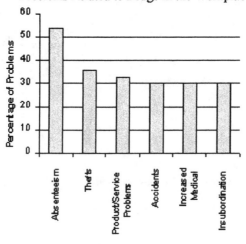

21.   Women and Low-Impact Fitness Activities

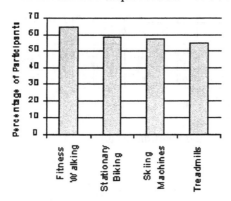

23.   The Satisfaction Americans Feel with Respect to Their Lives

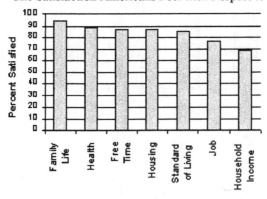

25. **(a)** Estimated populations used for answers:
1790 - 5 million,
1890 - 65 million,
1990 - 255 million

(b) 60 million   65 million - 5 million

(c) 190 million   255 million - 65 million

(d) 1200% $\dfrac{65 - 5}{5} \times 100\%$

(e) 292%. $\dfrac{255 - 65}{65} \times 100\%$

27. Note: Values are plotted between the vertical lines.

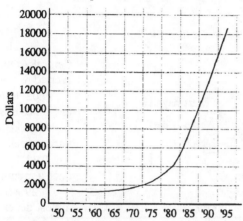

The United States Federal Debt on Per Capita basis

29. **(a)** Note: Values are plotted between the vertical lines.

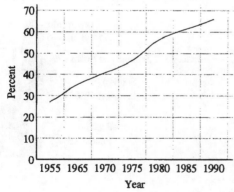

Women with Children Under 18 Participating in the Labor Force

(b) The percent increases from 1965 to 1975, and all fall on the line connecting the percents for 1965 and 1975.

31. Note: Values are plotted between the vertical lines.

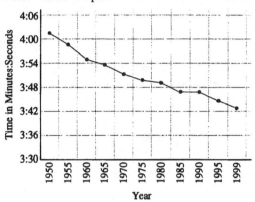

World Record Times for the Mile Run

33.

The Frequency of High School Students and Drinking

Central Angles:
every day
$$0.02 \times 360° = 7.2°$$
few times a week
$$0.12 \times 360° = 43.2°$$
once a week
$$0.31 \times 360° = 111.6°$$
once a month
$$0.29 \times 360° = 104.4°$$
almost never
$$0.26 \times 360° = 93.6°$$

35.

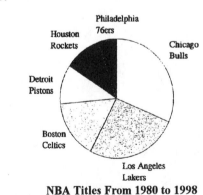

NBA Titles From 1980 to 1998

37.

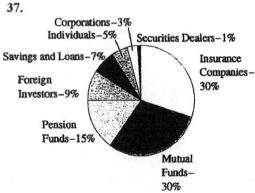

Owners of Junk Bonds

Section 2.2   Comparisons

Goals
1. Graph two or more sets of data using comparison graphs.

Key Ideas and Questions
1. How can the basic graphs be used to compare different, but related, data sets?

Vocabulary

Double Stem and Leaf Plot	Multiple Bar Graphs	Multiple Pie Charts
Comparison Histograms	Multiple Line Graphs	Proportional Bar Chart

· ·

Overview

We can use charts and graphs to make comparisons between different but related sets of data. Good visual presentation can show how sets of data are similar as well as how they are different, and can often help explain these similarities or differences.

Double Stem and Leaf Plot

Two data sets can be combined into the same plot in a **double stem and leaf plot.** The stem is placed in the middle and the two sets of leaves are placed to either side.

Comparison Histograms

Two data sets can be presented in two **comparison histograms** by putting both histograms on one graph and shading, coloring, or otherwise marking them differently so the comparison is visible.

Multiple Bar Graphs

Comparison bar graphs are also used to show relative strengths. When comparing two (or more) data sets on one graph, it is called a **double** (or multiple) **bar graph.**

Double and Multiple Line Graphs

Line graphs are used to show comparisons together with trends, much as bar graphs are. Often, the graph has two vertical scales enabling the visual presentation to show more information, such as amounts on one scale and percentage change on the other.

Multiple Pie Charts

One way to present trends in proportions is to give a series of pie charts.

Proportional Bar Graphs

Bar graphs can also be used to display relative amounts and trends in the same visual presentation. A **proportional bar graph** can be used. Each bar is the same height, and it corresponds to 100% of the total. Each bar is then divided into appropriate percentages.

· ·

Suggestions and Comments for Odd-numbered problems

General Comments:

When making comparison graphs, you sometimes have a choice as to which set of data comes first in the order, such as when you make a multiple bar graph for data that covers the same period of years or falls in the same categories or bins. Changing the order in which the bars are drawn may give a different perspective to the data. In other situations, the order may be dictated, but you still have choices such as the horizontal and vertical scales. If you are unsure as to how certain types of graphs are drawn, review the examples and problems from Section 2.1.

Solutions to Odd-numbered Problems

1. Grades on Sociology Midterm

Class 1		Class 2
	5	4
643	6	6
76664433	7	033445678
766554440	8	00223344556688
5221	9	46

Both classes have most scores in the 70s and 80s. The scores in Class 1 are also slightly more spread out than those in Class 2, with as many scores below 70 as Class 2, but more scores above 90.

3. Home Run Hitters for the Yankees

Babe Ruth		Mickey Mantle
	1	3589
52	2	12337
54	3	01457
9766611	4	02
944	5	24
0	6	

Babe Ruth consistently hit more home runs than Mickey Mantle. In only 4 out of 18 years did Mantle hit more than 40 home runs while Ruth hit less than 40 home runs in only 4 of his 15 years with the Yankees.

5.

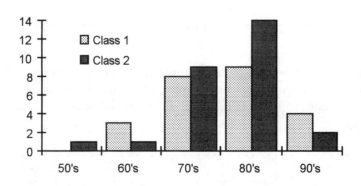

5. Alternate: Notice the difference if we reverse the order of the columns.

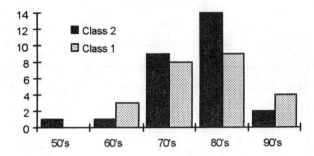

7.

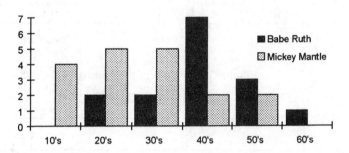

7. Alternate: Notice the difference if we reverse the order of the columns.

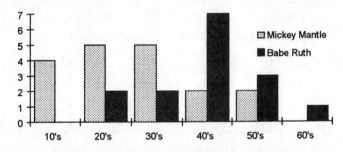

9. Comparison of Home Occupants with respect to Features

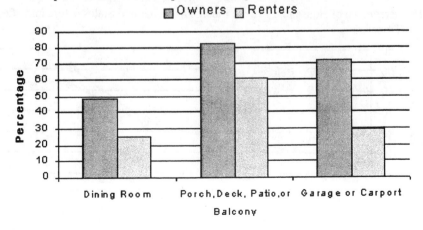

11. Changing Patterns in Meat Consumption (per person) in the U. S.

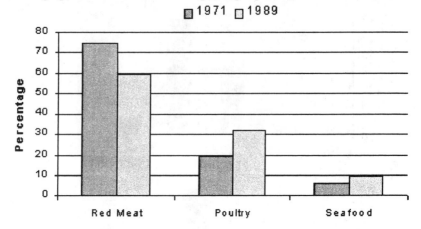

13.   Employed Workers (in thousands) 1975-1995

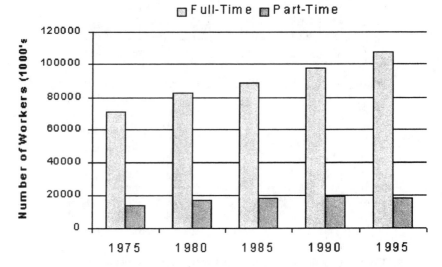

15. Percentage of Household Private Income in the United States before and after Taxes and Transfers

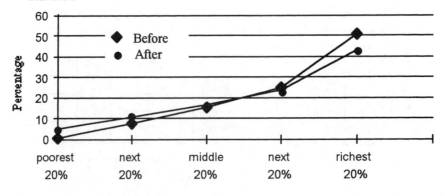

17. The effect of modifying the vertical axis:

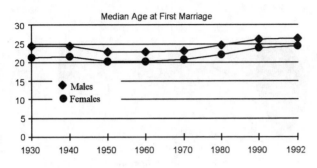

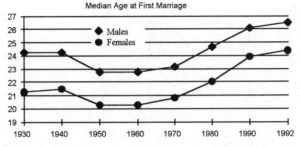

19.

21. Fertilizer consumption is much higher in the industrial world than in the non-industrialized regions. Asia, which is a mixture of both, is about in the middle. In North America the use of fertilizers is down slightly while in Europe and the USSR it is up.

23. (a) In 1955, an average of about 0.55 acres per person was devoted to grain production. By 1985 this average had dropped to about 0.35 acres.

    (b) The average amount of fertilizer used per person rose from about 13 pounds in 1955 to about 55 pounds in 1985.

    (c) The trend is a reflection of the fact that there is only a limited amount of arable land mass. In order to feed everyone, agricultural production has had to be increased through use of fertilizers.

25.

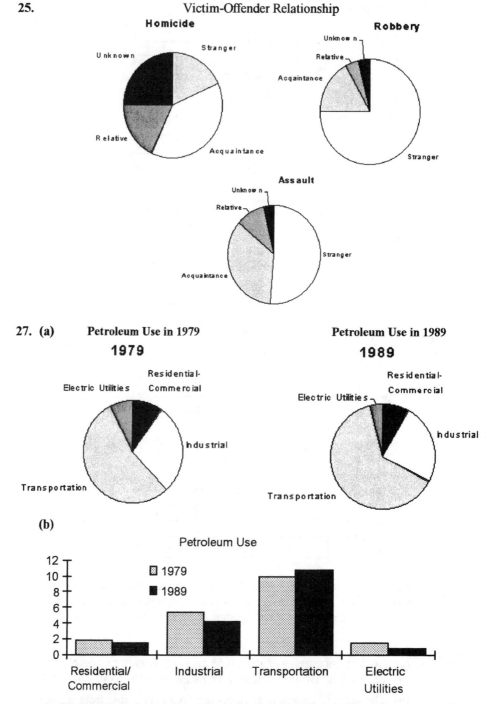

Victim-Offender Relationship

27. (a) Petroleum Use in 1979   Petroleum Use in 1989

(b)

(c) The graphs in (a) are better at showing proportions within a given year while the graphs in (b) are better at comparing the two years.

29. (a) People Living Alone

(b) People Living Alone

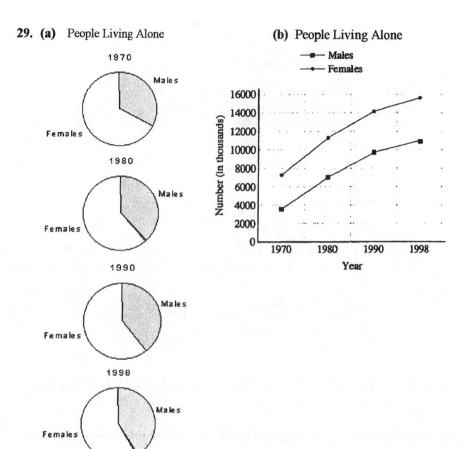

(c) The line graph shows trends more clearly, whereas pie charts show the gender shift more clearly.

31. (a)   **(b)**

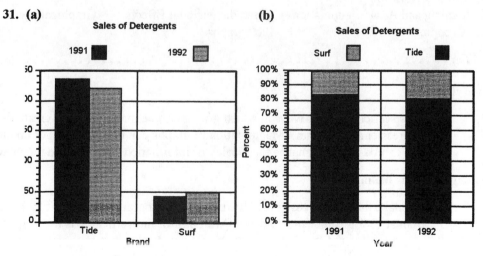

(c) The multiple bar graph allows for a clearer picture of the trends.

33. The Relative Changes in Costs of Selected Items

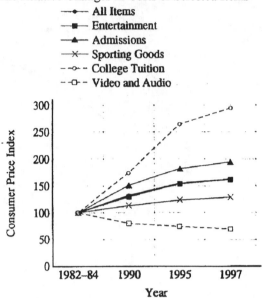

Section 2.3 Enhancement, Distraction and Distortion

Goals

1. Identify graphs that have been distorted and explain how to correct the distortions.

Key Ideas and Questions

1. Describe four ways that graphs may be altered to influence the perception of the viewer.

Vocabulary/Notation

Scaling and Axis Manipulation   Three-dimensional Effects   Graphical Maps
Cropping   Pictographs

. . . .

Overview

This section explains ways in which graphs can be manipulated to create different impressions of the data. Manipulation of graphs can be intentional or unintentional and can be used to enhance the display of information or to mislead the viewer.

Scaling and Axis Manipulation

To make the differences among the bars of a histogram or bar chart look more dramatic, you can display the chart with part of the vertical axis missing. You can intentionally omit the scale on the vertical axis, or begin it at one number or another that might make the view of the data more or less advantageous. You can also distort the nature of the data by reversing the axes or reversing the orientation of an axes.

Line Graphs and Cropping

Manipulating the scale of your visualization of the data includes a technique called **cropping**. Cropping refers to the choice of a "window" through which to frame and view the information. You can change this "window" by changing the scale of the vertical axis and using different increments. Graphs that crop the lower portion of the vertical scale emphasize the changes in the values. Graphs that rise above the edge of the vertical scale, or go to the edge of the scale, make the trend more dramatic.

Three-Dimensional Effects

Increased availability of computer graphing software makes it easier for anyone to create more attractive graphs. However, 3-D effects can obscure the true picture of the data. Pie charts can also be manipulated by making them 3-D, and rotating and exploding the image.

Pictographs

Graphs are embellished with pictures in many different ways to make them more interesting. Such enhancements can be confusing and some are designed to deceive.

Pie charts can also be misused. Distortions occur in not labeling percentages, having percentages that do not add to 100% or over-emphasis of one sector.

Graphical Maps

Maps can be used to summarize or to show patterns related to national, regional, or world concerns.

. . . .

Suggestions and Comments for Odd-numbered problems

5. Make four bins for the bar graph, one for each year. The years on the line graph should be evenly spaced. The center of the bin is considered to be the placement of the date when you compare your graphs. That is, from the center of the first bin in the bar graph to the center of the last bin can be used to indicate the period of time from 1950 to 1992.

7. Two different changes can be made to emphasize the changes in the tax burden and make the increases seem more dramatic. First, the scale can be started higher than zero. Second, the horizontal scale can be shortened.

11. and 13.
The ovals represent the amounts in the pension fund, but distort the values because the lengths of the ovals corresponds to the amounts in the fund. When you construct the bar charts, make sure the widths are the same so that the heights correctly represent the amounts. The ovals are used as pie charts. The oval are also deceptive in another way. Compare the 18% as represented in the oval with the 18% you construct in the graphs.

19. In order to make another version of the graph, you need to read values from the graph that is given. This is very hard to do with any accuracy because the graph is poorly done with little help in terms of vertical and horizontal reference lines. You should be able to approximate values within $10, however, and get a reasonable facsimile of the graph. Remember, we are mainly interested in the effect the change of scale produces.

21. When circles are used to represent quantities, the areas of the circles should be in the same ratio as the amounts they represent. Circles are two-dimensional, and the area of a circle is given by $A = \pi r^2$. Similarly, when spheres are being used the volumes of the spheres should be in the same ratio as the amounts they represent.

For a sphere, $V = \dfrac{4}{3} \pi r^3$.

Solutions to Odd-numbered Problems

1. Note: Values are plotted between the vertical lines.

Year

World Record Times for the Mile Run

(b) The change in the vertical scale makes the downward trend more apparent.

3. (a) and **(b)** When a bar graph represents a quantity for a given year, the center of the bar graph generally coincides with the place-holder for the date. The lower graph has been placed so that the markers on the line graph for 1950 and 1992 coincide with the centers of the corresponding bars.

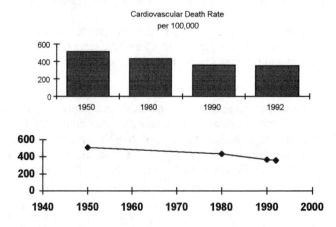

5.

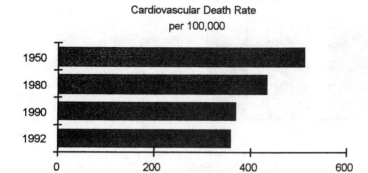

7. The Federal Tax Burden per Capita

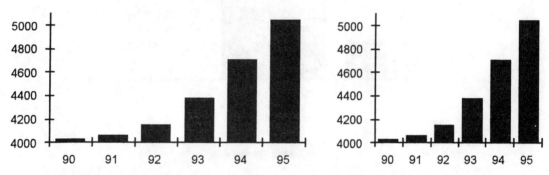

Two different changes can be made to emphasize the changes in the tax burden and make the increases seem more dramatic.

9. Annual Increase in the Atmosphere of Major Greenhouse Gases, 1957-1987

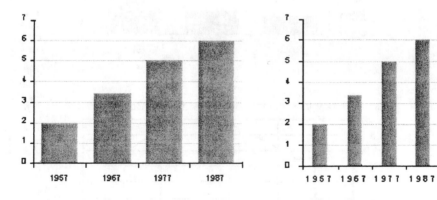

11.

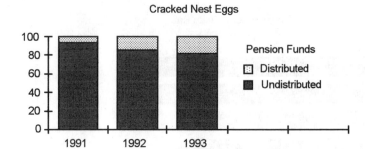

13.

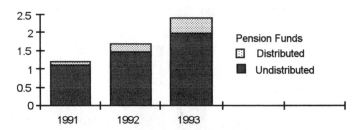

15.

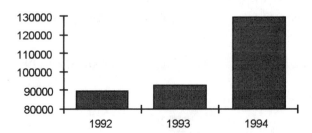

17.

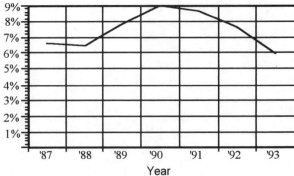

19.

21. (a) The area of the circles should be in the same ratio as the amounts they represent. That is, the area for the circle representing company B should be twice the area of the circle representing company A because its revenue is twice as much. Because the graphs are two-dimensional, the areas of the circles are proportional to the squares of their radii. In a circle, Area $= \pi r^2$. If the radius for the circle for company A is 1 inch, then the radius of the circle for company B should be $\sqrt{2}$, or about 1.4 in. For company A, if $r = 1$, then $A = \pi$. For company B, if $r = \sqrt{2}$, then $A = 2\pi$. $1^2 : (\sqrt{2})^2 = 1 : 2 = 5{,}000{,}000 : 10{,}000{,}000$

(b) The volumes of the two spheres should be in the same ratio as the amounts they represent. The volume of a sphere is given by $V = \dfrac{4}{3}\pi r^3$. Because the graphs are three-dimensional, the amounts represented by the spheres vary by the cube of the radius. If we want to show a volume that is twice as large as a sphere with a radius of 1, we need a sphere with a radius of $\sqrt[3]{2}$, or about 1.26. $1^3 : (\sqrt[3]{2})^3 = 1 : 2$

23.

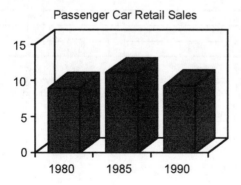

25.

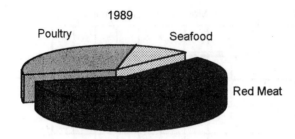

1989

27. Golf Balls: Combined Yardage for Driver, 5-iron, and 9-iron

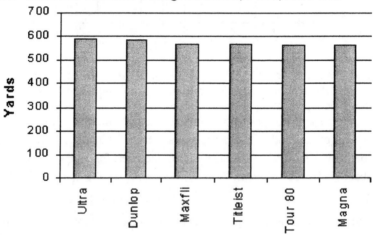

29. Smog Levels Above Standards in Selected Canadian and U.S. Cities

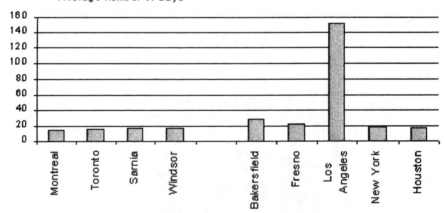

Section 2.4   Scatterplots

Goals
1. Learn to graphically represent and analyze the relationship between two variables through the use of a scatterplot.

Key Ideas and Questions
1. How does a scatterplot with a strong correlation allow you to make confident predictions?
2. Does a strong correlation between two variables allow you to infer that a change in one variable is the cause of a change in the other?

Vocabulary/Notation

Scatterplot	No Correlation	Strong and Weak Correlation
Outlier	Positive Correlation	Correlation Coefficient
Regression Line	Negative Correlation	

· ·

Overview

Scatterplots provide a visual display of the relationship between two variables. Data is often grouped into pairs of numbers that may have a relation to one another. Such pairs of numbers can be plotted as points on a portion of the (x, y)-plane to form what is called a **scatterplot**. We use the scatterplot to investigate the possible relationship between the data. With some data, there doesn't appear to be any relationship. In other cases, it often happens that you can see a pattern, and many times the data points appear to lie approximately on a line. These linear-type relationships will be emphasized.

You may recall that data points separated from others by gaps are called **outliers**. When outliers are ignored, the other data points may show a relationship where they appear to lie almost on a line. A line sketched roughly through the points reveals this relationship. This revealed relationship is called a **correlation**. The specific line that fits the data the best is called a **regression line**.

The data may or may not fit the regression line well. The fit of the data to the regression line is measured by the **correlation coefficient**. The correlation coefficient is a number between +1 and -1. When the correlation coefficient is +1, the points fit the line perfectly; as one variable increases, the other also increases proportionately. When the correlation coefficient is -1, the points also fit the line exactly, but as one variable increases the other decreases. A correlation coefficient of 0 means that the two variables are unrelated to each other. Even pairs of variables with a correlation coefficient or 0.5 or -0.5 will not show much of a linear relationship,

Instead of having to compute complex correlation coefficients to check the fit of data, we will use intuitive descriptions of the fit. These terms are illustrated next.

Correlation: Positive or negative; strong or weak

Strong Positive Correlation   **Weak Positive Correlation**   **No Correlation**

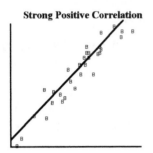   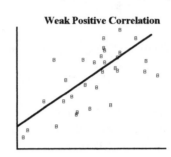   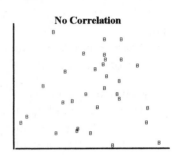

Weak Negative Correlation   **Strong Negative Correlation**

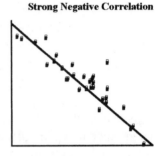

Correlation and Causation

Remember that "correlation is not causation." Two quantities can be highly correlated without one quantity being the cause of the other. There may be an underlying cause that affects both of the quantities, or it may simply be a statistical oddity.

The Regression Line

Since the regression line is used to make predictions, it is most useful to have a formula or equation to work with. The equation of the regression line is based on two basic principles:

1. It should pass through a point which represents the average value of x and the average value of y. This point is represented by $(\bar{x}, \bar{y})$.

2. It should be the "best fit" in the sense that the total of the distances between the points of the scatterplot and the regression line are minimized.

The equation of the regression line is $y - \bar{y} = m(x - \bar{x})$, where m is the slope of the regression line, and $m = \dfrac{n\Sigma xy - \Sigma x \Sigma y}{n\Sigma x^2 - (\Sigma x)^2}$. "$\Sigma$" is the "summation" symbol.

• •

Suggestions and Comments for Odd-numbered problems

General Comments

Either use graph paper or a carefully drawn grid for all your work. Good analysis or answers are dependent on a good graph. "Strong" and "weak" are relative terms. You might think in terms such as "tight" or "loose" when considering the way the points cluster around a line drawn through the middle of the scatter of points. Think of drawing an oval around all, or almost all, of the points. As the shape of the oval goes from long and narrow to circular, the correlation goes from very strong to no correlation at all. In a positive correlation, the variables increase or decrease together. In a negative correlation, one variable increases while the other one decreases.

9. through 15.

Making a reasonable prediction from a scatterplot depends on several factors, beginning with a carefully constructed scatterplot. Next comes the placement of the regression line; it should go through the "center" of the plotted points so that the points cluster about the line as closely as possible with nearly equal numbers of points on either side of the line. Finally, when making predictions, use a straight edge (ruler or folded paper) to get vertical and horizontal lines from the axes to the regression line.

25. and 27.

The regression line can be written in the form $y = mx + b$; where m is the slope, and b is the y-intercept (where the line crosses the y-axis). For a positive correlation, m is positive; and for a negative correlation, it is negative. Whether m is small or large tells us nothing about the strength of the correlation.

29. and 31.

The strength of the correlation is measured by the *correlation coefficient*.
The value of r can be either positive or negative, and will be between -1 and +1. A positive correlation will have a positive value for r. As the value of r gets closer to either -1 or +1, it indicates a stronger correlation. For strong correlations, the *absolute value* of r is generally greater than 0.75.

Solutions to Odd-numbered Problems

1. **(i)**

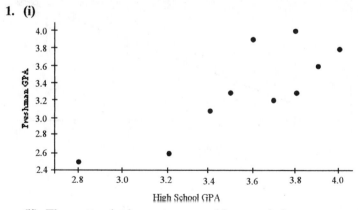

(ii) The scatterplot has a strong positive correlation, as opposed to a weak positive correlation. In general, as high school GPA increases, so does freshman GPA.

(iii) There are no outliers.

3. (i)

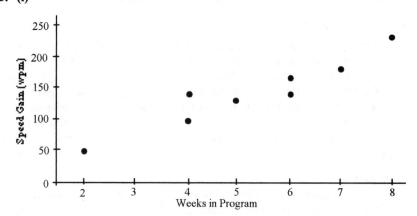

(ii)   The scatterplot has a strong positive correlation.
(iii) There are no outliers.

5. The scatterplot has a (fairly) strong negative correlation.

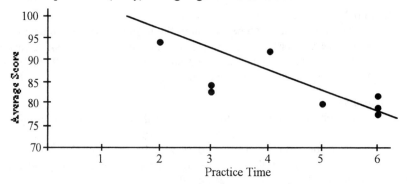

7. The scatterplot has a strong positive correlation.

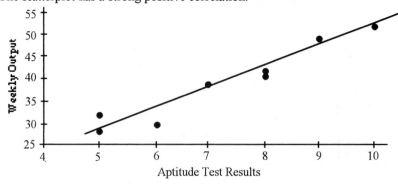

9.

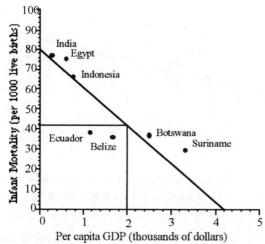

Based on the strong correlation and the regression line, we would predict that a country with a per capita gross domestic product of $2000 would have an infant mortality rate of 42 deaths per 1000 live births.

11. (a) Approximately 81 or 82   **(b)** Approximately 87

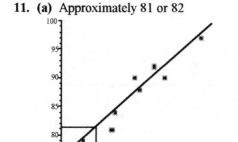 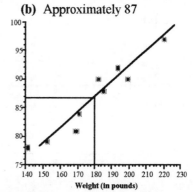

Answers will vary depending on where you estimate the regression line.

13. (a) and (b)

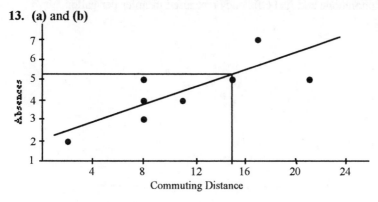

(c) We would estimate 5 or 6 absences per year for a worker who commutes 15 miles. Answers will vary depending on your scatterplot and where you draw the regression line.

15. (a) and (b)

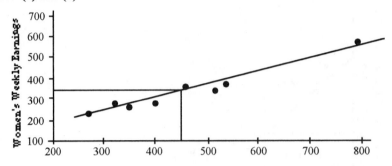

(c) We would predict a mean weekly salary of $350 for a woman if the mean weekly salary for a man was $450. Answers will vary depending on your scatterplot and where you draw the regression line.

17. The scatterplot shows a strong positive correlation. All the points can be enclosed in a long narrow oval.

19. The scatterplot shows no correlation between the variables. Any oval that contains the data points would be nearly circular in shape.

21. (a)

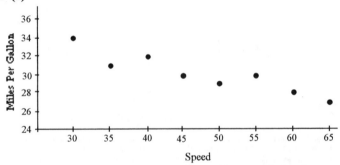

(b) The scatterplot shows a strong negative correlation between the speed of the automobile and fuel efficiency measured in miles per gallon.

23. (a)

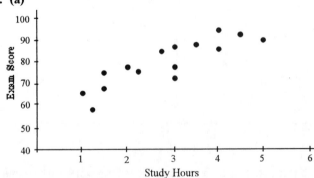

(b) The scatterplot shows a (fairly) strong positive correlation between the time spent studying and the exam score.

25. First, we set up a table to prepare for the necessary calculations.

x	y	x^2	xy
2.8	2.5	7.84	7.00
3.2	2.6	10.24	8.32
3.4	3.1	11.56	10.54
3.7	3.2	13.69	11.84
3.5	3.3	12.25	11.55
3.8	3.3	14.44	12.54
3.9	3.6	15.21	14.04
4.0	3.8	16.00	15.20
3.6	3.9	12.96	14.04
3.8	4.0	14.44	15.20
Sums: 35.7	33.3	128.63	120.27

The equation for the regression line is $y - \overline{y} = m(x - \overline{x})$, where

$$m = \frac{n\sum xy - \sum x \sum y}{n\sum x^2 - (\sum x)^2}$$

$\overline{y} = \frac{33.3}{10}$, $\overline{x} = \frac{35.7}{10}$, and $m = \frac{10(120.27) - (35.7)(33.3)}{10(128.63) - (35.7)^2} = 1.17612$

y - 3.33 = 1.18(x - 3.57)

y - 33.3 = 1.18x - 4.21

y = 1.18x - 0.87.

27. First, we set up a table to prepare for the necessary calculations.

x	y	x^2	xy
6	79	36	474
3	83	9	249
4	92	16	368
6	78	36	468
3	84	9	252
2	94	4	188
5	80	25	400
6	82	36	492
Sums: 35	672	171	2891

The equation for the regression line is $y - \overline{y} = m(x - \overline{x})$, where

$$m = \frac{n\sum xy - \sum x \sum y}{n\sum x^2 - (\sum x)^2}$$

$\overline{y} = \frac{672}{8}$, $\overline{x} = \frac{35}{8}$, and $m = \frac{8(2891) - (35)(672)}{8(171) - (35)^2} = -2.74126$

y - 84 = -2.74(x - 4.38)

y - 84 = -2.74x + 12   (12.0012 rounded)

y = -2.74x + 96.

29. First, we set up a table to prepare for the necessary calculations.

x	y	x^2	y^2	xy
2.8	2.5	7.84	6.25	7.00
3.2	2.6	10.24	6.76	8.32
3.4	3.1	11.56	9.61	10.54
3.7	3.2	13.69	10.24	11.84
3.5	3.3	12.25	10.89	11.55
3.8	3.3	14.44	10.89	12.54
3.9	3.6	15.21	12.96	14.04
4.0	3.8	16.00	14.44	15.20
3.6	3.9	12.96	15.21	14.04
3.8	4.0	14.44	16.00	15.20
Sums: 35.7	33.3	128.63	113.25	120.27

$$r = \frac{n\sum xy - (\sum x)(\sum y)}{\sqrt{(n\sum x^2 - (\sum x)^2)(n\sum y^2 - (\sum y)^2)}}$$

$$r = \frac{10(120.27) - (35.7)(33.3)}{\sqrt{(10(128.63) - (35.7)^2)(10(113.25) - (33.3)^2)}} = 0.83182$$

$r = 0.83$ This indicates a strong positive correlation.

31. First, we set up a table to prepare for the necessary calculations.

x	y	x^2	y^2	xy
6	79	36	6241	474
3	83	9	6889	249
4	92	16	8464	368
6	78	36	6084	468
3	84	9	7056	252
2	94	4	8836	188
5	80	25	6400	400
6	82	36	6724	492
Sums: 35	672	171	56694	2891

$$r = \frac{n\sum xy - (\sum x)(\sum y)}{\sqrt{(n\sum x^2 - (\sum x)^2)(n\sum y^2 - (\sum y)^2)}}$$

$$r = \frac{8(2891) - (35)(672)}{\sqrt{(8(171) - (35)^2)(8(56694) - (672)^2)}} = -0.73893$$

$r = -0.74$ This indicates a (fairly) strong negative correlation.

Chapter 2 Review Problems

Solutions to Odd-numbered Problems

1. **(a)** Rounded to the nearest 100, the values are:

$28,500	$26,100	$27,100	$24,900
$26,900	$28,200	$27,700	$25,800
$27,800	$27,500	$26,500	$25,900
$26,500	$28,100	$26,700	$27,200

(b) The dot plot:

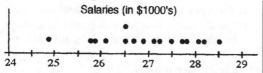

(c) The stem and leaf plot with stems as 1000's and leaves as 100's:

```
28 | 1   2   5
27 | 1   2   5   7   8
26 | 1   5   5   7   9
25 | 8   9
24 | 9
```

3.

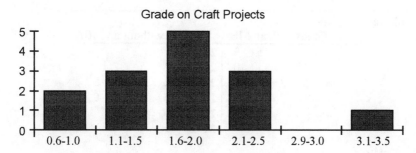

5. The table values are found by multiplying the height (as read on the vertical scale) by
   $100,000. The sales for 1988 are 20 ∞ $100,000 = $2,000,000.

Year	Sales
1988	2,000,000
1989	2,500,000
1990	3,500,000
1991	3,000,000
1992	2,000,000
1993	2,500,000
1994	1,500,000

The largest sales of $3,500,000 were in 1990

7. The percentage change is found by dividing the change in sales by the sales in the first of two years.

Years	Percentage Change	
1989 - 1990	+133%	$\dfrac{3500000 - 1500000}{1500000} = \dfrac{2000000}{1500000} \approx +1.33$
1990 - 1991	- 14.3%	$\dfrac{2500000 - 3500000}{3500000} = \dfrac{-1000000}{3500000} = -0.2857$

From 1989 to 1990 there was a 133% increase.
From 1990 to 1991 there was a 29.6% decrease.

9. A stem and leaf plot gives both complete information about the data and a graphical picture of it. The dot plot gives you good information about the values, and only a hint of what the graph will look like. The histogram is a good graphical picture of the distribution of the values, but the details about the values are lost.

11. The best choice would be a multiple bar graph.
For showing the amounts, the graph type of first choice should be a version of a bar graph. A proportional bar graph shows the proportions but not the amounts, so a multiple bar graph is preferred. A multiple bar graph allows for easy comparison of sales on a year-by-year basis, since the bars representing different types of appliances will be side-by-side. With adequate shading or other "coding" of the bars, sales trends can also be followed.

A multiple line graph could be used if you wanted to follow the trends for each appliance more easily, but this would definitely be the second choice.

13. **(a)**

Grads with an MBA							Grads without an MBA				
			5	3	**29**					
		1	1	0	**28**	1	2	5		
9	8	6	3	3	0	**27**	1	2	5	7	8
			7	5	4	**26**	1	5	5	7	9
				8	5	**25**	8	9		
					24	9				

(b)

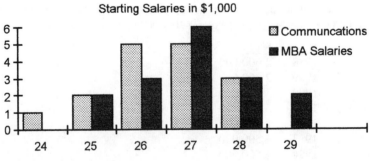

Continued on next page

13. (continued)

Notice the difference if the order of the columns is reversed.

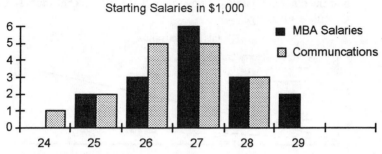

Starting Salaries in $1,000

■ MBA Salaries

▨ Communcations

(c) We would conclude that graduates with MBAs generally receive higher starting salaries than those without MBAs.

15. (a)

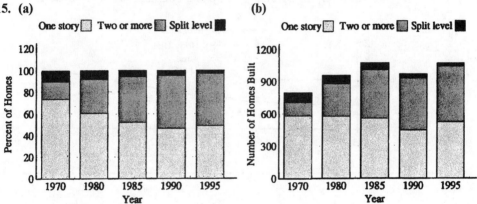

One story ▨ Two or more ▨ Split level ■

(b) One story ▨ Two or more ▨ Split level ■

17. (a)

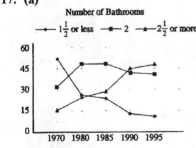

Number of Bathrooms

— $1\frac{1}{2}$ or less — 2 — $2\frac{1}{2}$ or more

(b)

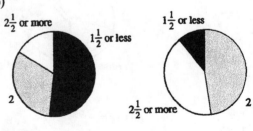

(c) (1)

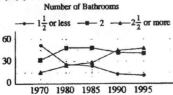

Number of Bathrooms

— $1\frac{1}{2}$ or less — 2 — $2\frac{1}{2}$ or more

(2)

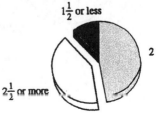

19. (a) Since Up and Coming has the sales advantage, cropping the vertical scale visually minimizes the actual totals and exaggerates the differences between Up and Coming and the other companies. Showing only the most recent years is also to Up and Coming's advantage.

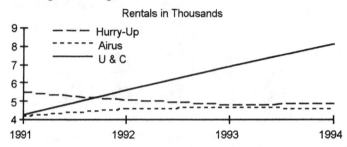

(b) To de-emphasize the differences in values, the graph should be long and low. However, the advantage of Up and Coming can't be eliminated.

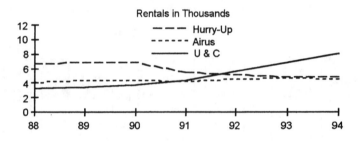

21.

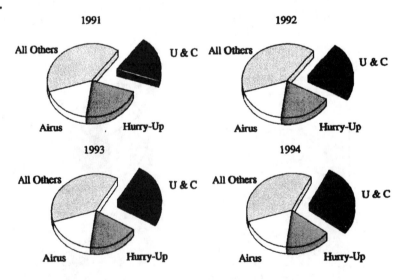

3 Collecting and Interpreting Data

Section 3.1   Populations, Samples and Data

Goals
1. Identify biased and unbiased samples.
2. Learn to select simple random samples.

Key Ideas and Questions
1. How can you sample so that there is no bias?
2. In what ways might bias be present in other types of sampling?

Vocabulary

Population	Quantitative Data	Bias
Variable	Qualitative Data	Simple Random Sample
Measurement	Ordinal Data	Random Number Table
Sample	Nominal Data	Random Number
	Representative Sample	Generator

· ·

Overview

One of the ways we become informed about a specific group or other subject (taxpayers, senior citizens, traffic accidents) is through the use of statistics. Statistical methods are used to gather information and help decision-makers draw conclusions about specific groups of people or other topics. Instead of analyzing the whole group in question--which is called the **population**--it is more common to choose a portion or **sample** from the population and analyze the sample instead. A carefully and properly chosen sample can be representative of the population. Results such as means or percentages obtained from the sample can be used as estimates for the values in the population.

Bias

Bias is a flaw in the sampling procedure that makes it more likely that the sample is not representative of the population. There are many different types of bias. Some are unintentional, but some are intended. Some are systematic. That is, there is a "structure" that is not taken into account, much like a deck of cards that hasn't been shuffled, the deck is "stacked". Late night call-in show surveys, for example, don't reflect the views of more than a small minority of the population - unless the audience after 11:00 p.m. is representative of the whole population, which is highly unlikely.

Simple Random Samples

A **simple random sample** is any sample chosen in such a way that all samples of the same size have the same chance of being chosen. A simple random sample is the only sample of a fixed size that has no bias. When you choose a sample one element at a time, then all unselected elements should have the same chance of being chosen at any step in the process.

A **random number generator** is a computer or calculator program designed to produce numbers that are as random as possible. This means that there is no apparent pattern to the numbers. A **random number table** is a table produced by a random number generator.

. . . .

Suggestions and Comments for Odd-numbered problems

General Comments

Sampling is generally done to answer a question. The population in a sample is the set of people, places, or things to which the answer will be applied.

For example, suppose the question is asked "What do registered voters in the United States think is the most important problem facing the country (out of a list of six identified problems)?" and a poll (a sample) is taken in *your* home state. The population being studied consists of only registered voters in the United States, not everyone in the United States. The fact that only people in your home state were included in the sample introduces a bias. In actuality, the results only apply to them.

Using the Random Number Table

Most of the problems involving the taking of a simple random sample don't allow you to make a random response. In general, the directions will tell you to start on a given line using specific digits from a specific column. The answers provided only have relevance if you follow the directions. When you start in a different position or follow a different pattern, you end up with a different sample. The directions for assigning digits to members of the population also need to be followed. The conventions are these:

For populations up to 100 in size assign 00 to 99. This is a little inconvenient some times, and in practice you could use another numbering scheme.

For independent sampling, you should begin with 01 for up to 99 and begin with 001 for populations up to 1000. In independent sampling, the population has generally been arranged in some order, and there is no "00" position.

Using a table of random numbers is easier if you have one you can write on and mark up. Make several photo copies of the table ahead of time.

Random Samples

When using the table of random numbers for simple random samples, each member of the population is assigned a number and the table is "searched" until the right size of sample has been chosen. The occurrence of an assigned number during the search means the member is selected.

1. through 7.

The "population" is the set of objects (people or cars, for example) in which we are interested. The "sample" is the actual subset from which we get the data.

13. through 19.

Identifying bias in a process is sometimes very difficult to detect, even when the results can have significant consequences. Is there any feature of the sampling design that will interfere with getting a representative sample? Do the people taking the sample have an interest in results that support their point of view or position?

21. through 27.

The only thing *random* about using a random number table is the position in the table that marks your beginning point. After that, a systematic method of moving through the table (decided on in advance) is used. For the exercises, specific starting points are given so that you can obtain results that can be checked.

29. and 31.

Sample Variability is a very important issue. Small samples, even simple random samples with no bias, can display a great deal of variability. In Chapter 4, the relationship between sample size and sample variability will be examined..

Solutions to Odd-numbered Problems

1. The population consists of all light bulbs manufactured by the company.
   The sample consists of the package of 8 bulbs selected for testing.

3. The population consists of all full time students enrolled at the university.
   The sample consists of the 100 students chosen to be interviewed.

5. The population is the 6000 cars that are produced.
   The sample is the 60 cars selected for detailed inspection.

7. The population consists of the 7140 registered voters in the city.
   The sample consists of the 420 people in the neighborhood that is canvassed.

9. Quantitative variables are "age" and "identification number".
   Qualitative variables are "location", "kind of tree", and "health".
   Of the qualitative variables, "location" and "kind of tree" are nominal, while "health" is ordinal.

11. Quantitative variables are "age" and "height".
    Qualitative variables are "country of origin" and "profession".

13. Local dentists may be more likely to use a local product than dentists across the country, or at least view it more favorably. However, there can be a reverse bias. To avoid the bias, local dentists should not know that it is a local product. Any results have validity in the Minneapolis-St. Paul area, because that's the only population being studied.

15. When people are being observed, their behavior often changes. If the customers are aware that the makers of the lemon-lime drink are doing the testing, they may be more likely to say they prefer the lemon-lime: perhaps to please the people making the commercial and be included, or just to be nice.

17. The population consists of all the fish in the lake, because that is what the biologist wants to know about. The sample is the 500 fish that are caught and examined for tags. The strategy goes like this:

The biologist catches, tags, and releases fish back into the lake anticipating that the tagged fish will circulate throughout the lake or in certain patterns. The percentage of tagged fish in any subsequent sample will be approximately the same as in the entire lake. Bias can result from the fact that some of the tagged fish may die from being handled before the sample is taken or that the fish might not redistribute throughout the lake. There is also the possibility that some of the tagged fish will be caught or otherwise removed from the lake, but the biologist could assume that a proportionate number of untagged fish were also included. In that case, the percentages in the lake would still be the same. Biologists and other investigators have developed sophisticated techniques to deal with their specific situations.

19. The population consists of all doctors, while the sample consists of the 20 doctors chosen for the study. Although the sample could be chosen in an unbiased manner, the company's plan is biased in their favor. The company is in control of the data and can choose to ignore it. They will commission studies until they get the result they want. This could happen: Perhaps they have a product with an active ingredient that is recommended by 75% of all doctors. In a samples of size 20, there is almost a 10% chance that 18 (90%) or more will recommend the active ingredient.

21. Beginning in row 115 and using the last two digits of the third column, we look for digits from 00 to 35, and avoid any duplicates. Only the first five rows are shown.

			xx			
115	18075	32457	500**11**	42175	41029	07733
116	52754	43382	02151	46182	40557	94157
117	05255	73603	15957	99738	62835	62959
118	76032	69846	633**16**	48201	11580	45699
119	97050	48883	178**28**	98601	74821	06605

The numbers we consider are:
11, 51, 57, **16**, **28**, 62, 41, 99, **18**, **32**

The students selected for the sample are 11, 16, 18, 28, 32

23. We begin in row 110 and use the first 3 digits of the second column, looking for numbers between 000 and 249

		xxx				
110	84281	57601	78425	36246	79348	41681
111	61589	93355	41310	17068	65700	54464
112	25318	28496	80120	31632	06746	90642
113	40113	91130	74270	27914	80511	70243
114	58420	96471	28464	72438	37667	16233

The numbers we consider are:
576, 933, 284, 911, 964, 324, 433, 736, 698, 488, 555, **121**, 554, **066**, 843, **146**, **060**, 939, **025**

The students selected for the sample are numbers 025, 060, 066, 121, 146

25. The students are numbered as follows:

00-Allen	05-Tom	10-Chris
01-Fred	06-Amy	11-Matt
02-Patty	07-Jane	12-Dan
03-Margaret	08-Mary	13-Jamie
04-Bill	09-John	14-Tyler

We begin in row 118 and use the first 2 digits of the second column, looking for numbers between 00 and 14

		xx				
118	76032	69846	63316	48201	11580	45699
119	97050	48883	17828	98601	74821	06605
120	29030	55519	63362	55720	15296	78787
121	45609	**12**114	36541	53609	09322	28694
122	07608	55455	49299	90355	35334	29000

The numbers we consider are:
69, 48, 55, **12**, 55, **06**, 84, **14**, <u>06</u>, 93, **02**

The members selected for the committee are: Dan, Amy, Tyler, and Patty.

27. First, we assign numbers to the letters of the alphabet as follows:

00-a	05-f	10-k	15-p	20-u	25-z
01-b	06-g	11-l	16-q	21-v	
02-c	07-h	12-m	17-r	22-w	
03-d	08-i	13-n	18-s	23-x	
04-e	09-j	14-o	19-t	24-y	

We begin in row 105 and use the third and fourth digits of the fourth column, looking for numbers between 00 and 25

				xx		
105	43857	49021	49026	93 60 8	51382	49238
106	91823	38333	37006	78 54 5	23827	39103
107	34017	00983	48659	39 44 5	90910	29087
108	49105	95041	94232	50 78 4	59181	44253
109	72479	24246	35932	33 35 8	34853	77573

The numbers we consider are:
60, 54, 44, 78, 35, **24**, **06**, 63, 91, 43, **17**, **18**, 73, **20**, 60, 72, 60, 35, 80, 30, 79, 79, 43, 70, **09**

The letters selected for the sample are g, j, r, s, u, and y.

29. The students are numbered as follows:

00-Allen	05-Tom	10-Chris
01-Fred	06-Amy	11-Matt
02-Patty	07-Jane	12-Dan
03-Margaret	08-Mary	13-Jamie
04-Bill	09-John	14-Tyler

Note: Rather than list all the digits, only those corresponding to the digits assigned to students (including any repeats) are listed in the order they occur.

Continued on next page.

29. Continued

    (a) Sample 1: 05, 07, 01, (05), 00, 12;
          Tom, Jane, Fred, Allen, Dan

        Sample 2: 09, 11, (09), 01, 03, (01), 00
          John, Matt, Fred, Jamie, Allen

        Sample 3: 02, 00, 12, 01, 07
          Patty, Allen, Dan, Fred, Jane

        Sample 4: 08, 14, 01, (01), 09, (09), 02
          Mary, Tyler, Fred, John, Patty

    (b) Margaret, Bill, Amy, and Chris were not chosen in any sample.

31. The students are numbered from 00 (Arnold) to 23 (Xia).

    (a) Sample 1: 00, 12, 06, 14, (06), 02

          Arnold, Molly, Glenda, Oliver, Chris

        Sample 2: 13, 01, 00, 21, (01), (21), 20

          Natalie, Bob, Arnold, Victor, Ursula

        Sample 3: 06, 17, 18, 20, 09

          Glenda, Raul, Sandra, Ursula, Jason

        Sample 4: 10, 00, 11, 21, 22

          Kelly, Arnold, Lester, Victor, Wesley

    (b) Percentages of people in the samples having high blood pressure
        Sample 1: 40%   Sample 2: 20%
        Sample 3: 40%   Sample 4: 0%

Note: As demonstrated here, small samples can be unrepresentative of the population from which they are taken. The incidence of high blood pressure in the population is 17% (4 out of 24), while the samples show from 0% to 40%

Section 3.2   Independent Sampling, Stratified Sampling and Cluster Sampling

Goals
1. Describe and use various sampling and surveying methods.

Key Ideas and Questions
1. What other kinds of sampling are possible if a simple random sample is not practical?
2. What are the advantages and disadvantages of the different methods of sampling?

Vocabulary

Independent Sampling	Strata	Sampling Unit
Systematic Sampling	Stratified Random	Frame
1-in-k Systematic Sampling	Sampling	Multistage Sampling
Quota Sampling	Cluster Sampling	

• •

Overview

Independent Sampling

Independent sampling occurs when each member of the population has a fixed chance of being selected for the sample regardless of whether other members of the population were selected or not.

Systematic Sampling

Systematic sampling first organizes (numbers) the population in an appropriate manner, and then goes through the population in a systematic way to select the sample. In **1-in-k sampling**, the population is grouped into subsets of size k (with the possibility of one small group at the end). For 1-in-6 sampling, this would be 01-06, 07-12, 13-18, etc. A starting position in the first group is selected, say 02, and then every 6th member is selected: 02, 08, 014, etc.

Quota Sampling

Quota sampling utilizes natural groupings within a population as the basis of sampling. The number of members from each group taken in the sample is proportional to the size of the group in the total population. The main goal of quota sampling is to ensure that minorities, in particular, are not underrepresented. That is, if an identifiable subgroup represents 15% of the population, then it should be represented by 15% of the sample.

Stratified Sampling

In **stratified random sampling,** the population is divided into **strata**, groups or units that are similar in some way that is important to the purposes of the sampling. A set of the strata are selected, and then a simple random sample is taken from each of the strata that is used.

Cluster Sampling

Cluster sampling is similar to stratified sampling in that the population is already organized into groups, called sampling units, on some basis, such as a neighborhood, a department, or floors of a residence hall. The set of all these groups is called the sampling frame. A simple random sample is used to select sampling units, and then all members of the sampling unit become part of the final sample.

Multistage sampling

In **multistage sampling**, successively smaller units within the population are selected. The main concern is that the objectives of randomness and representation are preserved. The U.S. Census Bureau uses a multistage sampling system. Different methods of sampling can be used at each stage.

. . . .

Suggestions and Comments for Odd-numbered problems

Independent Samples

In an independent sample, certain digits are randomly picked as "identifiers" or "selectors". Each member of the population is associated with a position in the table of random numbers. If the "identifier" occurs in a given position, the member associated with the position is selected. In a 10% independent sample, each member of the population has a 10% chance of being selected, independent of what happens to other members.

1-in-*k* Sampling

Only two things need to be considered: the value of k (the size of each group) and the position of selection within each group. Determine the value of k, and then randomly pick a value from 1 to k to begin the sample.

Quota Sampling

The size of the simple random sample taken in each group should be proportional to the size of the group with respect to the total population.

Solutions to Odd-numbered Problems

1. For the 10% independent sample, the digit 0 will be our identifier. The cars are numbered from 1 to 100 (not 00 to 99 for this application). The positions of digits in the portion of the random digits table we are using are also assigned the numbers from 1 to 100. The positions of the 0's in the table indicate the cars that are to be chosen.

> The numbers of the Vipers chosen to be tested are:
> 19, 23, 28, 29, 55, 57, 60, 65, 72, 73, 76

Note: Although the digit 0 is randomly distributed to 10% of the table, some strange patterns occur in the location of the digit. There is one string of 25 digits and another of 24 digits that contain no 0's, while one string of 22 digits contains 7 0's.

101	47772
102	63857
103	95237
104	12201
105	49026
106	37006
107	48659
108	94232
109	35932
110	78425
111	41310
112	80120
113	74270
114	28464
115	50011
116	02151
117	15957
118	63316
119	17828
120	63362

3. In the 20% independent sampling from the 24 office workers, Chris and Glenda are selected.

Note: You might expect that a 20% sample would select 4 or 5 from the group ($0.20 \times 24 = 4.8$).

However, *20%* refers to the chance of each one being selected, not the percentage being selected. Also, this is a small sample. If we were sampling from a very large group, the percentage selected would be closer 20%.

130	77175
131	80925
132	23629
133	77764
134	75867

5. Beginning in column 6, line 101, the states numbered 7, 11, 19, 29, 33 are selected in the sample.

State	Per Capita Income
Connecticut	28,110
Hawaii	23,354
Maine	18,895
New Hampshire	22,659
North Carolina	18,702

101	99445
102	20429
103	04755
104	33306
105	49238
106	39103
107	29087
108	44253
109	77573
110	41681

7. We use the third digit of column 2 on line 134 as the random digit; it is an 8.
   For 1-in-10 sampling that means that our numbers are in the form $r + 10i$, where $r = 8$ and i is a non-negative integer.
   The cars selected are numbered 8, 18, 28, 38, 48, 58, 68, 78, 88, 98

9. For a 1-in-5 sampling system, our selected numbers will have the form $r + 5i$, where r is a random digit from 1 through 5, and i is a non-negative integer.

   We begin with the fourth digit of column 4 on line 121; it is a 0, not a digit from 1 through 5. Therefore, we read down the column looking for one.

   The digit on line 122 is 5. The numbers selected are 5, 10, 15, 20. These correspond to Esther, Jason, Oliver, and Teresa.

11. For a 1-in-10 sampling system, our selected numbers will have the form $r + 10i$, where r is a random digit from 1 through 10 (0 as a digit), and i is a non-negative integer.

    We begin with the first digit of column 5 on line 137; it is a 5.

    The numbers selected are 5, 15, 25, 35, 45. These correspond to

State	per capita income
California	$21, 821
Iowa	18,315
Missouri	19,463
Ohio	19,688
Vermont	19,467

13. First, we will select the men, then the women. Since there are 80 of each, we can number them with two digits, either 00 through 79, or 01 through 80. We will choose the latter because it is more straight forward.

    For the men, we'll use the second and third digits of column 2 beginning on line 113. The digits are 11, which correspond to the first man selected. The men selected are numbered 11, 64, 24, 33, 36, 55, 21, 54, 66, 43.

    For the women, we'll use the second and third digits of column 3 beginning on line 113. The digits are 42, which correspond to the first woman selected. The women selected are numbered 42, 21, 59, 33, 78, 65, 46, 75, 29, 64.
    Note: the decision to number 01 through 80 eliminated 00 from selection; it was the third number in the column of digits we examined.

15. **(a)** Since there are 2000 students enrolled and 40 are being sampled, this represents one student out of every 50. There should be (950 ÷ 50 =) 19 freshmen and sophomores, (800 ÷ 50 =) 16 juniors and seniors, and (250 ÷ 50 =) 5 graduate students.

    (b) . Beginning with the first three digits on line 105 of column 1, the freshmen and sophomores are 438, 918, 340, 491, 724, 842, 615, 253, 401, 584, 180, 527, 052, 760, 290, 456, 076, 949, 505.
    Beginning with the first three digits on line 105 of column 3, the juniors and seniors are 490, 370, 486, 359, 784, 413, 742, 284, 500, 021, 159, 633, 178, 365, 492, 046.
    Beginning with the first three digits on line 105 of column 5, the graduate students are 238, 067, 115, 152, 093.

17. **(a)** The numbers 00 through 99 give you 100 numbers with two digits. If you use 100 as the last number, you have to use three digits and look for 100 numbers (001 to 100) out of 1000 numbers (000 to 999).

    (b) Each room represents a cluster, and you need 20 rooms to get a sample of size 60. Using the second and third digits of column 2 in Table 3.1, beginning on line 115, the rooms selected are 24, 33, 36, 98, 88, 55, 21, 54, 66, 43, 46, 60, 39, 25, 41, 71, 09, 77, 58, 29.

19. **(a)** The average size homeroom is about 24. Even with some absences, four home rooms will most likely have at least 80 students total. It's much simpler to select 4 out of 14 than it is to select 80 out of 335.

    (b) Using the second and third digits of column 3 of Table 3.1, beginning on line 108, the rooms selected are 13, 01, 05, 06.

21. 8's and 9's are eliminated from the portion of the table we use, and the remaining positions are "numbered" from 1 to 50, the same as the states (Alabama is 1 and Wyoming is 50).

State	Representatives		
		120	63362
		121	36541
Georgia	11	122	4*2**
Illinois	20	123	0461*
Kansas	4	124	17532
Kentucky	6	125	32*67
New Mexico	3	126	*6457
North Carolina	12	127	7*160
North Dakota	1	128	124*7
Texas	30	129	5**53
Washington	9	130	*6514
		131	64073
		132	7

Section 3.3   Measures of Central Tendency: Means, Medians, and Modes

Goals
1. Compare numerical summaries of data such as the mean and the median.

Key Ideas and Questions
1. What are the three main measures of central tendency?
2. How are the mean and median similar? How are they different?
3. How do the mean and median compare for symmetric data? For skewed data?

Vocabulary

Data Point	Median	Skewed Data
Central Tendency	Symmetric Data	Distribution
Sample Mean	Distribution	Weighted Mean
Population Mean	Mode	

. . . .

Overview

Common measures which represent an average, middle, or most frequent value for a distribution are the mean, median and mode. These **measures of central tendency** can be combined with visual displays to provide a more complete, meaningful, and effective presentation of the data.

Mean

The **mean** of a set of data is the arithmetic average of the values in the set. The mean represents one central ("center") value of the data. To find the mean, add the values of the data set and divide this total by the number of data points.

Median

The **median** is another measure of central ("center") tendency. The median is the value closest to the middle of the data set. To find the median, first arrange the data set in increasing order. If there are an odd number of data points, there is one in the exact middle. This data point is the median of the data set. If there are an even number of data points, then there are two data points in the middle. The median in this instance is the average of these two points. The median is not affected by a very large or very small value in the set.

Mode

The **mode** is the most frequently occurring value in the data, but can often be very unrepresentative of the data as a whole

Weighted Mean

If data points have different levels of importance, or if data points are grouped, a **weighted mean** may be used that reflects the importance of the different data points or the sizes of the groups.

. . . .

Suggestions and Comments for Odd-numbered problems

17. Think backwards on this problem. To find the mean, you first find the sum of the scores and then divide by the number of scores. If you know the mean and the number of scores, you can find the total for all the scores. How does this help you find the missing number?

19. Organize the data into a table. This will make the mode and mean apparent and facilitate calculation of the weighted mean.

21. The per capita figure for the countries taken as a group will be the total of the gross national products divided by the total of the populations.

23. First round the values, make a list of the data by tens (180, 190, etc.) and tally the data by category as you go through the list from top to bottom. When you have the frequency for each category, add these to make sure the total is the same as the number of players on the roster.

Solutions to Odd-numbered Problems

1. Mean $= \dfrac{3 + 7 + 12 + 9 + 10 + 15}{6} = \dfrac{56}{6} = 9.\overline{3}$

Median = 9.5   The data arranged in order: 3, 7, 9, 10, 12, 15;
                 we take the average of the two middle values

Mode: There is no mode

3. Mean $= \dfrac{2 + 4 + 7 + 10 + 11 + 12 + 15 + 21}{8} = \dfrac{82}{8} = 10.25$

Median = 10.5   The data are arranged in order; since there is no single middle value,
                we take the average of the two middle values.

Mode: There is no mode

5. Mean $= \dfrac{2 + 2 + 2 + 5 + 7 + 30}{6} = \dfrac{48}{8} = 8$

Median = 3.5   The data are arranged in order; since there is no single middle value,
                we take the average of the two middle values.

Mode = 2

7. Mean $= \dfrac{2 + 4 + 6 + 8 + 10 + 12 + 38 + 45}{8} = \dfrac{125}{8} = 15.625$

Median = 9   The data are arranged in order; since there is no single middle value,
                we take the average of the two middle values.

Mode   There is no mode

9. A value in the set that is much larger than the others can make the mean larger than most values in the set; in problem 5, only one of the six values was larger than the mean. This indicates that the mean is not always a good representative value for the values in a set. A similar situation occurs when one or two values are much smaller than the rest of the values in a set. For this reason, it is often a good idea to give both the mean and the median.

11. Mean winning score in the Rose Bowl 1980 - 1989
$$= \frac{17 + 23 + 28 + 24 + 45 + 20 + 45 + 22 + 20 + 22}{10} = \frac{266}{10} = 26.6$$

13. Mean Total Points in the Rose Bowl 1980-1989
$$= \frac{33 + 29 + 28 + 38 + 54 + 37 + 73 + 37 + 37 + 36}{10} = \frac{402}{10} = 40.2$$

15. The answer to problem 13 is the sum of the answers to problems 11 and 12.

17. Since the mean of the 10 scores is 80.7, the total of the scores is 10×80.7, or 807. The 9 remaining scores have a sum of 728. The missing score is
$$807 - 728 = 79.$$

19. The frequency distribution for the data is

Scores	5	6	7	8	9	10
Number	1	3	8	6	5	3

Grades

The mode is 7, since it has the greatest frequency.

The mean is 7.77 $\quad \dfrac{1 \times 5 + 3 \times 6 + 8 \times 7 + 6 \times 8 + 5 \times 9 + 3 \times 10}{1 + 3 + 8 + 6 + 5 + 3} = \dfrac{202}{26} \approx 7.7692$

The median is 8 Since there are 26 values, the median will be the average of the 13th and 14th values (in order) - both are 8s.

The data is not symmetric, but is not especially skewed.

21. To calculate the desired per capita gross domestic product, we need the total gross domestic product for the five countries and the total population.
To simplify the writing, we will use all data in millions.

Per capita gross domestic product $= \dfrac{700000 + 1000000 + 1200000 + 500000 + 900000}{12 + 16 + 22 + 8 + 13} = \dfrac{4300000}{71}$

Per capita gross domestic product $\approx \$60,563$

23. The rounded data fall into the following categories:

Weight	180	190	200	210	220	230	240	250	260	270	280	290	300	310	320	330
Frequency	6	9	2	4	7	6	2	6	1	3	2	2	1	3	2	2

Football Player Weights

The **mode** is 190   The **mean** is 235.7   The **median** is 230

Since the mode is to the left of the median and the median is to the left of the mean $(190 < 230 < 235.7)$, we say the data is skewed to the right. The histogram shows this as well, although the skewing is only moderate because of the high number of data points for 220, 230, and 250. Skewing is not always present in a distribution, and is most meaningful when there is only one set of high frequencies close together.

Section 3.4   Measures of Variability

Goals

1. Compute numerical summaries of the spread or variability of data, such as standard deviation and interquartile range.

Key Ideas and Questions

1. How can we express the amount of spread or variability in a data set?
2. What does a box and whisker plot show about a data set?
3. How are the mean and median influenced by unusually large or small individual values?

Vocabulary

Range	Five-number Summary	Sample Standard Deviation
First Quartile	Box and Whisker Plot	Population Variance
Third Quartile	Deviation from the Mean	Population Standard
Interquartile Range (IQR)	Sample Variance	Deviation

. . . .

Overview

It is important to understand the dispersion or spread for data sets. The spread, or deviation from the middle, or center, tells us how far away from center the bulk of the points are (on average) and , therefore, how representative the middle value is for the data set. If the spread is large, we can expect a greater amount of variation in the value of data points. The first of the measures of spread (the range and interquartile range) use the median as the center of the data. The second measure of spread (the standard deviation) uses the mean as the center of the data.

Range

The **range** is the crudest measurement of the spread of a data set. It is the difference of the largest and smallest values in the set.

Quartiles

Together with the low and high values in the data set, the quartiles break the data set into four subsets of relatively equal size.

The **first quartile, q_1,** is defined as the median of the lower half of the points. With an odd number of data points, q_1 is the median of the lower half of the points, not including the middle data point. With an even number of data points, q_1 is the median of the lower half of the points. The **third quartile, q_3,** is the median of the upper half of the points. The **second quartile, q_2,** is the median. The **interquartile range (IQR),** is $q_3 - q_1$. The IQR measures the amount of dispersion or spread in the data. The range is not as refined a measure of spread as the IQR, since a large or small outlier will affect the range and probably not the IQR.

Box and Whisker Plots

A **5-number summary** of a data set is the set: $\{s, q_1, m, q_3, L\}$ where s is the smallest data point, L is the largest data point, m is the median, and q_1 and q_3 are the first and third quartiles. These numbers can be graphed in a special way called a **box and whisker plot**. This plot gives a picture of the data while omitting the details. By drawing box and whisker plots for two data sets and displaying them side-by-side, you can compare data sets quickly and easily.

Deviation

The difference between a data point and the mean is called the **deviation from the mean**, or simply the **deviation**, of the point. Since the average of the deviations from the mean is always zero, it does not provide helpful information about the spread of the data.

The **variance**, which is the average of the *squares* of the deviations, is more helpful. The variance will always be positive if at least two values are different. In general, a data set that is more spread out has a larger variance. The population variance uses n (the size of the data set) as the divisor, while the sample variance uses $n - 1$ as a divisor. The square root of the variance is called the **standard deviation** of a data set. and standard deviation. The population and sample standard deviation are the most commonly used measures of spread in technical or scientific work.

Suggestions and Comments for Odd-numbered problems

1. through 8.
The first quartile is the median for the numbers that are less than the median for the entire set. Likewise, the third quartile is the median for the numbers in the set that are greater than the median of the entire set. The interquartile range ("between the quartiles") is the difference between these two values.

Drawing Box and Whisker Plots
The first step is finding the five-number summary and then making a number line that includes both the largest and smallest values. The line should be drawn carefully, with a straight edge and equally spaced intervals. Another line which will be the "backbone" of the box and whisker plot should be draw lightly, parallel to the number line, and the positions for the five number summary indicated.

13. The reference line should included the largest and smallest values of both sets combined.

17. Two ways you could use to compare are percentage and relative class standing.

27. through 37.
The process for finding the standard deviation and sample standard deviation is the same except for the divisor: n for standard deviation, $n-1$ for sample standard deviation.

33. and 35.
Modifying a data set in a specific way can have predictable results. These same type of modifications will be important when working with normal distributions in Chapter 4. Use the population variance and standard deviation on these problems.

Solutions to Odd-numbered Problems

1. (i) $\{1, 3, 4, 5, 6, 8, 9, 10, 11, 12, 15\}$ (ii) The range is 14 (15 - 1)
(iii) Because there are 11 values, the median will be in the middle with 5 scores above it and 5 below it. The median is 8.
(iv) First quartile = median of $\{1, 3, 4, 5, 6\} = 4$
Third quartile = median of $\{9, 10, 11, 12, 15\} = 11$
(v) Interquartile range = Third quartile - First quartile = 11 - 4 = 7

3. (i) $\{6, 9, 10, 12, 13, 14, 18, 21, 24, 26\}$ (ii) The range is 20 (26 - 6)
(iii) Because there are 10 values, the median will have 5 scores above it and 5 below it. Since there is no value directly in the middle, we take the average of the two scores on either side: $\frac{13 + 14}{2} = 13.5$
(iv) First quartile = median of $\{6, 9, 10, 12, 13\} = 10$
Third quartile = median of $\{14, 18, 21, 24, 26\} = 21$
(v) Interquartile range = Third quartile - First quartile = 21 - 10 = 11

5. (i) $\{2, 4, 5, 6, 8, 8, 9, 10, 10, 12, 15\}$ (ii) The range is 13 (15 - 2)
(iii) Because there are 11 values, the median will have 5 scores above it and 5 below it. The median is 8, even though there is another 8.
(iv) First quartile = median of $\{2, 4, 5, 6, 8\} = 5$
Third quartile = median of $\{9, 10, 10, 12, 15\} = 10$
(v) Interquartile range = Third quartile - First quartile = 10 - 5 = 5

7. (1) {1, 1, 2, 2, 3, 4, 4, 4, 5, 5, 6, 8, 9, 10}
    (ii) The range is 9 (10 - 1)
    (iii) Because there are 14 values, the median will have 7 scores above it and 7 below it.
         Since there is no value directly in the middle, we take the average of the two scores
         on either side. Since both scores are 4, the median is 4.
    (iv) First quartile = median of {1, 1, 2, 2, 3, 4, 4} = 2
         Third quartile = median of {4, 5, 5, 6, 8, 9, 10} = 6
    (v) Interquartile range = Third quartile - First quartile = 6 - 2 = 4

9. (a) Five number summary = {73, 77.5, 79.5, 82.5, 95}

    (b)

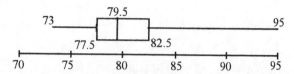

11. The home run totals arranged in order:
        {6, 11, 22, 25, 29, 34, 35, 41, 41, 46, 46, 47, 49, 54, 54, 59, 60}
    (a) The mean = 38.76   Total = 659; 659 ÷ 17 = 38.7647
        The median = 41

        First quartile = median of {6, 11, 22, 25, 29, 34, 35, 41} = $\frac{25 + 29}{2}$

        Third quartile = median of {46, 46, 47, 49, 54, 54, 59, 60} = $\frac{49 + 54}{2}$

        5 -number summary = {6, 27, 41, 51.5, 60}

    (b)

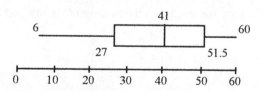

13. (a) 5-number summary for male doctors = {20, 27, 34, 50, 86}
        5-number summary for female doctors = {5, 10, 18.5, 29, 33}

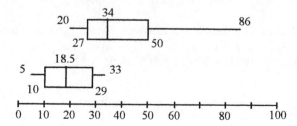

15. First, the scores must be arranged in order;
$\{66,67,68,70,71,73,74,75,76,78,80,81,82,82,83,84,84,87,88,90,91,92,93,94\}$

    (a) The median = 81.5 (the average of the 12th and 13th numbers, 81 and 82)

        First quartile = median of $\{66,67,68,70,71,73,74,75,76,78,80,81\}$ = $\dfrac{73+74}{2}$

        Third quartile = median of $\{82,82,83,84,84,87,88,90,91,92,93,94\}$ = $\dfrac{87+88}{2}$

        5 -number summary = $\{66, 73.5, 81.5, 87.4, 94\}$

    (b)

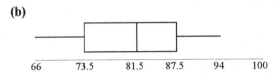

      66   73.5   81.5   87.5   94   100

17. The scores should first be arranged in order.
Economics class:
$\{66,67,68,70,71,73,74,75,76,78,80,81,82,82,83,84,84,87,\mathbf{88},90,91,92,93,94\}$

Sociology class:
$\{27,28,34,35,36,37,37,37,38,\mathbf{38},40,40,41,41,42,43,45,48\}$

The score of 88 was a better score because it was just above the third quartile of 87.5, while the score of 38 was the median in that set of scores.

19. The mean = 18.72   Total = 224.6   224.6 ÷ 12 = 18.71666...
The values arranged in order are:
        $\{0.4, 1.2, 1.8, 2.3, 6.4, 9.6, 11.6, 13.9, 19.9, 23.6, 37.5, 96.4\}$
Since there are 12 values, the median is the average of the 6th and 7th values.
Median = $\dfrac{9.6+11.6}{2}$ = 10.6
The first quartile is the median of $\{0.4, 1.2, 1.8, 2.3, 6.4, 9.6\}$
      = $\dfrac{1.8+2.3}{2}$ = 2.05
The third quartile is the median of $\{11.6, 13.9, 19.9, 23.6, 37.5, 96.4\}$
      = $\dfrac{19.9+23.6}{2}$ = 21.75
5-number summary = $\{0.4, 2.05, 10.6, 21.75, 96.4\}$

21. When the United Kingdom is excluded from the set, the values arranged in order are
$\{0.4, 1.2, 1.8, 2.3, 6.4, 9.6, 11.6, 13.9, 19.9, 23.6, 37.5\}$
The 5-number summary = $\{0.4, 1.8, 9.6, 19.9, 37.5\}$

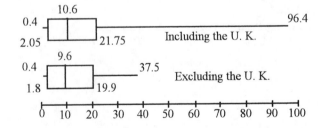

23. The values arranged in order are
  {4.3, 4.5, 4.7, 4.7, 5.7, 5.7, 5.8, 6.1, 6.2, 6.4, 6.4, 9.2}
  (a) The mean = 5.81   Total = 69.7   69.7 ÷ 12 = 5.808333...
  The median = $\dfrac{5.7 + 5.8}{2}$ = 5.75

  The first quartile is the median of {4.3, 4.5, 4.7, 4.7, 5.7, 5.7}
    = 4.7 (since the middle values are the same).
  The third quartile is the median of {5.8, 6.1, 6.2, 6.4, 6.4, 9.2}
    = $\dfrac{6.2 + 6.4}{2}$ = 6.3

  5-number summary = {4.3, 4.7, 5.75, 6.3, 9.2}

  (b)

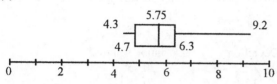

25. The deviation from the mean of a data point x is d = x - mean.

Data point	Deviation
3	3 - 7 = - 4
7	7 - 7 = 0
12	12 - 7 = 5
9	9 - 7 = 2
4	4 - 7 = - 3

A data point below the mean will have a negative value as the deviation.
A data point above the mean will have a positive deviation.
A data point equal to the mean will have 0 deviation.

27. The sample variance is the mean of the squared deviations, dividing by n - 1.
The sample standard deviation is the square root of the sample variance.

Data point	Deviation	Deviation squared
3	3 - 7 = - 4	16
7	7 - 7 = 0	0
12	12 - 7 = 5	25
9	9 - 7 = 2	4
4	4 - 7 = - 3	9 Sum of deviations squared = 54

Sample Variance = $\dfrac{54}{4}$ = 13.5

Sample Standard Deviation = $\sqrt{\text{sample variance}}$ = $\sqrt{13.5} \approx 3.67$

29. (a) Mean $= \dfrac{4 + 6 + 7 + 10 + 13}{5} = \dfrac{40}{5} = 8$

Data point	Deviation	Deviation squared
4	4 - 8 = - 4	16
6	6 - 8 = - 2	4
7	7 - 8 = - 1	1
10	10 - 8 = 2	4
13	13 - 8 = 5	25 Sum of deviations squared = 50

Sample Variance $= \dfrac{50}{4} = 12.5$

Sample Standard Deviation $= \sqrt{\text{sample variance}} = \sqrt{12.5} \approx 3.54$

(b) Mean $= \dfrac{-2 + -2 + 1 + 2 + 4 + 12}{6} = \dfrac{15}{6} = 2.5$

Data point	Deviation	Deviation squared
-2	-2 - 2.5 = - 4.5	20.25
-2	-2 - 2.5 = - 4.5	20.25
1	1 - 2.5 = - 1.5	2.25
2	2 - 2.5 = - 0.5	0.25
4	4 - 2.5 = 1.5	2.25
12	12 - 2.5 = 9.5	90.25 Sum of deviations squared = 135.5

Sample Variance $= \dfrac{135.5}{5} = 27.1$ Sample Standard Deviation $= \sqrt{27.1} \approx 5.21$

(c) Mean $= \dfrac{3 + 4 + 4 + 4 + 5 + 5 + 5 + 6}{8} = \dfrac{36}{8} = 4.5$

Data point	Deviation	Deviation squared
3	3 - 4.5 = - 1.5	2.25
4	4 - 4.5 = - 0.5	0.25
4	4 - 4.5 = - 0.5	0.25
4	4 - 4.5 = - 0.5	0.25
5	5 - 4.5 = 0.5	0.25
5	5 - 4.5 = 0.5	0.25
5	5 - 4.5 = 0.5	0.25
6	6 - 4.5 = 1.5	2.25 Sum of deviations squared = 6

Sample Variance $= \dfrac{6}{7} \approx 0.8571$ Sample Standard Deviation $= \sqrt{0.8571} \approx 0.926$

31. Mean $= \dfrac{2.72 + 3.84 + 4.07 + 4.80 + 5.61 + 6.78}{6} = \dfrac{27.82}{6} \approx 4.64$

Data point	Deviation	Deviation squared
2.72	2.72 - 4.64 = - 1.92	3.69
3.84	3.84 - 4.64 = - 0.80	0.64
4.07	4.07 - 4.64 = - 0.57	0.32
4.80	4.80 - 4.64 = 0.16	0.03
5.61	5.61 - 4.64 = 0.97	0.94
6.78	6.78 - 4.64 = 2.14	4.58 Sum of deviations squared = 10.2

Sample Variance $= \dfrac{10.2}{5} = 2.04$ Sample Standard Deviation $= \sqrt{2.04} \approx 1.43$

33. Original data set: {3, 10, 9, 7, 15}

$$\text{Mean} = \frac{3 + 10 + 9 + 7 + 15}{5} = \frac{44}{5} = 8.8$$

Modified data set: {8, 15, 14, 12, 20}

$$\text{Mean} = \frac{8 + 15 + 14 + 12 + 20}{5} = \frac{69}{5} = 13.8$$

Variance and Standard Deviation

	Original Data			Modified Data	
Data	Deviation	$(\text{Deviation})^2$	Data	Deviation	$(\text{Deviation})^2$
3	3 - 8.8	$(-5.8)^2 = 33.64$	8	8 - 13.8	$(-5.8)^2 = 33.64$
10	10 - 8.8	$(1.2)^2 = 1.44$	15	15 - 13.8	$(1.2)^2 = 1.44$
9	9 - 8.8	$(0.2)^2 = 0.04$	14	14 - 13.8	$(0.2)^2 = 0.04$
7	7 - 8.8	$(-1.8)^2 = 3.24$	12	12 - 13.8	$(-1.8)^2 = 3.24$
15	15 - 8.8	$(6.2)^2 = 38.44$	20	20 - 13.8	$(6.2)^2 = 38.44$

Because the squares of the deviations are the same for both sets, the variance and standard deviation will be the same. Sum of deviations squared = 76.8

$$\text{Population Variance} = \frac{76.8}{5} = 15.36$$

$$\text{Standard Deviation} = \sqrt{\text{variance}} \approx 3.92$$

When all data points are increased by the same amount, the mean changes by that amount as well. However, the "spread" between the data points <u>does not</u> change. Therefore, the standard deviation (which measures the spread of the data) <u>does not</u> change when a constant is added (or subtracted) to all data points.

35. Modified data set: {27, 21, 9, 30, 45}

$$\text{Mean} = \frac{27 + 21 + 9 + 30 + 45}{5} = \frac{132}{5} = 26.4$$

The new mean is 26.4 (= 3 × 8.8)

Data point	Deviation		Deviation squared
27	27 - 26.4 =	0.6	0.36
21	21 - 26.4 =	- 5.4	29.16
9	9 - 26.4 =	- 17.4	302.76
30	30 - 26.4 =	3.6	12.96
45	45 - 26.4 =	18.6	345.96 Sum of deviations squared = 691.20

$$\text{Population Variance} = \frac{691.20}{5} = 138.24$$

$$\text{Standard Deviation} = \sqrt{\text{variance}} \approx 11.76$$

The new standard deviation is approximately 11.76 (= 3 × 3.92)

When all the data points are multiplied by the same factor, the mean is <u>also multiplied</u> by that factor as well. In addition, the "spread" between the data is <u>also changed</u> by that factor (9 - 7 = 2; 27 - 21 = 6). Therefore, the standard deviation <u>also changes</u> by that same factor.

37. We will make the calculations for each unique value and then multiply the results by the number of data points having each value (this is a weighted average).

Data point	Number of points
13	1
14	4
15	7
16	6
17	0
18	2

Total points = 20   (Always double check the total.)

(a) Mean $= \dfrac{13 + 4 \times 14 + 7 \times 15 + 6 \times 16 + 2 \times 18}{20} = \dfrac{306}{20} = 15.3$

Data point	Deviation	(Deviation)^2	× "points"	= Subtotal
13	13 - 15.3 = - 2.3	5.29	1	5.29
14	14 - 15.3 = - 1.3	1.69	4	6.76
15	15 - 15.3 = - 0.3	0.09	7	0.63
16	16 - 15.3 = 0.7	0.49	6	2.94
18	18 - 15.3 = 2.7	7.29	2	14.58

Sum of squared deviations = 30.20

Sample variance $= \dfrac{30.20}{19} \approx 1.59$ Sample standard deviation $\approx \sqrt{1.59} \approx 1.26$

(b)

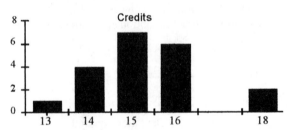

39. The most likely candidate for an outlier is 95.

If 95 is an outlier, the difference between it and the third quartile must be 1.5 times the interquartile range (IQR).

The 5-number summary for the golf scores is {73, 77.5, 79.5, 82.5, 95}

1.5 × IQR = 1.5 × (82.5 - 77.5) = 1.5 × 5 = 7.5

Difference between 95 and third quartile = 95 - 82.5 = 12.5 > 7.5.

Therefore, 95 is an outlier. 73 is not an outlier since it is only 4.5 less than the first quartile.

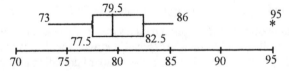

41. 5-number summary for Ruth's home runs as an outfielder (problem 11)
$\{6, 27, 41, 51.5, 60\}$

$1.5 \times IQR = 1.5 \times (51.5 - 27) = 1.5 \times 24.5 = 36.75$

Difference between low value and first quartile $= 27 - 6 = 21 < 36.75$

Difference between high value and third quartile $= 60 - 51.5 = 8.5 < 36.75$

There are no outliers. The box-and-whiskers plot is unchanged.

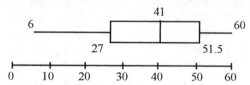

43. **(a)** 5-number summary for U.S. investment in the ECC (problem 19)
$\{0.4, 2.05, 10.6, 21.75, 96.4\}$

$1.5 \times IQR = 1.5 \times (21.75 - 2.05) = 1.5 \times 19.7 = 29.55$

Difference between high and third quartile $= 96.4 - 21.75 = 74.65$

Since this is greater than 29.55, we see that the U.K. is an outlier.

(b) The second highest value is 37.5. The difference between this value and the third quartile is $15.75 < 29.55$. So, 37.5 is not an outlier.

The difference between the low value and the first quartile is 1.65. The low value is not an outlier.

96.4 is the only outlier.

(c)

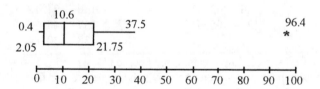

Chapter 3 Review Problems

Solutions to Odd-numbered Problems

1. The population consists of the voters in the town.

The sample consists of the people who were surveyed.

Sources of bias: The people walking or otherwise passing by the high school are likely to be high school students and parents or friends. These people may already have accepted rollerbladers as "part of their world".

3. Standard Use of a Random Number Table

Label the people 0, 1, 2, 3, 4. Choose a starting point in a random number table and a pattern to follow in the table. Record the first three numbers that are distinct and one of 0, 1, 2, 3, 4. These numbers tell which people to put in the sample.

The basic idea in selecting an unbiased sample from a group is that at each step in the selection process, each of the remaining choices has an equally likely chance of being selected. There may be an overall pattern involved in the process, even if it is what we refer to as being "random". The random number table is not strictly "random", but was generated by a process that is very close to being random. Other methods for selecting unbiased samples are also available, but tend to be more sophisticated and complex.

5. The first sample uses the first two digits in column 5, beginning on line 109.
The second sample uses the first two digits in column 4, beginning on line 117.

	Sample 1			Sample 2	
109	34853	*	117	99738	
110	79348		118	48201	
111	65700		119	98601	
112	06746		120	55720	
113	80511		121	53609	
114	37667		122	90355	
115	41029	*	123	82809	
116	40557		124	57302	
117	62835		125	10792	
118	11580		126	78791	*
119	74821		127	19436	
120	15296	*	128	(99)703	
121	09322		129	83094	
122	35334		130	61955	
123	76952		131	70415	
124	81752		132	24642	
125	(81)713		133	33455	
126	44380		134	(53)490	
127	00813		135	50892	
128	08901		136	04212	*
129	05261		137	40076	
			138	30584	

In the first sample, there were three people (34, 41, 15; 15%) who develop heart disease. In the second sample, there are only two people (78 and 04; 10%). We would expect these numbers to often be different because of sample variability.

7. Answers may vary. One correct answer is as follows:
For a 1-in-5 sampling system, our selected numbers will have the form $r + 5i$, where r is a random digit from 1 through 5, and i is a non-negative integer.

We begin with the first digit of column 5 on line 130; it is 5, $r = 5$. The fifth person will be selected as will every fifth person after that. Here, we have to deal with the consequences of the numbering system we chose; we numbered them 00 to 99 so that we could work with two digits. The fifth person is numbered 04. The people selected are numbered

04*, 09, 14, 19, 24, 29, 34*, 39*, 44, 49*, 54, 59, 64, 69, 74, 79, 84, 89*, 94, 99
Five of the people in this sample (04, 34, 39, 49, 89) developed heart disease.

To get a second sample we need to get a new digit, and go to column 6, row 121, and read from right to left (we could have read down the column, we just need to determine which way in advance). The first number in the range of 1 through 5 is a 2; the second person will be selected, and that person is numbered 01. The people selected are numbered

01, 06, 11, 16*, 21, 26, 31*, 36, 41*, 46*, 51, 56, 61, 66*, 71, 76, 81, 86*, 91, 96,
Six of the people in this sample (16, 31, 41, 46, 66, 86) developed heart disease.

9. First, we organize our data into a convenient format:

Credit hours	Mean Grade	Number of courses
two	3.2	13
three	3.6	22
four	3.5	21

$$\text{Weighted mean} = \frac{13 \times 3.2 \times 2 + 22 \times 3.6 \times 3 + 21 \times 3.5 \times 4}{13 \times 2 + 22 \times 3 + 21 \times 4} = \frac{614.8}{176} \approx 3.49$$

11. We organize the data and find the relevant means using a weighted mean.

Data point	Number of points
6	4
7	9
8	6
9	3
10	1

$$\text{Weighted mean} = \frac{6 \times 4 + 7 \times 9 + 8 \times 6 + 9 \times 3 + 10}{23} = \frac{172}{23} \approx 7.478$$

Data point	Deviation	(Deviation)^2	$\times$ "points" =	Subtotal
6	6 - 7.478 = - 1.478	2.1845	4	8.7380
7	7 - 7.478 = - 0.478	0.2285	9	2.0565
8	8 - 7.478 = 0.522	0.2725	6	1.6350
9	9 - 7.478 = 1.522	2.3165	3	6.9495
10	10 - 7.478 = 2.522	6.3605	1	6.3605

Sum of squared deviations = 25.7395

$$\text{Sample variance} = \frac{25.7395}{22} \approx 1.16997$$

$$\text{Sample standard deviation} \approx \sqrt{1.16997} \approx 1.082$$

Inferential Statistics

Section 4.1   Normal Distributions

Goals
1. Compute the fraction of measurements from a normal distribution that fall in a given interval..
2. Understand the concept of z-scores, and be able to compute and use z-scores.

Key Ideas and Questions
1. Describe the characteristics of a normal distribution.
2. How does knowing (or assuming) that a distribution is normal allow you to compute percentages of the distribution that lies between two values?

Vocabulary
Statistical Inference	Standard Normal Distribution
Normal Distribution	Population z-score

· ·

Overview

This chapter looks at making inferences about a population based on information from a sample taken from that population. Variations in characteristics such as the mean and proportion between the sample and the population are systematic. Information from a sample can provide meaningful information about the entire population.

Deviation

The difference between a data point and the mean is called the **deviation from the mean**, or simply the **deviation**, of the point. Since the average of the deviations from the mean is always zero, it does not provide helpful information about the spread of the data.

The **variance**, which is the average of the squares of the deviations, is more helpful. The square root of the variance is called the **standard deviation** of a data set. In general, the data set that is more spread out has the larger variance and standard deviation.

Z-Scores

The mean and the standard deviation can be used to compare data points from different data sets. Each data point has a **z-score** which measures the number of standard deviations between the data point and the mean.

$$\text{z-score} = \frac{\text{data point - mean}}{\text{standard deviation}}$$

The data points below the mean have negative z-scores; the ones above the mean have positive z-scores. The z-score of the mean (when it is a data point) is zero, and the sum of the z-scores is zero.

Every data set has an associated set of z-scores for each data point. This new set of data has a mean of 0 and standard deviation of 1. You can picture this by using two different horizontal scales: one for the numbers in the data set and one for the z-scores. The data set {1, 2, 3, 5, 9} has a mean of 4 and a standard deviation of 2.83. The z-score for 9 is 1.77. The z-score for 2 is -0.71.

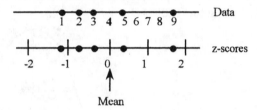

Standard Normal Distribution

A data set represented by an ideal bell-shaped curve is called a **normal distribution**. When the bell-shaped curve represents a data set that has a mean of 0 and a standard deviation of 1, it is called a **standard normal distribution**.

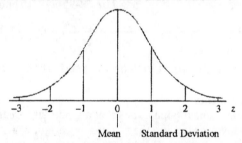

Computing with the Standard Normal Distribution

The area under the entire standard normal curve is 100% of the data, with 50% of the data above the mean, and 50% of the data below the mean. The area that lies between any two vertical lines corresponding to z-scores of the standard normal distribution can be expressed as a percentage of the entire area under the curve.

Computing with Normal Distributions

Many populations have a normal distribution (or one that is approximately normal) but not a standard normal distribution. The mean is usually different from 0 and the standard deviation different from 1. A normal (or nearly normal) distribution is very common. The standard normal distribution is a model to which all others can be compared. To do this, we assign a z-score to each data point.

To compute percentages using the normal distribution most efficiently, you can first add a **data axis** below the horizontal axis of the standard normal distribution. The z-axis for the standard normal distribution has 0 at the center and 1 at the standard deviation. The mean of your data goes below the 0. The value for the mean of the data points plus one standard deviation goes below the 1 on the z-axis, and the value of the mean of the data points minus one standard deviation goes below the -1 on the z-axis. Other data values relating to z-scores from -3 to 3 are added in the same way.

Although the areas related to the standard normal distribution can be computed using a (complicated) formula, they are usually found in a table of values (some of which can be very detailed) or using the 68-95-99.7 property. This is illustrated below; the values given here are more precise than in the usual statement of the property, and should be verified using the Table 4.1.

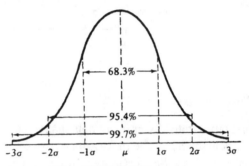

The 68-95-99.7 Property

. . . .

Suggestions and Comments for Odd-numbered problems

General Comments

In questions about populations that are normally distributed (or nearly so), the answer usually involves finding the z-scores of certain values. From the z-score of a value from the population, you can determine what percentage of the population is either above or below the given value. These percentages can be multiplied by the size of the population to find out how many in the population are represented in a given range of values.

What are the z-scores for the numbers involved in the question? How do the z-scores relate to the percentages associated with the standard normal curve?

Using a reasonably good sketch of the bell-shaped curve is usually a good idea. The places where the curve seems to "flex" on each side of the mean coincide with z-scores of 1 and -1.

Solutions to Odd-numbered Problems

1. 20%   Each of the intervals represents 20% 0f the distribution. Note that the intervals are not of the same length.

3. 40%   The interval from 57 to 67 spans two of the intervals. The basic principle is that the percentages of adjacent intervals are added.

5. The set with the highest mean is Data Set III, since its mean is located farthest to the right. The set with the lowest mean is Data Set I (farthest to the left).

7. The set with the highest standard deviation is Data Set II, since it has a wider spread than the others. The set with the lowest standard deviation is Data Set I (narrowest).

9.

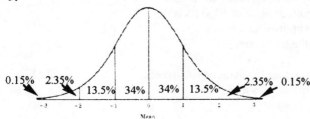

(a) 13.5% + 2.35% = 15.85%   **(b)** 2.35% + 0.15% = 2.5%   **(c)** 100% - 2(34%) = 32%

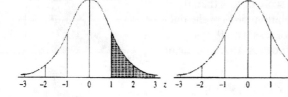
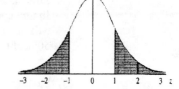

11.

    (a) 2.35%   **(b)** 100% - 2.5% = 97.5%   **(c)** 2 × 2.5% = 5%

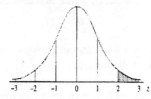
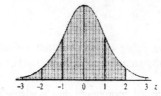
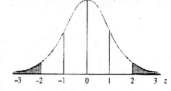

13. In order to find percentages in a normal population, we have to convert the data to z-scores.

Data point	z-score
10	$(10 - 12) \div 2 = -1$
14	$(14 - 12) \div 2 = 1$

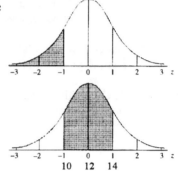

(a) Weighing "less than 10 pounds" is the same as having a z-score of less than -1. 16% of the turkeys weigh less than 10 pounds.

(b) Weighing "between 10 pounds and 14 pounds" is the same as having a z-score between -1 and 1. 68% of the turkeys weigh between 10 and 14 pounds.

(c) In any large number of turkeys from the ranch, 16% should be expected to weigh less than 10 pounds. Out of 1000 turkeys, there would be approximately 160 weighing less than 10 pounds.

15. First, the given data must be changed to z-scores.

Data point	z-score
35	$(53 - 40) \div 5 = -1$
40	$(40 - 40) \div 5 = 0$

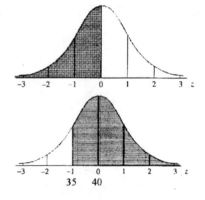

(a) Lasting "less than 40,000 miles" is the same as having a z-score of less than 0. In any normal population, 50% of the data will have a z-score of less than 0. 50% of the data will be less than the mean.

(b) Lasting "at least 35,000 miles" is the same as having a z-score greater than -1. 84% of the tires will last at least 35,000 miles.

17. 44.52%   From Table 4.1, we see that a z-score of 1.6 is associated with an area of 0.4452 that is under the curve between 0.0 and 1.6.

19. 25.80%   Table 4.1 represents the "positive" z-scores of a symmetric distribution. The area under the curve between 0.0 and -0.7 is the same as the area under the curve between 0.0 and +0.7. From Table 4.1, we see that a z-score of 0.7 is associated with an area of 0.2580 that is under the curve between 0.0 and 0.7.

21. 78.35%   When the z-scores are on opposite sides of the mean and we want the area between the z-scores, we add the respective areas.

z-score	relevant area
-1.1	0.3643
1.4	0.4192
	0.7835

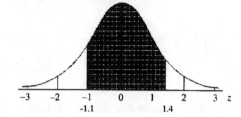

23. 12.75%   When the z-scores are on the same side of the mean and we want the area between the z-scores, we subtract the respective areas.

z-score	relevant area
2.4	0.4918
1.1	0.3643
	0.1275

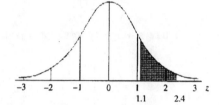

25. 3.59%   When we want the area for one of the "tips" of the distribution, we subtract the relevant area from 0.5000 because half of the distribution is on each side of the mean.

z-score	relevant area
z > 0.0	0.5000
1.8	0.4641
	0.0359

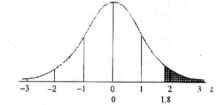

27. 96.41%   When we want the area that is before the right hand "tip" of the distribution or after the left hand "tip" of the distribution, we add the relevant area to 0.5000 because half of the distribution is on each side of the mean, and will be included.

z-score	relevant area
z < 0.0	0.5000
1.8	0.4641
	0.9641

Note: Contrast this problem to problem 25.

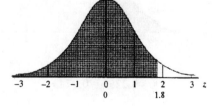

29. Scores   x   9   10   11   14   17

z-scores $\dfrac{x-10}{2}$ -0.5   0.0   0.5   2.0   3.5

31. Scores   x   64   80   96   111   136   145

z-scores $\dfrac{x-100}{15}$ -2.400 -1.333 -0.267 0.733 2.400 3.000

33. Scores   x   45.16 49.82 55.20 58.63

z-scores $\dfrac{x-55.6}{11.3}$ -0.924 -0.512 -0.035 0.268

35. First, we will convert the data points to z-scores. Then we find the relevant areas related to each z-score.

Scores	z-scores	Area (between 0 and z)
9	-1.5	0.4332
13	0.5	0.1915
16	2.0	0.4772

Turkeys weighing less than 9 pounds   Turkeys weighing between 13 and 16 pounds

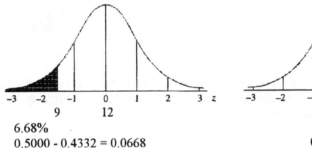 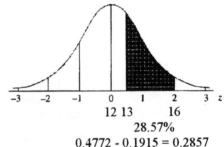

6.68%   28.57%

0.5000 - 0.4332 = 0.0668   0.4772 - 0.1915 = 0.2857

37. First, we will convert the data points to z-scores. Then we find the relevant areas related to each z-score.

Scores	z-scores	Area (between 0 and z)
37000	-0.6	0.2257
53000	2.6	0.4953

Tires lasting less than 37000 mi.   Tires lasting more than 53000 mi.

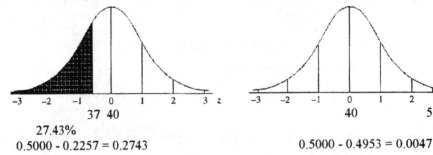

27.43%   0.47%

0.5000 - 0.2257 = 0.2743   0.5000 - 0.4953 = 0.0047

39. The distribution has a mean of 40,000 mi. and a standard deviation of 5,000 mi.
A value of 30,000 mi. has a z-score of -2.0.
The related area under the standard normal curve is 0.4772 (this represents the interval from 30,000 to 40,000). The area under the curve for values less than 30,000 is 0.0228.

2.28% of the tires will last for less than 30,000 mi. and may have to be replaced.
For each lot of 1000 tires, the company can expect to replace 22.8 tires (on average).
At a cost of $86 each, the company can expect to pay (22.8 × $86) = $1960.80.

Section 4.2   Confidence Intervals and Reliable Estimation

Goals

1. Learn to compute and interpret the mean and standard deviation of the set of sample proportions.
2. Be able to use the standard error to compute confidence intervals and margins of error.

Key Ideas and Questions

1. Explain the meaning of "95% confidence interval"..
2. Why does sampling allow for meaningful results to come from surveys of relatively small numbers of people or objects from a population?
2. How does the size of the sample affect the confidence interval?

Vocabulary

Population Proportion   Standard Error
Sample Proportion   95% Confidence Interval

. . . .

Overview

In this section you will use information about how the means or proportions of samples are distributed, together with information about the standard deviations of the samples, to make conclusions about the populations. You will also learn to assign a level of confidence to your results.

Sample Proportion

A sample proportion compares a portion of a sample with the entire sample. A sample proportion is represented by the symbol $\hat{p}$, called "*p hat*". For a sample of size *n*, we compute the sample proportion as follows.

Sample Proportions

If a sample of size *n* is selected from a population, then the sample proportion of a particular group in the sample is given by

$$\hat{p} = \frac{\text{number sampled that belong to the group}}{n}$$

For a population of even relatively small size, say $N = 100$, and a relatively small sample size, say $n = 20$, the number of possible samples is in excess of 5×10^{20}. The amazing thing is, as the sample size increases, the variability in sample proportions decreases, and the histogram of the sample proportions begins to approach the bell-shaped curve.

Distribution of Sample Proportions

By using an unbiased sample of large enough size, you can ensure that it is very likely the sample proportion will be close to the population proportion. The two main facts about the distribution of sample proportions are:

> If samples of size n are taken from a population having a population proportion p, then the set of all sample proportions has a mean of p and
>
> a standard deviation of $\sqrt{\dfrac{p(1-p)}{n}}$.

It is useful to know the mean and standard deviation for the sample proportion because once the sample size is at least 30, then the set of all proportions for samples of that size taken from the population is represented by a nearly bell-shaped curve. When you know this, then you can estimate percentages that lie under any part of the curve.

The Standard Error

When we have figured the sample proportion, we need to estimate how close it is to the true population proportion. The following formula helps determine how close our results are:

> #### Standard Error
>
> Given a sample of size n with sample proportion $\hat{p}$, the standard deviation of the set of all sample proportions is approximately
>
> $$\hat{s} = \sqrt{\dfrac{\hat{p}(1-\hat{p})}{n}}$$
>
> which is known as the **standard error** of the sample.

Confidence Intervals

In any normal distribution, 95% of the data must be within two standard deviations of the mean. If the standard deviation is not known, we use the standard error and conclude that in 95% of the cases, the population proportion is within two standard errors of the sample proportion. So, a **95% confidence interval** is the interval of numbers from $\hat{p} - 2\hat{s}$ to $\hat{p} + 2\hat{s}$, as shown in the figure.

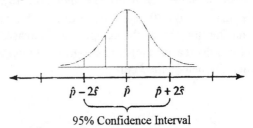

95% Confidence Interval

Any value in the 95% confidence interval is a reasonable estimate for the population proportion p.

The value $2\hat{s}$ is called the **margin of error** for the estimate of the population proportion. As the sample size increases, the margin of error decreases.

. . . .

Suggestions and Comments for Odd-numbered problems

5. Calculate the proportion of humanities courses in each of the possible selections. Treat these fractions as a population and calculate the mean in the usual manner.

11. through 17.
The standard error uses the same formula as the one used to calculate the standard deviation in the set of all sample proportions. However, it uses the proportion from the sample instead of the proportion from the population (which is generally unknown).

19. through 25.
Finding the values for confidence intervals.
The answers are based on the following guidelines for calculations:
1. Population proportion, sample proportion, and confidence intervals will be given to the nearest 0.1%.
2. Results from one step will be rounded before going to the next.
3. Calculations of standard error will use proportions rounded to the nearest percent.

If you want to carry more precision, the answers may vary by 0.1% for the 95% confidence interval.

If there is a choice between rounding up and rounding down in the calculation of the standard error, technically you should always round "up". This will increase the width of the interval and increase the confidence level (if ever so slightly). We won't require such precision, however.

27. and 29.
Confidence intervals can be calculated to correspond to any level of confidence. The structure of the calculation is always the same. What changes is the number of standard errors involved. The wider the interval, the higher the level of confidence that is involved. In which of the following two statements would you have more confidence, in the ordinary sense?

> The true answer is between 10 and 11.
> The true answer is between 0 and 21.

Obviously, whenever the first one is true, the second is automatically true. Conversely, the second can be true while the first one is false.

31. First, you can to get a good estimate of the number of gigaxes that might be returned. Find the 99.7% confidence interval for the *proportion* of gigaxes that will be defective. Then, knowing the company's production level, you can estimate the *number* of defective gigaxes.

Solutions to Odd-numbered Problems

1. The population is the 6000 cars produced. Population proportion $= \frac{300}{6000} = 0.05$.

   The sample is the 600 cars selected. Sample proportion $= \frac{5}{60} = 0.083$.

3. The population is the registered voters. Population proportion $= \frac{3250}{7140} = 0.455$.

   The sample is the voters in the area canvassed. Sample proportion $= \frac{210}{420} = 0.5$.

5. **(a)** The population proportion for Humanities courses $= \frac{3}{5} = 0.6$

   (b) The proportions in each of the samples are:
   $$\frac{1}{3}, \frac{1}{3}, \frac{1}{3}, \frac{2}{3}, \frac{2}{3}, \frac{2}{3}, \frac{2}{3}, \frac{2}{3}, \frac{2}{3}, 1$$

   (c) The distribution of the sample proportions is:

   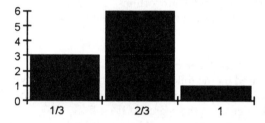

   (d) The mean of the sample proportions $= \dfrac{3 \times \frac{1}{3} + 6 \times \frac{2}{3} + 1}{10} = \dfrac{1 + 4 + 1}{10} = 0.6$

7. The population proportion $= \frac{300}{6000} = 0.05$.

   The mean of the sample proportions equals the population proportion $= 0.05$.

   The standard deviation of sample proportion $= \sqrt{\dfrac{(0.05)(0.95)}{60}} = 0.028$

9. The population proportion $= \frac{3250}{7140} = 0.455$.

   The mean of the sample proportions equals the population proportion $= 0.455$.

   The standard deviation of sample proportion $= \sqrt{\dfrac{(0.455)(0.545)}{200}} = 0.035$

11. Standard error $= \sqrt{\dfrac{(0.45)(0.55)}{600}} = 0.0203$

    Compare this to the calculation in problem 9. The proportions are about the same, but the larger sample size produces a lower standard error.

13. Standard error $= \sqrt{\dfrac{(0.65)(0.35)}{640}} = 0.0189$

15. The sample proportion is $\hat{p} = \frac{265}{500} = 0.53$

    Standard error $= \sqrt{\dfrac{(0.53)(0.47)}{500}} = 0.0223$

17. The sample proportion is $\hat{p} = \frac{128}{220} = 0.58181818...$

Since the calculation is not "sensitive" to small differences in $\hat{p}$, we will round sample proportions to two decimal places.

Standard error $= \sqrt{\frac{(0.58)(0.42)}{220}} = 0.033276$ or 0.0333

If we use 0.582 in the calculation, we get 0.033254.

19. (a) Sample Proportion $= \frac{209}{431} = 0.485$ or 48.5%

Standard error $= \sqrt{\frac{(0.48)(0.52)}{431}} = 0.02406$ or 2.4%

(b) 95% confidence interval = sample proportion $\pm 2 \times$ standard error

48.5% $\pm 2 \times$ (2.4%) = 48.5% $\pm$ 4.8% or $\boxed{43.7\% \text{ to } 53.3\%}$.

21. (a) Sample Proportion $= \frac{172}{280} = 0.6143$ or 61.4%

Standard error $= \sqrt{\frac{(0.61)(0.39)}{280}} = 0.02915$ or 2.9%

(b) 95% confidence interval = sample proportion $\pm 2 \times$ standard error

61.4% $\pm 2 \times$ (2.9%) = 61.4% $\pm$ 5.8% or $\boxed{55.6\% \text{ to } 67.2\%}$.

23. (a) Sample Proportion $= \frac{442}{535} = 0.82617$ or 82.6%

Standard error $= \sqrt{\frac{(0.83)(0.17)}{535}} = 0.01624$ or 1.6%

(b) 95% confidence interval = sample proportion $\pm 2 \times$ standard error

82.6% $\pm 2 \times$ (1.6%) = 82.6% $\pm$ 3.2% or $\boxed{79.4\% \text{ to } 85.8\%}$.

If as many digits as possible had been carried in the calculator, the 95% confidence interval would be 79.34001% to 85.89364%.
See the discussion under "Hints."

25. (a) Sample Proportion $= \frac{105}{240} = 0.4375$ or 43.8%

Standard error $= \sqrt{\frac{(0.44)(0.56)}{240}} = 0.03204$ or 3.2%

(b) 95% confidence interval = sample proportion $\pm 2 \times$ standard error

43.8% $\pm 2 \times$ (3.2%) = 43.8% $\pm$ 6.4% or $\boxed{37.4\% \text{ to } 50.2\%}$.

27. From problem 21: Population proportion = 61.4%; standard error = 2.9%

(a) 90% confidence interval = sample proportion $\pm 1.65 \times$ standard error

61.4% $\pm 1.65 \times$ (2.9%) = 61.4% $\pm$ 4.8% or $\boxed{56.6\% \text{ to } 66.2\%}$.

(b) 99.7% confidence interval = sample proportion $\pm 3 \times$ standard error

61.4% $+ 3 \times$ (2.9%) = 61.4% $\pm$ 8.7% or $\boxed{52.7\% \text{ to } 70.1\%}$.

29. From problem 23: Population proportion = 82.6%; standard error = 1.6%

    (a) 90% confidence interval = sample proportion ± 1.65 × standard error

        82.6% ± 1.65 × (1.6%) = 82.6% ± 2.6% or $\boxed{80.0\% \text{ to } 85.2\%}$.

    (b) 99% confidence interval = sample proportion ± 2.58 × standard error

        82.6% ± 2.58 × (1.6%) = 82.6% ± 4.1% or $\boxed{78.5\% \text{ to } 86.7\%}$.

31. First, we need to find a 99.7% confidence interval for the proportion of defective gigaxes.

$$\text{Sample proportion} = \frac{28}{800} = 0.035 \text{ or } 3.5\%$$

$$\text{Standard error} = \sqrt{\frac{(0.035)(0.965)}{800}} = 0.0065 \text{ or } 0.65\%$$

99.7% confidence interval = sample proportion ± 3 × standard error

3.5% ± 3 × (0.65%) = 3.5% ± 1.95% or 1.55% to 5.45%

We interpret this to mean that from 1.55% to 5.45% of all the gigaxes manufactured will be defective. If the company produces 100,000 gigaxes, then from 1,550 to 5,540 will be defective. At $30 per gigax, the company can anticipate spending from $46,500 to $163,500 to cover guarantees (proportions were rounded to 4 places).

Chapter 4 Review Problems

Solutions to Odd-numbered Problems

1. 83.85%, 0.15%, and 16% or 84%, 0.13%, and 15.87% (depending on source)
If we use the approximate values from Figure 4.9, we have the following:
For $-1 < z < 3$, the percentage is 83.85% (0.34 + 0.34 + 0.135 + 0.0235 = 0.8385)
For $3 < z$, the percentage is 0.15% (0.0015)
For $z < -1$, the percentage is 16% (0.135 + 0.0235 + 0.0015 = 0.1600)

If we use Table 4.1, the values are:
For $-1 < z < 3$, the percentage is 84% (0.3413 + 0.4987)
For $3 < z$, the percentage is 0.13% (0.0013)
For $z < -1$, the percentage is 15.87%% (0.5000 - 0.3413)

These values sum to 100% because they cover ALL possible values with no duplication.

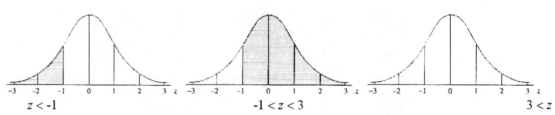

$$z < -1 \qquad\qquad\qquad -1 < z < 3 \qquad\qquad\qquad 3 < z$$

3. 2.28% and 15.87%
   First, we compute the z-scores for each of the values in question; then we use Table 4.1

   $$x = 210, z = \frac{210 - 250}{20} = -2; \qquad x = 270, z = \frac{270 - 250}{20} = +1$$

   If x < 210, then z < -2
   The percentage of a standard normal population less than -2 is 2.28% (0.5 - 0.4772)

   If x > 270, then z > +1
   The percentage of a standard normal population greater than 1 is 15.87% (0.5 - 0.3413)

5. 4.46%

   First, we compute the z-score for 40 tests: $z = \frac{40 - 54.5}{8.4} \approx -1.726$

   We round this to the nearest 0.1, since Table 4.1 is in tenths.
   From Table 4.1, the percentage related to z = 1.7 is 0.4554 (area from 0 to 1.7)
   Since the region we're interested in is "outside" this value of z, we subtract:
   0.5 - 0.4554 = 0.0446 or 4.46%

7. The sample proportion of chocolates is 30% (9 ÷ 30 = 0.30), so the best guess for the
   population proportion is also 30%. This inference is based on the assumption that the

   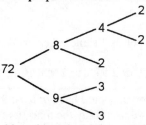

   method is unbiased and the sample is representative of the population. However, a
   different sample would probably produce another estimate. We can eliminate bias, but
   we can't eliminate variability.

   The best guess for the total number of chocolates is 30% of 200, or 60.

9. First, the population proportion is 25% (5,000 ÷ 20,000).
   For samples of size 48, the mean of the sample proportions will be 25%, and the

   standard deviation of the sample proportion is $\sqrt{\frac{(0.25)(0.75)}{48}}$ = 0.0625.

   95% of the sample proportions will fall within 2 standard deviations of the mean of the
   proportions: 25% ± 2 × 6.25% or $\boxed{12.5\% \text{ to } 37.5\%}$.

   Note: We have not found a 95% confidence interval for the population proportion. Our
   answer is an exact description of how the sample proportions are distributed. There is a
   subtle, but very important, difference.

11. **(a)** mean = 0.63 (or 63%)
The mean of all sample proportions is the same as the population proportion

standard deviation = 0.05 $SD = \sqrt{\dfrac{(0.63)(0.37)}{93}} = 0.05007$

(b) For p = 0.5, we have $z = \dfrac{0.50 - 0.63}{0.05} = -2.6$
The area under the curve for z < -2.6 is 0.0047 (0.5000 - 0.4953)

(c) For p = 0.7, we have $z = \dfrac{0.70 - 0.63}{0.05} = 1.4$
The area under the curve for z > 1.1 is 0.0808 (0.5000 - 0.4192)

13. **(a)** We begin with $s = \sqrt{\dfrac{p(1-p)}{n}}$ and square both sides: $s^2 = \dfrac{p(1-p)}{n}$

Multiply by n: $ns^2 = p(1-p)$ and divide by s^2: $n = \dfrac{p(1-p)}{s^2}$

(b) For $s = 0.01$, we have $n = \dfrac{(0.05)(0.95)}{(0.01)^2} = 475$
The sample size must be at least 475.

15. **(a)** The margin of error (ME) for the poll is 2 standard errors

$ME = 2 \times \sqrt{\dfrac{(0.66)(0.34)}{1039}} \approx 0.02939$ or 2.94% or 2.9% using standard rounding.

Note: Since a 95% confidence interval should have *at least* 95%, rounding should actually be "outward". Therefore we would say the margin of error was 3.0%.
It is also a commonly accepted practice to carry one more decimal place in the standard error or margin of error than in the sample proportion..

(b) 95% confidence interval: $\hat{p} \pm 2 \times \sqrt{\dfrac{(0.66)(0.34)}{1039}} \approx \hat{p} \pm 3.0\%$

95% confidence interval: 63.0% to 69.0%

(c) 99.7% confidence interval: $\hat{p} \pm 3 \times \sqrt{\dfrac{(0.66)(0.34)}{1039}} \approx \hat{p} \pm 0.044,$ or $\hat{p} \pm 4.4\%$

99.7% confidence interval: 61.6% to 70.4%

If standard rounding is used, the 95% confidence interval would be 63.1% to 68.9%, and the 99.7% confidence interval would be 61.6% to 70.4%

17. In each situation, the sample proportion is 60% (120/200 and 480/800)

(a) For $\hat{p} = 0.6$ and n = 200, ME = $2 \times \sqrt{\dfrac{(0.6)(0.4)}{200}} \approx 6.928\%$

For $\hat{p} = 0.6$ and n = 800, ME = $2 \times \sqrt{\dfrac{(0.6)(0.4)}{800}} \approx 3.464\%$

(b) The margin of error in part (2) is one-half the margin of error in part (1).

(c) The sample size in part (2) was 4 times the sample size in part (1), and the margin of error was $\dfrac{1}{2}$ as large. As sample size gets larger, the margin of error gets reduced by a factor that is the square root of the factor by which the sample size increases. For a sample of size 1800 (9 times as large) we would expect the margin of error to be reduced by a factor of 3, or, ME $\approx 6.928\% \div 3 \approx 2.309\%$

Confirmation: ME = $2 \times \sqrt{\dfrac{(0.6)(0.4)}{1800}} \approx 0.023094$

5 Probability

Section 5.1   Computing Probabilities in Simple Experiments

Goals
1. Compute probabilities of events in situations where all outcomes have the same chance of occurring.

Key Ideas and Questions
1. Describe the main features of a probability model with equally likely outcomes, including the sample space, events, outcomes and how to assign probabilities to events.
2. Illustrate the main features of a probability model with an experiment such as tossing two dice or spinning a roulette wheel with 38 slots.
3. Compare theoretical and experimental probability.
4. Compare and contrast "Odds" and "Probability"

Vocabulary

Experiment	Equally Likely Outcomes	Mutually Exclusive
Outcome	Theoretical Probability	Complement of an Event
Sample Space	Fair Coin/Fair Die	Properties of Probability
Event	Union	Odds For/Odds Against
Experimental Probability	Intersection	

· ·

Overview

Probability is the mathematics of chance. This section introduces the language and general concepts of probability and the rules governing it.

If we repeat an experiment over and over, the fraction of times the event occurs should be the probability of the event. The probability is always between zero and one. Probabilities can be given as fractions, decimals, or percents since we can readily change from one form to the other.

An **experiment** is making an observation, or taking a measurement of some act. An **outcome** is one of the possible results of an experiment. The set of all possible outcomes is called a **sample space.** An **event** is any collection of the possible outcomes. An event is a subset of the sample space; that is, an event is a collection of specific outcomes.

Equally Likely Outcomes

One way to find the probability of event E is to make many repetitions of the experiment and determine the frequency with which E occurs. The relative frequency of E occurring is called its **experimental probability.**

Probability of an Event with Equally Likely Outcomes

Suppose the outcomes in the sample space S are equally likely to occur. Let E be an event. Then the **probability of event E**, denoted **P(E)**, is

$$P(E) = \frac{\text{number of outcomes in E}}{\text{number of outcomes in S}}$$

If you consider just one outcome from a set of equally likely outcomes, its probability is 1 divided by the number of outcomes in the entire sample space. Recall that the probability of any event is a number from 0 to 1. The event containing no outcomes has probability zero, and the event containing all the outcomes in the sample space has probability one.

Since we are never really sure that a die or coin in the real world is perfectly balanced, we are never really sure that the outcomes in the sample space are *equally* likely. When we apply the definition, we are computing **theoretical probabilities** which work quite well for real world problems. The convention used to indicate the theoretical probabilities of an ideal coin or die is to refer to them as a **fair coin** or a **fair die.**

The **union** of two events $(A \cup B)$ refers to all outcomes that are in one or the other or both events. The **intersection** $(A \cap B)$ refers to outcomes that are in both events. Events with no outcomes in common are called **mutually exclusive**.

Probability of Mutually Exclusive Events

If L and M are mutually exclusive events, then $P(L \cup M) = P(L) + P(M)$.

The set of outcomes in the sample space S, but not in event E, is called the **complement of event E**, written $\overline{E}$.

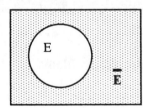 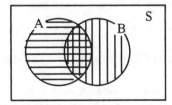

In the second figure above, notice that the region $A \cap B$ is shaded twice--once from A and once from B. It follows that for any sets A and B, to find the number of elements in $A \cup B$, we find the sum of the numbers of elements in set A and set B, but we then subtract the number of elements in $A \cap B$ so they are not counted twice. Hence,

$$P(A \cup B) = P(A) + P(B) - P(A \cap B)$$

Properties of Probability

1. For any event A, $0 \leq P(A) \leq 1$.
2. $P(\varnothing) = 0$.
3. $P(S) = 1$, where S is the sample space.
4. For mutually exclusive events A and B, $P(A \cup B) = P(A) + P(B)$.
5. For any events A and B, $P(A \cup B) = P(A) + P(B) - P(A \cap B)$.
6. If $\overline{A}$ denotes the complement of event A, then $P(A) = 1 - P(\overline{A})$.

The Odds For and Against

A concept related to probability that is used as a way to express the chances of an event happening or not happening is "Odds". In general, the "**Odds for**" (or the "Odds in favor of") an event is expressed as a ratio of the probability the event *will* happen to the probability that the event *won't* happen (the complement). The "**Odds against**" an event is the ratio of the probability that the event *won't* happen to the probability that it *will* happen. For example, if the probability of the event is $\frac{3}{5}$, then the probability of

the complement is $\frac{2}{5}$. The ratio is usually expressed with whole numbers, so we would

say the odds *for* the event are 3 : 2 (or 3 to 2) while the odds *against* the event are 2 : 3. If the odds *for* (or *in favor of*) an event are given, the probability of the event can be calculated as follows:

Computing Probability from the Odds

If the Odds for an event are $a : b$, then the probability of the event is given by

$$P(E) = \frac{a}{a + b}$$

When an event and its complement are equally likely (that is, $P(E) = 0.5$), we most often say the odds are 1:1 (note: 50:50 is often used). In common use, the terms "odds" is often used incorrectly in place of "probability". The ratios used are most often rounded to whole numbers that are easy to use. For example, if the probability of an event is 0.617, we would say the odds for are 3 : 2 rather than 617 : 383.

. . . .

Suggestions and Comments for Odd-numbered problems

5. When 4 coins are tossed, and each coin is considered individually, there are 16 different sequences of heads and tails that can occur. They can be written down in a pattern based on the number of ways to get from all heads to no heads.

11. There are eight different ways the coins can come up, if you distinguish one coin from another. Consider tossing a penny, a nickel, and a dime.

13. Since each of the pairs is equally likely, carefully count the pairs that match the description of the event.

15. If part of a name is in a sector, then the sector has that color.

17. Review Example 5.2(d). Renumber the faces of the dice 1 through 4.

19. Modify Example 5.2(d) for dice whose faces are numbered {1, 2, 3, 4, 5, 6}. The faces of these dice are numbered {2, 2, 3, 3, 3, 5}.

25. List all possible pairs of marbles in the order in which they are drawn. That is, (R,G) is different than (G,R). Read the description of the complement of the event in the problem carefully.

29. Review Example 5.2(b).

Solutions to Odd-numbered Problems

1. (iii) To say there is a 20% chance of snow in the county tomorrow is to say that in the past, when conditions have been similar to those now, it has snowed 20% of the time.

3. (a) {H, T}

   (b) {A, B, C, D, E, F}

   (c) {1, 2, 3, 4}

5. (a) When 4 coins are tossed, and each coin is considered individually, there are 16 different sequences of heads and tails that can occur. They can be written down in a pattern.

       HHHH
       HHHT, HHTH, HTHH, THHH
       HHTT, HTHT, HTTH, TTHH, THTH, THHT
       HTTT, THTT, TTHT, TTTH
       TTTT

   (b) A head appears on the first coin in 8 sequences:
       {HHHH, HHHT, HHTH, HTHH, HHTT, HTHT, HTTH, HTTT}

   (c) There are 4 sequences that contain 3 heads:
       {HHHT, HHTH, HTHH, THHH}

   (d) All sequence have to end with either a head or a tail, so the event is the entire sample space.

   (e) There are 4 sequences with a head on the second coin and a tail on the third coin:
       {HHTH, HHTT, THTH, THTT}

7. The sample space can be constructed in the following way:

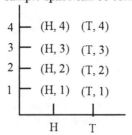

9. The experimental probability of an event is the relative frequency of the event based on the number of trials taken. $P(E) = \dfrac{\text{"successes"}}{\text{trials}}$, where a success is a trial in which the event occurs.

(a) $\dfrac{12 \text{ successes}}{60 \text{ trials}}$: $P(E) = \dfrac{1}{5}$

(b) $\dfrac{(10 + 10 + 8) \text{ successes}}{60 \text{ trials}}$: $P(E) = \dfrac{28}{60} = \dfrac{7}{15}$

(c) $\dfrac{(12 + 8 + 11) \text{ successes}}{60 \text{ trials}}$: $P(E) = \dfrac{31}{60}$

11. (a)

outcome	0	1	2	3
Probability	$\dfrac{1}{8}$	$\dfrac{3}{8}$	$\dfrac{3}{8}$	$\dfrac{1}{8}$

(b) $P(E) = \dfrac{3}{8} + \dfrac{3}{8} + \dfrac{1}{8} = \dfrac{7}{8}$; or

$P(\text{not getting a head}) = \dfrac{1}{8}$

$P(\text{getting at least one head}) = 1 - P(\text{not getting a head}) = 1 - \dfrac{1}{8} = \dfrac{7}{8}$.

13. (a) $\dfrac{6}{36} = \dfrac{1}{6}$ (b) $\dfrac{9}{36} = \dfrac{1}{4}$

(1, 1)	(1, 2)	(1, 3)	**(1, 4)**	(1, 5)	(1, 6)
(2, 1)	(2, 2)	(2, 3)	**(2, 4)**	(2, 5)	(2, 6)
(3, 1)	(3, 2)	(3, 3)	**(3, 4)**	(3, 5)	(3, 6)
(4, 1)	(4, 2)	(4, 3)	**(4, 4)**	(4, 5)	(4, 6)
(5, 1)	(5, 2)	(5, 3)	**(5, 4)**	(5, 5)	(5, 6)
(6, 1)	(6, 2)	(6, 3)	**(6, 4)**	(6, 5)	(6, 6)

(1, 1)	(1, 2)	(1, 3)	(1, 4)	(1, 5)	(1, 6)
(2, 1)	**(2, 2)**	(2, 3)	**(2, 4)**	(2, 5)	**(2, 6)**
(3, 1)	(3, 2)	(3, 3)	(3, 4)	(3, 5)	(3, 6)
(4, 1)	**(4, 2)**	(4, 3)	**(4, 4)**	(4, 5)	**(4, 6)**
(5, 1)	(5, 2)	(5, 3)	(5, 4)	(5, 5)	(5, 6)
(6, 1)	**(6, 2)**	(6, 3)	**(6, 4)**	(6, 5)	**(6, 6)**

(c) $\dfrac{21}{36} = \dfrac{7}{12}$ (d) 0

(1, 1)	(1, 2)	(1, 3)	(1, 4)	(1, 5)	**(1, 6)**
(2, 1)	(2, 2)	(2, 3)	(2, 4)	**(2, 5)**	**(2, 6)**
(3, 1)	(3, 2)	(3, 3)	(3, 4)	**(3, 5)**	**(3, 6)**
(4, 1)	(4, 2)	**(4, 3)**	**(4, 4)**	**(4, 5)**	**(4, 6)**
(5, 1)	**(5, 2)**	**(5, 3)**	**(5, 4)**	**(5, 5)**	**(5, 6)**
(6, 1)	**(6, 2)**	**(6, 3)**	**(6, 4)**	**(6, 5)**	**(6, 6)**

(1, 1)	(1, 2)	(1, 3)	(1, 4)	(1, 5)	(1, 6)
(2, 1)	(2, 2)	(2, 3)	(2, 4)	(2, 5)	(2, 6)
(3, 1)	(3, 2)	(3, 3)	(3, 4)	(3, 5)	(3, 6)
(4, 1)	(4, 2)	(4, 3)	(4, 4)	(4, 5)	(4, 6)
(5, 1)	(5, 2)	(5, 3)	(5, 4)	(5, 5)	(5, 6)
(6, 1)	(6, 2)	(6, 3)	(6, 4)	(6, 5)	(6, 6)

15. Assuming that all sectors (portions of the circle) are equally likely, the probability that the spinner will stop on yellow is;

$$P(Y) = \dfrac{3 \text{ yellow sectors}}{8 \text{ sectors total}} = \dfrac{3}{8}$$

17. (a) {2, 3, 4, 5, 6, 7, 8}

(b)

(i) $P = \frac{1}{4}$	(ii) $P = \frac{1}{2}$	(iii) $P = \frac{13}{16}$

(1,1)	(1,2)	(1,3)	**(1,4)**	**(1,1)**	(1,2)	**(1,3)**	(1,4)	(1,1)	(1,2)	**(1,3)**	**(1,4)**
(2,1)	(2,2)	**(2,3)**	(2,4)	(2,1)	**(2,2)**	(2,3)	**(2,4)**	(2,1)	**(2,2)**	**(2,3)**	**(2,4)**
(3,1)	**(3,2)**	(3,3)	(3,4)	**(3,1)**	(3,2)	**(3,3)**	(3,4)	**(3,1)**	**(3,2)**	**(3,3)**	**(3,4)**
(4,1)	(4,2)	(4,3)	(4,4)	(4,1)	**(4,2)**	(4,3)	**(4,4)**	**(4,1)**	**(4,2)**	**(4,3)**	**(4,4)**

19. Sample space = {2,2,3,3,3,5}

(a) $\dfrac{2 \text{ sides with a "2"}}{6 \text{ sides}} = \dfrac{2}{6} = \dfrac{1}{3}$ **(b)** $\dfrac{4 \text{ sides without a "2"}}{6 \text{ sides}} = \dfrac{4}{6} = \dfrac{2}{3}$

(c) $\dfrac{4 \text{ odd-numbered sides}}{6 \text{ sides}} = \dfrac{4}{6} = \dfrac{2}{3}$ **(d)** $\dfrac{2 \text{ even-numbered sides}}{6 \text{ sides}} = \dfrac{2}{6} = \dfrac{1}{3}$

Note: The answers to parts (b) and (d) could have been found from parts (a) and (c), respectively, by using the complement.

21. (a) The sample space for two spins of the spinner events and the events A, B, A ∩ B, and A ∪ B.

Event A: $P(A) = \frac{1}{4}$	Event B: $P(B) = \frac{1}{4}$

BB	BG	BR	BY		BB	BG	BR	**BY**
GB	**GG**	**GR**	**GY**		GB	GG	GR	**GY**
RB	RG	RR	RY		RB	RG	RR	**RY**
YB	YG	YR	YY		YB	YG	YR	**YY**

Event A ∩ B: $P(A \cap B) = \frac{1}{16}$	Event A ∪ B: $P(A \cup B) = \frac{7}{16}$

BB	BG	BR	BY		BB	BG	BR	**BY**
GB	GG	GR	**GY**		**GB**	**GG**	**GR**	GY
RB	RG	RR	RY		RB	RG	RR	**RY**
YB	YG	YR	YY		YB	YG	YR	**YY**

(b) $P(A \cup B) = P(A) + P(B) - P(A \cap B) = \frac{1}{4} + \frac{1}{4} - \frac{1}{16} = \frac{7}{16}$

23. (a) and (b) The sample space for tossing two 4-sided dice and the related events A, B, A ∩ B, and A ∪ B.

Event A: $P(A) = \frac{1}{4}$	Event B: $P(B) = \frac{1}{2}$

(1,1)	(1,2)	(1,3)	(1,4)		(1,1)	**(1,2)**	(1,3)	**(1,4)**
(2,1)	(2,2)	(2,3)	(2,4)		(2,1)	**(2,2)**	(2,3)	**(2,4)**
(3,1)	**(3,2)**	**(3,3)**	**(3,4)**		(3,1)	**(3,2)**	(3,3)	**(3,4)**
(4,1)	(4,2)	(4,3)	(4,4)		(4,1)	**(4,2)**	(4,3)	**(4,4)**

Event A ∩ B: $P(A \cap B) = \frac{1}{8}$	Event A ∪ B: $P(A \cup B) = \frac{5}{8}$

(1,1)	(1,2)	(1,3)	(1,4)		(1,1)	**(1,2)**	(1,3)	**(1,4)**
(2,1)	(2,2)	(2,3)	(2,4)		(2,1)	**(2,2)**	(2,3)	**(2,4)**
(3,1)	**(3,2)**	(3,3)	**(3,4)**		**(3,1)**	**(3,2)**	**(3,3)**	**(3,4)**
(4,1)	(4,2)	(4,3)	(4,4)		(4,1)	**(4,2)**	(4,3)	**(4,4)**

(c) $P(A \cup B) = P(A) + P(B) - P(A \cap B) = \frac{1}{4} + \frac{1}{2} - \frac{1}{8} = \frac{5}{8}$

25. (a) The sample space and the related events.

Event A: $P(A) = \frac{5}{12}$ Event B: $P(\overline{A}) = \frac{7}{12}$

RG	RY	**RW**	**RG**	**RY**	RW
GR	**GY**	**GW**	GR	GY	GW
YR	YG	**YW**	**YR**	**YG**	YW
WR	WG	WY	**WR**	**WG**	**WY**

(b) $P(\overline{A}) = 1 - P(A) = 1 - \frac{5}{12} = \frac{7}{12}$

27. If $P(A) = \frac{1}{5}$, then $P(\overline{A}) = \frac{4}{5}$
  (a) Odds in favor = 1 : 4
  (b) Odds against = 4 : 1

29. (a) In problem 11, we found that the probability of getting all three heads was $\frac{1}{8}$. This means the probability of not getting three heads is $\frac{7}{8}$. The odds *in favor* of getting all heads is 1 : 7.
  (b) In problem 11, we also found that the probability of getting exactly one head was $\frac{3}{8}$. The probability of *not* getting one head is $\frac{5}{8}$. The odds *against* getting only one head is 5 : 3.

31 (a) m(S) is the measure of the sample space, which is the roadway.
    m(S) = 100 miles
  (b) m(A) = 20 miles
  (c) $P(A) = \frac{m(A)}{m(S)} = \frac{20}{100} = \frac{1}{5}$

Section 5.2 Computing Probabilities in Complex Experiments

Goals
1. Use tree diagrams to make probability computations.

Key Ideas and Questions
1. How does a probability tree diagram simplify computations in comparison with a tree diagram?
2. What are the additive and multiplicative properties of probability tree diagrams?
3. Illustrate the main features of a probability tree diagram with an example.

Vocabulary
Tree Diagram	Fundamental Counting Principle	Drawing With Replacement
One Stage Tree	Probability Tree Diagram	Drawing Without Replacement
Two Stage Tree	Additive Property of Trees	Multiplicative Property of Trees
Primary/Secondary Branches		

Overview

Sometimes it is difficult to make a list of all possible outcomes in an experiment. A **tree diagram** is a graphic device that can be used to represent the outcomes of an experiment. The simplest tree diagrams have **one stage** and are used when the experiment involves only one action. **Two stage trees** are used to represent experiments that are a sequence of two experiments.

When you look at a figure showing a two stage tree, you can see that we can compute the number of outcomes from looking at the tree diagram, rather than having to count all of them. This counting procedure leads to the following principle:

Fundamental Counting Principle

If an event A can occur in r ways, and (for each of these r ways) an event B can occur in s ways, then the number of ways events A and B can occur, in succession, is $r \times s$.

Probability Tree Diagrams

Tree diagrams can be used to determine probabilities in complex experiments. Tree diagrams that are labeled using the probabilities of the events are called **probability tree diagrams**. The idea that you can add the probabilities at the ends of branches in a probability tree diagram is summarized in the following property:

Additive Property of Probability Tree Diagrams

If a complex event E is the union of simple events $E_1, E_2, \ldots, E_n$, where each pair of the events is mutually exclusive, then

$$P(E) = P(E_1) + P(E_2) + \ldots + P(E_n).$$

The probabilities of the events $E_1, E_2, \ldots, E_n$ can be viewed as those associated with the ends of the branches in a probability tree diagram.

This property is an extension of the property $P(A \cup B) = P(A) + P(B)$ where A and B are mutually exclusive events.

When making experiments where items are drawn from a set, there are two options. If you draw object A and replace it, this is referred to as **drawing with replacement**. When you draw an object and do not replace it, the process is called **drawing without replacement**.

You can multiply the probabilities along a series of branches in a probability tree diagram to find the probability at the end of a branch. This property is based on the fundamental counting principle and is stated below.

> ### Multiplicative Property of Probability Tree Diagrams
>
> Suppose an experiment consists of a sequence of simpler experiments that are represented by branches of a probability tree diagram. Then the probability of any of the simpler experiments is the product of all the probabilities on its branch.

To summarize, the probability of a complex event can be found as follows:

1. Construct the appropriate probability tree diagram.
2. Assign probabilities to each branch in the diagram.
3. Multiply the probabilities along individual branches to find the probability of the outcome at the end of each branch.
4. Add the probabilities of the relevant branches, depending on the event.

. . . .

Suggestions and Comments for Odd-numbered problems

3. Typically, the serial number on a dollar bill (Federal Reserve note) begins with a letter, followed by 8 digits, and ending with a letter. However, the number of digits in the serial number is not important. How many choices are there for the <u>last digit</u> in the serial number?

7. and beyond
   Think of each problem as describing a sequence of activities. Draw the tree showing the possible results at each step in the process. Learn to interpret each branch of the tree as one set of choices for the sequence of activities.

15. The basic question when constructing a tree is: "How many choices are there for each activity at each stage in the process?"

25. After the first card has been drawn, how many cards are left? How many of these will make a pair with the one you've drawn?

29. How many ways are there to draw two cards in succession from the deck?
    How many ways are there to draw two hearts in succession from the deck?

33. There are 360° in a circle. How many degrees are accounted for?

39. Make a model of the sample space similar to the sample space for two regular dice, except list the sides as 1, 2, 2, 3, 3, 3 instead of 1, 2, 3, 4, 5, 6.

42. and beyond
    Many complex problems and relationships can be understood using a step-by-step approach and carefully analyzing what happens at each step. Follow the directions as they are given.

Solutions to Odd-numbered Problems

1.

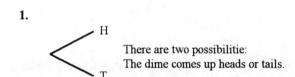

There are two possibilitie:
The dime comes up heads or tails.

3.

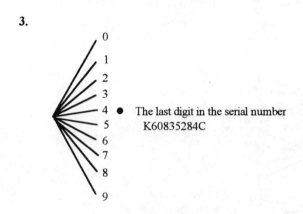

The last digit in the serial number
K60835284C

5.

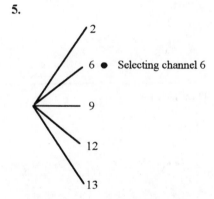

Selecting channel 6

7. The first stage of the tree diagram   The second stage of the tree diagram

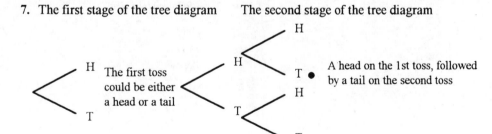

The first toss
could be either
a head or a tail

A head on the 1st toss, followed
by a tail on the second toss

9.

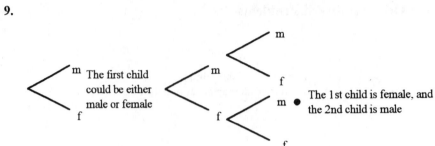

11.

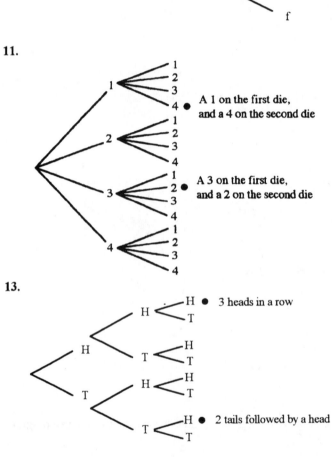

13.

15.

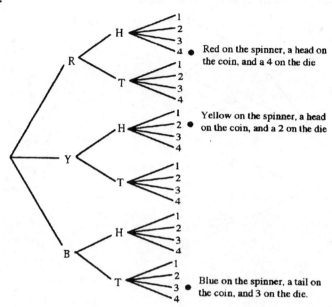

17.   (a) - (d)

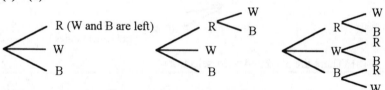

(e) There are a total of 6 possible outcomes.

19.

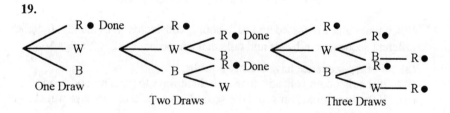

21. (a) 2   The coin can come up either heads or tails.

   (b) 6   We assume these are regular 6-sided dice

   (c) 6

   (d) 72 By the Fundamental Counting Principle,
         we have $2 \times 6 \times 6 = 72$

23. (a)   This is a multi-stage tree with three levels.

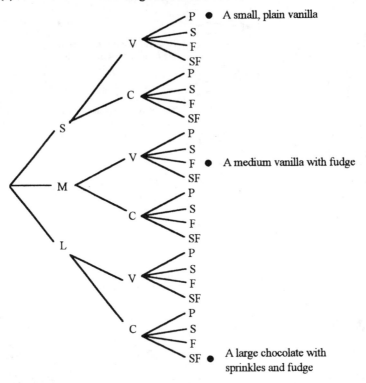

● A small, plain vanilla

● A medium vanilla with fudge

● A large chocolate with sprinkles and fudge

(b) There are $3 \times 2 \times 4 = 24$ selections

25. $P(\text{pair}) = \dfrac{7}{47}$ After the first card has been selected, there are 47 cards left, and 7 of the cards would provide a match.

27. When four coins are tossed and each coin is considered individually, there are 16 different sequences of heads and tails that can occur: $2 \times 2 \times 2 \times 2 = 16$.

(a)   6 of the sequence have exactly 2 heads.
(i)   If the first coin is heads, there are three ways to get a second head.
(ii)   If the first coin is tails and the second is heads, there are two ways to get a second head.
(iii)   If the first and second coins are tails, there is only one way to get two heads.
Note: The 6 sequences could also be obtained from problem 5, Section 5.1.

(b) $P(2 \text{ heads}) = \dfrac{6}{16} = \dfrac{3}{8}$ All sequences are equally likely

29. (a) The first card can be drawn 52 ways, leaving 51 cards for the second draw. There are 52×51 ways to draw the two cards in a given order, but since order doesn't matter in this case, only half that number will be distinct pairs (the ace of diamonds followed by the 7 of hearts is the same pair as the 7 of hearts followed by the ace of diamonds).

The number of possible outcomes is $\dfrac{52 \times 51}{2} = 1326$

(b) The first heart could be drawn 13 ways, leaving 12 hearts for the second draw.

The number of distinct outcomes for the event that both cards are hearts is $\dfrac{13 \times 12}{2} = 78$.

(c) $P(2 \text{ hearts}) = \dfrac{78}{1326} = \dfrac{39}{663} = \dfrac{13}{221} = \dfrac{1}{17}$

31. (a)   **(b)**   **(c)**

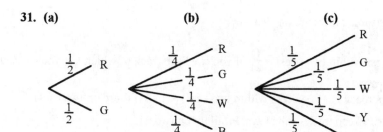

33. (a) $P(A) = \dfrac{120}{360} = \dfrac{1}{3}$; $P(B) = \dfrac{60}{360} = \dfrac{1}{6}$; $P(C) = \dfrac{180}{360} = \dfrac{1}{2}$

```
                              1/3   A  1/9
                        A ────1/6── B  1/18
                 1/3 ╱        1/2   C  1/6

                              1/3   A  1/18
                 1/6    B ────1/6── B  1/36
              ◁               1/2   C  1/12

                 1/2          1/3   A  1/6
                        C ────1/6── B  1/12
                              1/2   C  1/4
```

(b) $\dfrac{1}{6} \times \dfrac{1}{6} = \dfrac{1}{36}$ **(c)** $\dfrac{1}{2} \times \dfrac{1}{3} = \dfrac{1}{6}$

35. On the first draw, $P(W) = \frac{2}{6} = \frac{1}{3}$ and $P(B) = \frac{2}{3}$

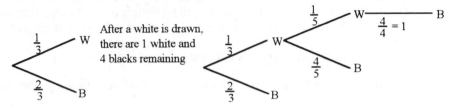

After a white is drawn, there are 1 white and 4 blacks remaining

   (a) P(one draw is needed) $= \frac{2}{3}$

   (b) P(two draws are needed) $= \frac{4}{15}$

   (c) P(three draws are needed) $= \frac{1}{15}$

37. P(success within two tries) $= \frac{11}{21}$ if you can keep track of which key you've used and $\frac{24}{49}$ if you can't.

Our approach is: P(success within two tries) = 1 - P(failure on two tries).

If you can keep track of the keys, P(failure on two tries) $= \frac{5}{7} \times \frac{4}{6} = \frac{20}{42} = \frac{10}{21}$.

$$P(\text{success within two tries}) = 1 - \frac{10}{21} = \frac{11}{21}$$

If you can't keep track of the keys, P(failure on two tries) $= \frac{5}{7} \times \frac{5}{7} = \frac{25}{49}$.

$$P(\text{success within two tries}) = 1 - \frac{25}{49} = \frac{24}{49}$$

Note: We could also do this problem directly using tree diagrams and adding up the appropriate branches.

39.   **(a)** P(two 3's) $= \frac{9}{36} = \frac{1}{4}$ **(b)** (same number) $= \frac{14}{36} = \frac{7}{18}$

(1, 1)	(1, 2)	(1, 2)	(1, 3)	(1, 3)	(1, 3)		**(1, 1)**	(1, 2)	(1, 2)	(1, 3)	(1, 3)	(1, 3)
(2, 1)	(2, 2)	(2, 2)	(2, 3)	(2, 3)	(2, 3)		(2, 1)	**(2, 2)**	**(2, 2)**	(2, 3)	(2, 3)	(2, 3)
(2, 1)	(2, 2)	(2, 2)	(2, 3)	(2, 3)	(2, 3)		(2, 1)	**(2, 2)**	**(2, 2)**	(2, 3)	(2, 3)	(2, 3)
(3, 1)	(3, 2)	(3 2)	**(3,3)**	**(3,3)**	**(3,3)**		(3, 1)	(3, 2)	(3 2)	**(3,3)**	**(3,3)**	**(3,3)**
(3, 1)	(3, 2)	(3 2)	**(3,3)**	**(3,3)**	**(3,3)**		(3, 1)	(3, 2)	(3 2)	**(3,3)**	**(3,3)**	**(3,3)**
(3, 1)	(3, 2)	(3 2)	**(3,3)**	**(3,3)**	**(3,3)**		(3, 1)	(3, 2)	(3 2)	**(3,3)**	**(3,3)**	**(3,3)**

   (c) P(two odd numbers) $= \frac{16}{36} = \frac{4}{9}$

(1, 1)	(1, 2)	(1, 2)	**(1, 3)**	**(1, 3)**	**(1, 3)**
(2, 1)	(2, 2)	(2, 2)	(2, 3)	(2, 3)	(2, 3)
(2, 1)	(2, 2)	(2, 2)	(2, 3)	(2, 3)	(2, 3)
(3, 1)	(3, 2)	(3 2)	**(3,3)**	**(3,3)**	**(3,3)**
(3, 1)	(3, 2)	(3 2)	**(3,3)**	**(3,3)**	**(3,3)**
(3, 1)	(3, 2)	(3 2)	**(3,3)**	**(3,3)**	**(3,3)**

43. (a) $\frac{1}{36}$ **(b)** $\frac{35}{36}$ **(c)** $\left(\frac{35}{36}\right)^{24} \approx 0.509$ **(d)** appr. 1 - 0.509 or 0.491

   (d) Rolling at least one pair of sixes is the complement of not rolling any pairs.

45. Note: The tree diagram is found in the solution for Problem 5, Section 5.1.

Section 5.3   Conditional Probability, Independence, and Expected Value

Goals
1. Compute probabilities when there is partial information available.
2. Use expected values to find the true cost of lottery tickets, insurance premiums and similar items.

Key Ideas and Questions
1. What is the formula for the conditional probability of an event A given that the event B has occurred?
2. Illustrate the formula for conditional probability with an example.
3. What does the independence of two events mean? Write a description.
4. What is the formula that independent events must satisfy?
5. If A and B are independent, then what is the conditional probability of A given B?
6. How do you find the expected value of an experiment?
7. What is the interpretation of the expected value of an experiment?

Vocabulary

Conditional Sample Space	Independent Events
Conditional Probability	Expected Value

. . . .

Overview

In this section, we will develop several additional properties of probability that will be used in analyzing complicated events, determining probable origins for certain sequences, and relating values to the outcomes of events.

Conditional Probability

Sometimes conditions make it necessary to focus on a portion of the sample space called the **conditional sample space**. When certain information is known about the experiment it can affect the possible outcomes. The original sample space can be reduced to those outcomes having the given condition. The following describes this **conditional probability.**

Conditional Probability

Suppose A and B are events in a sample space S such that $P(B) > 0$.
The **conditional probability** that the event A occurs, given that
the event B occurs, denoted $P(A \mid B)$, is

$$P(A \mid B) = \frac{P(A \cap B)}{P(R)} .$$

Independent Events

Two events are **independent** when one event does not influence the other.

Probability of Independent Events

When two events are independent, the probability of both happening equals the product of their probabilities. For independent events A and B
$$P(A \cap B) = P(A) \times P(B).$$

If two events are independent, then the occurrence of one of the events does not affect the probability that the other will occur.

Expected Value

When an experiment has numerical outcomes, it is often quite useful to know what the average should be for many repetitions of the experiment. This average is called the **expected value.** In some experiments, the possible outcomes may not actually be numbers; however, if an "average outcome" is desired, numbers can be associated with the outcomes. To find the expected value of an experiment with a numerical outcome or an associated numerical outcome, multiply each possible numerical outcome by its probability and add all of the products. **Expected value** is defined as follows:

Expected Value

Suppose that the outcomes of an experiment are numbers (values) called v_1, v_2, $\ldots$, v_n, and the outcomes have the probabilities $p_1, p_2, \ldots, p_n$, respectively. The expected value, E, of the experiment is the sum
$$E = (v_1 \times p_1) + (v_2 \times p_2) + \ldots + (v_n \times p_n)$$

• •

Suggestions and Comments for Odd-numbered problems

9. Complete the probability tree diagram. What is the probability of a, b, etc?

15. Draw a probability tree diagram. What are the choices for a person's condition?

17. Does the selection in one of the classes influence the selection in the other? If not, what does that mean with respect to the selections and how you compute the probability of the event?

19. If in doubt about the probabilities involved, refer to the sample space in Example 5.2(d).

25. Assign p as the probability for each of the other five faces. Then, the probability of the face with the 6 is $4p$. When the probabilities for all the faces are added, the total must be 1. What does that mean as far the probability for each face?

29. Make a table of all possible values that can be associated with any ticket. What is the one outcome that isn't given in the problem statement, and what is its probability?

Solutions to Odd-numbered Problems

1. {(2,1), (2,2), (2,3), (2,4), (2,5), (2,6), (4,1), (4,2), (4,3), (4,4), (4,5), (4,6), (6,1), (6,2), (6,3), (6,4), (6,5), (6,6) }

3. (a) $P(\text{at least one } 5) = \frac{11}{36}$ **(b)** $P(\text{sum} = 8) = \frac{5}{36}$

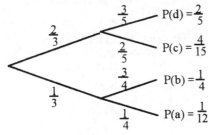

(1,5)
(2,5)   (2,6)
(3,5)   (3,5)
(4,5)   (4,4)
(5,1) (5,2) (5,3) (5,4) (5,5) (5,6)   (5,3)
(6,5)   (6,2)

(c) $P(\text{a die is a } 5 \text{ given the sum is } 8) = \frac{2}{5}$

(2,6)
(3,5)
(4,4)
(5,3)
(6,2)

5. (a) $P(A) = \frac{8}{15}$ **(b)** $P(B) = \frac{6}{15} = \frac{2}{5}$

(c) $P(A|B) = \frac{3}{6} = \frac{1}{2}$ **(d)** $P(B|A) = \frac{3}{8}$

7. (a) Literal translation: The probability that a person committed aggravated assault, given that the person was a drug dealer.
Liberal Translation: The probability a drug dealer committed aggravated assault.

(b) Liberal translation: The probability that someone who committed aggravated assault is a drug dealer.

(c) Liberal translation: The probability that a drug dealer has not committed aggravated assault.

(d) Liberal translation: The probability that someone who is not a drug dealer did not commit aggravated assault.

9. First, we need to complete the probability tree diagram.

$\frac{2}{3}$ — $\frac{3}{5}$ — $P(d) = \frac{2}{5}$

$\frac{2}{5}$ — $P(c) = \frac{4}{15}$

$\frac{1}{3}$ — $\frac{3}{4}$ — $P(b) = \frac{1}{4}$

$\frac{1}{4}$ — $P(a) = \frac{1}{12}$

(a) $P(A) = P(\{a, b, c\}) = \frac{1}{12} + \frac{1}{4} + \frac{4}{15} = \frac{5}{60} + \frac{15}{60} + \frac{16}{60} = \frac{36}{60} = \frac{3}{5}$

(b) $P(B) = P(\{b, c, d\}) = \frac{1}{4} + \frac{4}{15} + \frac{2}{5} = \frac{15}{60} + \frac{16}{60} + \frac{24}{60} = \frac{55}{60} = \frac{11}{12}$

9. **(c)** $P(A \cap B) = P(\{b, c\}) = \frac{1}{4} + \frac{4}{15} = \frac{15}{60} + \frac{16}{60} = \frac{31}{60}$ **(d)** $P(A \cup B) = P(\{a, b, c, d\}) = 1$

(e) $P(A|B) = \dfrac{P(A \cap B)}{P(B)} = \dfrac{31/60}{55/60} = \dfrac{31}{55}$ **(f)** $P(B|A) = \dfrac{P(B \cap A)}{P(A)} = \dfrac{31/60}{36/60} = \dfrac{31}{36}$

11. Answers are based on the assumption that each portion of the spinner face is an equally likely stopping place.

(a) $P(4) = \frac{1}{6}$ There are 6 equally likely sectors.

(b) $P(4 \mid \text{even}) = \frac{1}{3}$ There are 3 sectors labeled with even numbers.

(c) $P(4 \mid \text{odd}) = 0$ 4 is not in the (restricted) sample space.

13. **(a)** $P(\text{female} \mid \text{grad}) = \frac{14}{30} = \frac{7}{15}$ 14 females out of 30 graduate degrees.

(b) $P(\text{male} \mid \text{diploma}) = \frac{87}{186} = \frac{29}{62}$

(c) $P(\text{no degree} \mid \text{female}) = \frac{163}{205}$

The ones included have finished either only elementary school or high school.

15. First, we draw a tree diagram showing the probabilities related to each of the primary and secondary branches.

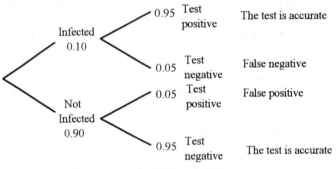

(a) $P(\text{false positive}) = 0.90 \times 0.05 = 0.045$

(b) $P(\text{false negative}) = 0.10 \times 0.05 = 0.005$

(c) $P(\text{infected} \mid \text{tested positive}) = \dfrac{\text{infected and tested positive}}{\text{tested positive}}$

$P(\text{infected and tested positive}) = 0.10 \times 0.95 = 0.095$

$P(\text{not infected and tested positive}) = 0.90 \times 0.05 = 0.045$

$P(\text{tested positive}) = 0.095 + 0.045 = 0.140$

$P(\text{infected} \mid \text{tested positive}) = \dfrac{0.095}{0.140} = 0.68$

17. Let A be the event: The person in the first class speaks Spanish.
Let B be the event: The person in the second class speaks Spanish.
$P(A) = \dfrac{20}{25} = \dfrac{4}{5}$, and $P(B) = \dfrac{12}{18} = \dfrac{2}{3}$
Since the selections are made randomly from each class, the events are independent.
We want the probability that both students speak Spanish.

$$P(A\cap B) = P(A) \times P(B) = \dfrac{4}{5} \times \dfrac{2}{3} = \dfrac{8}{15}$$

19. From previous work with the sample space for two dice, we have:
$$P(A) = \dfrac{1}{6}, \; P(B) = \dfrac{1}{6}, \text{ and } P(C) = \dfrac{1}{6}$$

(a) $P(A\cap B) = \dfrac{1}{36}$. Only one pair with a 3 on the first die has a sum of 7.

$$P(A) \times P(B) = \dfrac{1}{6} \times \dfrac{1}{6} = \dfrac{1}{36} = P(A\cap B)$$

A and B are independent.

(b) $P(A\cap C) = \dfrac{1}{36}$. Only one pair with a 3 on the first die has the same number on both dice.

$$P(A) \times P(C) = \dfrac{1}{6} \times \dfrac{1}{6} = \dfrac{1}{36} = P(A\cap C)$$

A and C are independent.

(c) $P(B\cap C) = 0$. No pairs have a sum of 7 and the same number on both dice.
$$P(B) \times P(C) = \dfrac{1}{6} \times \dfrac{1}{6} = \dfrac{1}{36} \neq P(B\cap C)$$

B and C are not independent.

21. First, we will make a tree diagram and note the number of girls in each branch of the tree. All branches are equally likely.

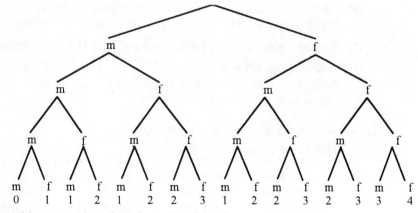

The table summarizes the information from the tree.

value	0	1	2	3	4
probability	$\dfrac{1}{16}$	$\dfrac{4}{16}$	$\dfrac{6}{16}$	$\dfrac{4}{16}$	$\dfrac{1}{16}$

23. The expected value is found by multiplying each of the values by the corresponding probability and adding the products.

$$\text{Expected value} = 0 \times \frac{1}{16} + 1 \times \frac{4}{16} + 2 \times \frac{6}{16} + 3 \times \frac{4}{16} + 4 \times \frac{1}{16} = \frac{32}{16} = 2$$

This can be interpreted as meaning that if there is a very large number of families with four children, the average number of girls is expected to be two, although you shouldn't expect a given family with four children to have two girls. In fact, the probability is $\frac{5}{8}$ that they *won't* have two girls.

25. If a 6 occurs four times as often as each of the other faces, then the probability of each of the other numbers is $\frac{1}{9}$ and the probability of a 6 is $\frac{4}{9}$. To see this, let p be the probability of each of the other faces and 4p be the probability of 6. The sum of the probabilities for the six faces must be 1, so we have $p + p + p + p + p + 4p = 9p = 1$. We make a table of values:

number	1	2	3	4	5	6
probability	$\frac{1}{9}$	$\frac{1}{9}$	$\frac{1}{9}$	$\frac{1}{9}$	$\frac{1}{9}$	$\frac{4}{9}$

$$\text{Expected value} = 1 \times \frac{1}{9} + 2 \times \frac{1}{9} + 3 \times \frac{1}{9} + 4 \times \frac{1}{9} + 5 \times \frac{1}{9} + 6 \times \frac{4}{9} = \frac{39}{9} = 4\frac{1}{3}$$

27. Since we have all the values and their probabilities, we can make the calculation for expected value. (The probabilities add up to 1.)

$$\text{Expected value} = 0\ 1 \times 0.52 + 1 \times 0.21 + 2 \times 0.18 + 3 \times 0.09$$
$$= 0 + 0.21 + 0.36 + 0.27 = 0.84$$

29. We'll make a table to show the prizes and probabilities.

Prize	10	15	30	50	0
Probability	$\frac{50}{1000}$	$\frac{10}{1000}$	$\frac{5}{1000}$	$\frac{1}{1000}$	$\frac{934}{1000}$

(a) The fair price of the ticket should be the expected value of the ticket.
Exp. value $= 10 \times 0.05 + 15 \times 0.01 + 30 \times 0.005 + 50 \times 0.001 + 0 = \$0.85$

(b) The ticket should be $1.35 if the average loss is $0.50 per person.

(c) Total Revenue = $2000
Total prizes $= 50 \times \$10 + 10 \times \$15 + 5 \times \$30 + \$50 = \$850$
Lottery Gain = $2000 - $850 = $1150

31. We organize the information in a table:

Gift - value	Probability
A - 9272.00	1/52000
B - 44.95	25736/52000
C - 2500.00	1/52000
D - 729.95	3/52000
E - 26.99	25736/52000
F - 1000.00	3/52000
G - 44.99	180/52000
H - 63.98	180/52000
I - 25.00	160/52000

Note: Since all the probabilities have a denominator of 52,000, the calculations can be made with just the prize values and the numerators, and dividing by 52,000 at the end.

Expected value = $36.39

33. (a) Expected claim for male drivers = $2420

$E = 2000 \times 0.23 + 4000 \times 0.16 + 6000 \times 0.09 + 8000 \times 0.06 + 10000 \times 0.03$

(b) Expected claim for female drivers = $1440

$E = 2000 \times 0.20 + 4000 \times 0.09 + 6000 \times 0.07 + 8000 \times 0.02 + 10000 \times 0.01$

(c) Male drivers should be charged a higher premium since the average claim is substantially higher.

Section 5.4   Systematic Counting

Goals
1. Use factorials, permutations, and combinations to perform complex counting tasks.

Key Ideas and Questions
1. How do you find the number of ordered arrangements of objects that are selected from a group of objects?
2. How do you find the number of unordered collections of objects that are selected from a group of objects?
3. Compare and contrast permutations and combinations.

Vocabulary

Factorial	With Replacement	Permutation
Ordered/unordered Samples	Without Replacement	Combination

· ·

Overview

When we began this chapter we defined the probability of an event E, in a sample space S, as $P(E) = \dfrac{\text{number of outcomes in E}}{\text{number of outcomes in S}}$, with the condition that all of the outcomes in S are equally likely. That makes the main task one of counting the number of outcomes. Although counting is the most fundamental task in mathematics, it is sometimes quite difficult.

There are two sets of conditions that allow us to use a systematic method in counting for certain complex, but frequently occurring situations, where subsets of a given set are to be selected, such as selecting 5 cards from a deck of cards or 5 people from an organization.
1. The elements are selected one at a time from the set, and either replaced or not replaced on a regular basis.
2. The order in which the elements are selected is important, or the elements can be rearranged in any manner.

In this section, we develop the mathematical tools based on of the Fundamental Principle of Counting for counting the number of outcomes in three situations.
1. Ordered samples with replacement,
2. Ordered samples without replacement, and
3. Unordered samples without replacement.

Ordered Samples with Replacement

If we have a set with n elements from which we are going to select k of them one at a time and each element can be selected repeatedly, then we say the selection is *with replacement* since the element is not removed from the set. We might think of the process of selection as filling a predetermined number of "slots".

$$\underset{1}{\rule{1cm}{0.4pt}} \;\; \underset{2}{\rule{1cm}{0.4pt}} \;\; \underset{3}{\rule{1cm}{0.4pt}} \;\; \underset{4}{\rule{1cm}{0.4pt}} \;\; \cdots\cdots \;\; \underset{k}{\rule{1cm}{0.4pt}} \qquad (k \text{ slots to be filled})$$

If each of the slots can be filled with any of the n elements, then the Fundamental Principle of Counting tells us the number of ways to fill all the slots is

$$n \times n \times n \times n \times \ldots\ldots \times n \;\; (k \text{ factors of } n) \quad \text{or} \quad n^k$$

As an example, suppose that a single die is tossed 3 times. How many possible outcomes are there if you record the results of each toss in order? Since any of the numbers can occur on any of the tosses, the number of outcomes is 6^3.

Ordered Arrangements of a Set - Permutations and Factorial Notation

If we assume that each element in a set can be used only once, then each successive trial has one fewer possibilities. Suppose 5 names are put in a container and drawn one at a time. There are 5 possibilities on the first draw, but only 4 on the second draw, then 3 on the third draw, and so on. The number of possible sequences in which these names can be drawn is $5 \times 4 \times 3 \times 2 \times 1$. In mathematical terminology, the ordered arrangement of all the elements in a set is known as a **permutation**, and a special notation, known as *factorial* notation is used to express the number of all possible permutations in the set. with this notation, $5 \times 4 \times 3 \times 2 \times 1 = 5!$ (read "5 factorial").

Permutations of k Objects

The number of ways to arrange k distinct objects, using each exactly once is
$$k \times (k-1) \times (k-2) \times \cdots \times 2 \times 1 = k!$$

Ordered Samples without Replacement

Factorial notation is helpful in counting the number of different ordered arrangements of objects form a set, even when all the objects are not used. Suppose 5 names are put in a container as before, and three names are drawn one at a time. There are 5 possibilities on the first draw, 4 on the second draw, and 3 on the third draw. The number of possible sequences in which these names can be drawn is $5 \times 4 \times 3$. This *partial* permutation can be expressed as the quotient of two factorials, one for the number of permutations for all the objects and one for the number of permutations of the objects *not* used.

$$5 \times 4 \times 3 = 5 \times 4 \times 3 \times \frac{2 \times 1}{2 \times 1} = \frac{5 \times 4 \times 3 \times 2 \times 1}{2 \times 1} = \frac{5!}{2!}$$

Ordered Samples Without Replacement of *k* Objects
Selected from Among *n* Objects

The number of ways to arrange *k* objects chosen from a set of *n* distinct objects using each object at most once is

$$\frac{n!}{(n-k)!}$$ This number is sometimes written $_nP_k$

Combinations: Choosing *k* Objects from among *n* Objects

When order is important and objects must be arranged, we use a permutation. However, in many situations, such as being dealt a hand in poker or bridge, or in selecting the general membership of a committee, the order has no bearing on the result, although object can still be used at most once. These *unordered samples without replacement* are known as **combinations**.

To find the number of combinations of k objects taken from among n objects, we first find the number of ordered samples of the given size and then divide by the number of ways to order samples of this size. This removes all the duplication. In practical terms, we divide the permutations of k objects from among n objects by the permutations of k objects. That is, we have $_nP_k \div k!$, which can be simplified as follows.

Number of Ways of Choosing *k* Objects
from Among *n* Objects

The number of ways to choose *k* objects from a set of *n* distinct objects is

$$\frac{n!}{(n-k)!} \div k! \text{ or } \frac{n!}{k!(n-k)!}$$

This number is sometimes written $_nC_k$ (read "from *n* choose *k*")

Summary

When you are choosing *k* objects from among *n* distinct objects, there are three situations for which we have counting tools:

- The objects can be repeated, and order makes a difference: use n^k
- The objects can't be repeated, and order makes a difference: use $_nP_k$
- The objects can't be repeated, and order doesn't make a difference: use $_nC_k$

· ·

Suggestions and Comments for Odd-numbered problems

1. through 11.

First think of the number of choices that are to be made. Visualize the steps in the actual physical process if you were doing the choosing, awarding, or whatever. If you are to "choose" 3 from a set of 8, then there are 3 steps. Then you should consider if the objects can be selected more than once and if order makes a difference.

21. and 23.
   Many calculators have permutations and combinations as built in functions. However, you should write these out as well to make sure you understand the calculation. These are counting techniques that you can apply outside the classroom, and you won't usually be carrying your calculator.

25. through 31.
   The concept of "order" should be generalized; are selections made for different purposes, or are all selections interchangeable?

33. through 37.
   How many separate tasks are involved? In how many ways can each task be completed?

39. When working with any collections, think of all "objects" being displayed in front of you. For card hands, visualize all the cards being spread out on a table and then picking the cards that you need: How many choices do you have to make? How many cards are there of the given type?

41. For picking three in a row, you have to match either the first three or the last three. In how many ways can you pick a digit that's different from the one that isn't matched? Can digits be repeated?

43. In how many ways can any 5 of the 6 numbers be matched? Think of the number of ways you can *not* match one of the numbers.

47. Use the result of problem 43 and the definition of expected value from section 5.3. The relevant payoffs include "- $1" for not matching either 5 or 6. What is the average payoff when spending $1 for a ticket?

49. (a) How many ways are there to choose 5 computers? How many ways are there to choose 5 "good" computers?
   (b) How many ways are there to choose 4 good computers and 1 defective computer?
   (c) How many good computers have to be included?

Solutions to Odd-numbered problems

1. There are 12 ways to choose a male student and 8 ways to choose a female; $12 \times 8 = 96$

3. 72 $3 \times 4 \times 6 = 72$

5. 1296   There are 6 letters, and they can be repeated; $6^4 = 1296$

7. 4096   Including the award for best overall time, there are four awards that can each be made to any of the eight competitors; $8^4 = 4096$

9. 24   She won't repeat any classes; and order matters; $4! = 24$; This is also $_4P_4$.

11. 720   Each report is given once; and order makes a difference; $6! = 720$ ($_6P_6$)

13. $7! = 7 \times 6! = 7 \times 720 = 5040$

15. $5! \times 6! = 120 \times 720 = 86,400$

17. 1320 $12 \times 11 \times 10 = 1320$

19. 495 $\frac{12\times11\times10\times9\times8\times7\times6\times5\times4\times3\times2\times1}{(8\times7\times6\times5\times4\times3\times2\times1)(4\times3\times2\times1)} = \frac{12\times11\times10\times9}{4\times3\times2\times1}$

Note: Since the terms in the denominator can be interchanged without affecting the result of the computation, this is the value for either $_{12}C_4$ or $_{12}C_8$.

21. 336 $_8P_3 = 8 \times 7 \times 6$

23. 56 $_8C_3 = \frac{8\times7\times6\times5\times4\times3\times2\times1}{(3\times2\times1)(5\times4\times3\times2\times1)} = \frac{8\times7\times6}{3\times2\times1}$

25. $_{15}P_3 = 2730$ First, the selections can't be repeated.
Since the tasks are different, they can be given a *natural order*.
That is, there's a first task to assign, then a second task, etc.

27. $_{12}P_3 = 1320$ "Selections" can't be repeated (each horse finishes in one place only).
Order makes a difference.

29. Since there is nothing to indicate order, this is a combination; $_{18}C_5 = 8568$
Note: If one or more students were selected to fill specific roles, then it would no longer be a combination. That doesn't mean, however, that it would be a permutation. Some counting problems require more than one principle or technique.

31. $_{52}C_{13} = 635,013,559,600 \approx 6.35 \times 10^{11}$ (635 Billion)
When playing bridge (or poker) the dealer is selecting the cards; they can be rearranged once they have been dealt. Note: If you could look at one hand *every second*, it would take more than 20,000 years to look at them all (talk about boring!).

33. $_8C_2 \times _5C_2 = 280$ This is a two-stage problem using the Fundamental Principle of Counting: "from 8 choose 2" and then "from 5 choose 2".
There is nothing to indicate that the order in which the books are chosen or read makes a difference.

35. $_{18}P_2 \times _{12}P_2 = 40,392$ This is a two-stage problem.
For each of the parties, the specific roles of leader and whip provide a natural order for the selections.

37. $_6C_2 \times _8C_4 \times _5C_2 = 10,500$ This is a three-stage problem.
1. Choosing 2 appetizers from among 6.
2. Choosing 4 main courses from among 8
3. Choosing 2 desserts from among 5
Note: If we had asked for possible menus listing the selections, then permutations would be used, and the answer would be 1,008,000.

39. Think of the cards laid out on the table in front of you.
  (**a**) $_{12}C_5 = 792$ There are 12 face cards; you can choose 5 and rearrange them
  (**b**) $_{12}C_4 \times {_{40}C_1} = 19{,}800$ Choose 4 of the 12 face cards and 1 of the other 40
  (**c**) $_{12}C_3 \times {_{40}C_2} = 171{,}600$ Choose 3 of the 12 face cards and 2 of the other 40
  (**d**) $_4C_3 \times {_8C_1} \times {_{40}C_1} = 1280$ This is a three-stage problem.
               1. Choose 3 of the four kings
               2. Choose 1 of the other 8 face cards
               3. Choose 1 of the other 40 cards

41. First, there are $9^4 = 6561$ possible ordered drawings
  (**a**) 17 There is only 1 way to match all 4 digits and 8 ways to match either the first 3 or the last 3.
  Suppose the numbers drawn were 7, 5, 7, and 4. How many ways are there to match the first three digits and not the last? Any digit other than 4 will do.
  (**b**) $\dfrac{17}{6561} \approx 0.0026$

43. (**a**) 193 There is only 1 way to match all 6 numbers, and $_6C_5 \times 32 = 192$ ways to match any 5 of the numbers. For matching any 5, you choose any 5 of the 6 numbers that were selected by the lottery and then one of the 32 that weren't selected.
  (**b**) $\dfrac{193}{2{,}760{,}681} \approx 0.00007$

45. $\dfrac{54{,}912}{2{,}598{,}960} \approx 0.02113$

In the paragraph preceding this problem, it is shown that there are 4,224 ways to have 3 kings and no other matching cards. There are 13 different types of "3 of a kind". Therefore, there are $4224 \times 13 = 54{,}912$ ways to have 3 of a kind.

47. Expected Value $\approx -\$0.29$ (for all the people playing the lottery, the expected result is an average loss of \$0.29)
There is 1 way to win the big prize, and 192 ways to win the smaller prize (Problem 43) There are 2,760,488 ways to lose. The *net* payoffs for winners are \$999,999 and \$4999.

$$EV = 999999 \times \tfrac{1}{2{,}760{,}681} + 4999 \times \tfrac{192}{2{,}760{,}681} + (-1) \times \tfrac{2{,}760{,}488}{2{,}760{,}681} \approx -0.29$$

Another way to calculate: $EV = 1000000 \times \tfrac{1}{2{,}760{,}681} + 5000 \times \tfrac{192}{2{,}760{,}681} - 1 \approx -0.29$

Note: If you already have a \$1 ticket, it is only "worth" approximately \$0.71.

49. (**a**) $_{17}C_5 \div {_{20}C_5} \approx 0.399$ The numerator represents the number of ways to choose 5 of the 17 "good" computers, and the denominator represents the number of way to choose any five of the 20 computers.
  (**b**) $(_3C_1 \times {_{17}C_4}) \div {_{20}C_5} \approx 0.461$ The numerator represents the number of ways to choose 1 of the 3 defective computers and 4 of the 17 good computers.
  (**c**) $(_3C_3 \times {_{17}C_2}) \div {_{20}C_5} \approx 0.009$ The numerator represents the number of ways of choosing all 3 of the defective computers and 2 of the 17 good computers.

51. $_4C_4 = 1$, $_4C_3 = 4$, $_4C_2 = 6$, $_4C_1 = 4$, $_4C_0 = 1$

53. $(_4C_4 + {_4C_3} + {_4C_2}) \div 16 = \dfrac{11}{16}$

Chapter 5 Review Problems

Solutions to Odd-numbered Problems

1. Sample space = {$10000, $20, 0}

$P(\$1000) = \frac{1}{100000}$, $P(\$20) = \frac{1}{1000}$,

$P(\text{at least } \$20) = \frac{1}{100000} + \frac{1}{1000} = \frac{1}{100000} + \frac{100}{100000} = \frac{101}{100000}$

3. Sample space = {PN, PD, PQ, ND, NQ, DQ}   D = dime, etc.

   (a) A = {PN, PD}, $P(A) = \frac{1}{3}$

   (b) B = {PQ, NQ, DQ}, $P(B) = \frac{1}{2}$

   (c) C = {PD, NQ, DQ}, $P(C) = \frac{1}{2}$

   (d) A and B are mutually exclusive; they have no common elements.

   (e) $P(A \cup B) = P(A) + P(B) - P(A \cap B) = \frac{1}{3} + \frac{1}{2} - 0 = \frac{5}{6}$

   $P(B \cup C) = P(B) + P(C) - P(B \cap C) = \frac{1}{2} + \frac{1}{2} - \frac{1}{6} = \frac{5}{6}$

   $P(B \cap C) = P(\{DQ\}) = \frac{1}{6}$

5. $P(\text{sum} = 4) = P(\{(1,3), (2,2), (3,1)\}) = \frac{3}{36} = \frac{1}{12}$

7. First, we draw the tree diagram, keeping track of the suit on the first draw and the matches of the suits on the second draw.

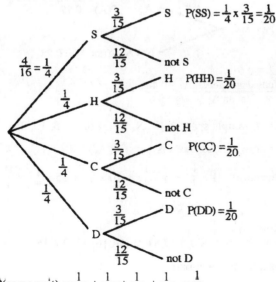

$P(\text{same suit}) = \frac{1}{20} + \frac{1}{20} + \frac{1}{20} + \frac{1}{20} = \frac{1}{5}$

9. Probability tree diagram with the probability of each sequence listed at the end of each branch.

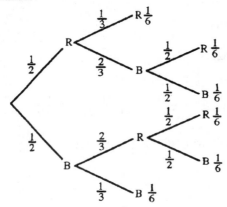

P(first marble is blue) $= \dfrac{1}{2}$

P(second marble is blue) $= (\dfrac{1}{2} \times \dfrac{2}{3}) + (\dfrac{1}{2} \times \dfrac{1}{3}) = \dfrac{1}{3} + \dfrac{1}{6} = \dfrac{1}{2}$

P(3 drawings ending in blue) $= \dfrac{1}{6} + \dfrac{1}{6} = \dfrac{1}{3}$

P(2 drawings needed) $= \dfrac{1}{6} + \dfrac{1}{6} = \dfrac{1}{3}$

11. The tree diagram:

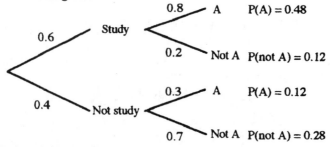

$P(A) = 0.48 + 0.12 = 0.6$

$P(\text{studied} \mid A) = \dfrac{\text{studied and got an A}}{\text{got an A}} = \dfrac{0.48}{0.6} = 0.8$

13. For drawing one coin, sample space = {P, N, D, Q} or {1, 5, 10, 25}

The probability for drawing each coin is $\dfrac{1}{4}$

Expected value (one coin) $= 1 \times \dfrac{1}{4} + 5 \times \dfrac{1}{4} + 10 \times \dfrac{1}{4} + 25 \times \dfrac{1}{4} = \dfrac{41}{4}$

$= 10\dfrac{1}{4}$ cents

For drawing two coins,
sample space = {PN, PD, PQ, ND, NQ, DQ} OR {6, 11, 26, 15, 30, 35}

The probability for drawing each pair is $\dfrac{1}{6}$

Expected value $= 6 \times \dfrac{1}{6} + 11 \times \dfrac{1}{6} + 26 \times \dfrac{1}{6} + 15 \times \dfrac{1}{6} + 30 \times \dfrac{1}{6} + 35 \times \dfrac{1}{6} = 20\dfrac{1}{2}$ cents

15. (a) 288   In this problem, there is the idea of order: first an ace, then another ace, then a third ace, then a king, and then another king. There are 4 ways to get the first ace, but only 3 ways to get the second ace, etc. There will be 4 ways to get the first king.

Using the Fundamental Principle of counting, we have $4 \times 3 \times 2 \times 4 \times 3$

(b) $288 \div {}_{52}P_5 \approx 0.00000092$ ${}_{52}P_5$ counts the ways to deal 5 cards in order

17. $1 - 0.1594 = 0.8406$

The probability that none of the computers is defective is ${}_{16}C_4 \div {}_{20}C_4 \approx 0.37564$
The probability that none of the printers is defective is ${}_{10}C_4 \div {}_{12}C_4 \approx 0.42424$
The probability that none of the computers and printers are defective is
$$0.37564 \times 0.42424 \approx 0.1594 \ \text{(allow for rounding)}$$

19. (a) $\frac{1}{495} \approx 0.002$ There is only 1 way to choose all seniors; there are ${}_{12}C_4 = 495$ ways to choose 4 students from the 12 nominees.

(b) $\frac{72}{495} \approx 0.145$ ${}_2C_1 \times {}_3C_1 \times {}_3C_1 \times {}_4C_1 = 72$ In simpler terms, there are 2 ways to choose a freshman, 3 ways to choose a sophomore, 3 ways to choose a junior, and 4 ways to choose a senior.

(c) $\frac{18}{495} \approx 0.036$ ${}_3C_2 \times {}_4C_2 = 3 \times 6 = 18$

6 Consumer Mathematics

Section 6.1   Interest

Goals
1. Calculate simple and compound interest.

Key Ideas and Questions
1. How does simple interest differ from compound interest?
2. What seems to be the more fair method of paying interest?
3. How can you compare different compound interest programs?
4. How would you compare borrowing money at 8% compounded annually with 7.6% compounded monthly?

Vocabulary
Simple Interest	Ordinary Interest	Balance
Principal	Compound Interest	Effective Annual Rate
Interest Rate	Compounding Period	Continuously Compounded
Time Period		

....

Overview

In this section, we discuss the two basic methods of lending or borrowing money, the differences between them, and ways to compare different interest payment situations. In the sections that follow, we will discuss different financial applications in which interest is charged in a variety of ways. The common questions are:

What is the amount of money and the period of time involved?

What is the stated interest rate, and how often is interest to be paid?

Simple Interest

When you pay **simple interest**, you are, in effect, paying rent on the use of someone else's money. The amount of money being used is called the **principal**. The amount you pay in interest is based on a stated percentage, called the **interest rate**, of the principal. The interest due is directly proportional to the amount of time the money is borrowed. To compute simple interest, you need to be sure that the interest rate and **time period** are in the same terms. That is, the interest rate is usually given as an annual percentage rate (APR). If the time period is in either months or days, then the interest rate needs to be changed appropriately: either dividing by twelve if the time period is in months, or by 365 or 366 when the time period is in days.

Simple interest (I) is the product of the principal (P), the rate of interest (r) expressed as a decimal, and the time period (t). This is expressed by the formula $I = Prt$. The final amount due (F), including the principal and interest is called the **future value**, or final balance. The formula that represents this is $F = P + Prt$.

Future Value for Simple Interest

If F represents the future value, P represents the principal, r the interest rate expressed as a decimal, and t the time in years, then
$$F = P + I = P + Prt = P(1 + rt)$$

When the time period for interest goes from the beginning of one month to the beginning of another month, then the term **ordinary interest** is used, and all months are assumed to be of equal length. When the time period is one day, the daily rate will depend on whether it is a leap year or not. The difference seems small, but when many interest charges are involved, even very small differences make for big changes in totals. The interest on most credit, such as credit cards, is charged on a daily basis.

Compound Interest

When **compound interest** is involved, simple interest is charged on the current balance at regular intervals, called the **compounding period**. This interest is then added to the existing balance to become the new balance for the next payment period. Since the balance increases from one compounding period to the next, the amount of interest for that period also increases. Over a long period of time, the effects of compounding can be very significant.

Calculating the amount of interest due for each of the time periods is repetitive. The total due at the end of a given period can be found as the sum of the principal and the interest due for each of the time periods by repeated application of the formula. This sum can be expressed compactly by the following formula.

Compound Interest Formula

If P represents the principal, r the interest rate expressed as a decimal, m the number of equal compounding periods (in a year), and t the time in years, then the final balance, F, is given as

$$F = P \times (1 + \frac{r}{m})^{mt}$$

The value of $\dfrac{r}{m}$ is the periodic rate corresponding to the compounding period. If the annual percentage rate is 12% and interest is paid quarterly (4 times a year), then the periodic rate is 3% quarterly.

Effective Annual Rate

For any given combination of principal, annual percentage rate, and time, if the compounding period is shortened, the final amount will be increased. The greatest increase due to compounding comes when the change is made from annual to semi-annual compounding. After that, the amount will still increase, but by smaller and smaller amounts. There is a limit to this growth, in which the compounding period is instantaneous and there are an infinite number of compounding periods.

The **Effective Annual Rate** (EAR) is used to compare the results of different compound interest arrangements. The Effective Annual Rate is defined as the simple interest rate that produces the same amount of interest in one year.

A comparison of the simple interest formula using the annual yield and the compound interest formula using the annual percentage rate (r) and n compounding periods per year for one year produces the formula for the annual yield. This does not depend on either the principal or the length of the loan.

Effective Annual Rate Formula

If EAR represents the Effective Annual Rate, r the interest rate expressed as a decimal, and m the number of equal compounding periods (in a year), then

$$EAR = (1 + \frac{r}{m})^m - 1$$

The maximum value for compounded interest occurs when the number of periods in a year is infinite and the interest is said to be **continuously compounded**.

. . . .

Suggestions and Comments for Odd-numbered problems

5. Change to a daily rate by dividing the annual rate by 365. The count of the number of days should include 21 days in January and 16 days in November.

7. Change to a daily rate using 366 days. Since it is a leap year, the count of days should include 17 days in February.

9. Remember that the first month and last month are both included. In doing the calculation, you can change the annual rate to a monthly rate by dividing by 12, or you can change the number of months to years by dividing by 12.

11. Change the annual rate to a quarterly rate by dividing by 4.

21. To compare these deals, calculate the Effective Annual Rate (EAR) for each. When compounding is done annually, the annual rate and the EAR are the same.

23. If interest is compounded every 2 months, there are 6 compounding periods in the year.

25. To find the rate, you must first find the amount of interest paid, and then use the formula $I = Prt$ to solve for the value of r.

29. The key question is: "How much should be deposited in September to produce a balance of $1500 in a period of 4 months (Sept 1 to Jan 1) at 6% interest?
    Balance $= P + Prt$ or Balance $= P \times (1 + rt)$.

31. Use the formula $EAR = e^r - 1$, where $e \approx 2.72$.

33. Use the formula $A = \$1200 \times (1 + 0.08)^n$, where n is the number of years.
    Use trial and error to find the value of n that produces approximately $1925.

35. Use the same approach as problem 31. The rate involved is the semi-annual rate (half the annual rate) and n is the number of 6-month periods.

37. Use the formula $\$40,000 = P \times (1 + 0.08)^{15}$ and solve for P.

39. Use the formula $I = Prt$ where $I = 25000$ and $t = 1$ to solve for P.

41. Use the formula $r = \left(\frac{P_1}{P_0}\right)^{1/n} - 1$.

47. and 49.
    The time that is required for an amount to double when it is being compounded continuously is approximately $t = \frac{0.6932}{r}$, where r is the annual rate. A rough approximation can also be obtained using the Rule of 72.

Solutions to Odd-numbered Problems

1. The (annual) interest rates are given and the time periods are in years, so the interest formula can be applied directly.

   (a) $126 $I = Prt = (600)(0.07)(3)$

   (b) $240 $I = (400)(0.12)(5)$

   (c) $864.50 $I = (1235)(0.07)(10)$

3. The (annual) interest rates are given, but since the time periods are in months, they have to be changed to years before using the interest formula.

   (a) $144 $I = Prt = (800)(0.06)(\frac{36}{12})$

   (b) $420 $I = (1400)(0.12)(\frac{30}{12})$

   (c) $154.38 $I = (1235)(0.075)(\frac{20}{12})$

5. In order to use the interest formula, we have to know how many days are involved. We list the months and count the days.

$$
\begin{array}{ll}
\text{January} & \text{21 days (31 - 10)} \\
\text{February} & 28 \\
\text{March} & 31 \\
\text{April} & 30 \\
\text{May} & 31 \\
\text{June} & 30 \\
\text{July} & 31 \\
\text{August} & 31 \\
\text{September} & 30 \\
\text{October} & 31 \\
\text{November} & 16 \\
\end{array}
$$

We see there are a total of 310 days involved. Since this is a non-leap year, our time is $t = \frac{310}{365}$

$I = (650)(0.06)(\frac{310}{365}) = \$33.12$

7. As in problem 5, we must count the days in the loan period. Since the loan is during a leap year, there will be 29 days in February and 366 days in the year. For the loan period, there are 17 days in February and 268 days total.

$I = (1600)(0.075)(\frac{268}{366}) = \$87.87$

9. In counting the number of months involved, we assume that the loan begins on the first day of the first month and ends on the last day of the last month. Since ordinary interest is being used, all months are considered of equal length.

   (a)
   $$
   \begin{array}{lll}
   1992 & \text{June-December} & \text{7 months} \\
   1993 & & \text{12 months} \\
   1994 & \text{January-February} & \text{2 months} \\
   \end{array}
   $$
   There are 21 months in the loan period.

   $I = (800)(0.06)(\frac{21}{12}) = \$84.00$

   (b)
   $$
   \begin{array}{lll}
   1992 & \text{October-December} & \text{3 months} \\
   1993 & & \text{12 months} \\
   1994 & & \text{12 months} \\
   1995 & \text{January-March} & \text{3 months} \\
   \end{array}
   $$
   There are 30 months in the loan period

   $I = (950)(0.075)(\frac{30}{12}) = \$178.13$

11. The same approach is used as in problem 9.
    (a) There are 39 months in the loan period.

    $I = (2350)(0.056)(\frac{39}{12}) = \$427.70$

    (b) There are 39 months in the loan period.

    $I = (7200)(0.05)(\frac{39}{12}) = \$1170.00$

13. $2300 $F = 2000 + (2000)(0.50)(3) = 2300.00$

15. $3975 \quad F = 3000 + (3000)(0.065)(5) = 3975$

17. Use the simple interest formula, but change the time period to years. $3068.54
$$F = 2575 + (2575)(0.0575)(\tfrac{40}{12}) = 3068.54$$

19. For each of the interest paying periods, we will use the compound interest formula. The periodic rate is 3% (12% ÷ 4)

3 months - $1545	$(1500)(1.03)$
6 months - $1591.35	$(1500)(1.03)^2$
9 months - $1639.09	$(1500)(1.03)^3$
1 year - $1688.26	$(1500)(1.03)^4$

21. In order to determine which offer is the best deal, we need to find the one with the highest Effective Annual Rate because this is an investment. If we were taking out a loan, we would want the one with the lowest Effective Annual Rate.

(a) $EAR = 0.05$

(b) $EAR = (1 + \tfrac{0.0495}{2})^2 - 1 = 0.050112563$

(c) $EAR = (1 + \tfrac{0.049}{12})^{12} - 1 = 0.050115575$

  4.9% compounded monthly is the best deal.

23. In order to find the Effective Annual Rate, we need the number of compounding periods per year. Since compounding is done every 2 months, $m = 6$.

$$EAR = 8.27\% \qquad EAR = (1 + \tfrac{0.08}{6}) - 1$$

25. The interest that John paid on the loan is $1000. To solve for the rate, we begin with the simple interest formula $I = Prt$ and solve for r. We then substitute the known information and calculate r.

$$4\% \qquad r = \frac{I}{Pt} = \frac{1000}{(5000)(5)} = 0.04$$

27. The increase in the value of the stock is to be treated as interest. First, we note that the investment of $16 per share produced a total value of $23 per share. Then we solve for r in the simple interest formula. See problem 23.

$$43.75\% \qquad r = \frac{23.00 - 16.00}{(16.00)(1)} = 0.4375$$

29. To find the amount that the student needs to deposit in order to have $1500 available on January 1, we will use 1500 as the balance at the end of the four-month period and the formula $F = P(1 + rt)$, and then solve for P.

$$1500 = P(1 + (0.06)(\tfrac{4}{12})) = P(1.02)$$

$$P = \tfrac{1500}{1.02} = 1470.59$$

Total needed on Sept. 1 = $1500 + $1470.49 = $2970.59

31. We use the Continuously Compounded Rate formula with $r = 0.08$ and $n = 12$.

$$8.33\% \qquad EAR = e^{.08} - 1 = 0.083287$$

33. To estimate the number of years, we use trial and error for different values of n and find that it takes approximately 6 years.

$$1200 \times (1.08)^5 = 1763.19$$
$$1200 \times (1.08)^6 = 1904.25$$
$$1200 \times (1.08)^7 = 2056.59$$

Note: To produce a total of *at least* $1925 would require 7 years, since no interest would be paid until the end of the year.

35. To estimate the number of years needed, we must find the number of half-year periods involved since interest is paid on a semi-annual basis. We use the semiannual rate of 7.5% (15% ÷ 2). Trial and error with several values of n shows that it will take approximately 11 years (to the nearest half-year).

$$500 \times (1 + 0.075)^{20} = 2123.93$$
$$500 \times (1 + 0.075)^{21} = 2283.22$$
$$500 \times (1 + 0.075)^{22} = 2454.46$$
$$500 \times (1 + 0.075)^{23} = 2638.54$$

37. To find the principal when the other information is known, we make the appropriate substitutions and solve for P.

$$40,000 = P \times (1.08)^{15}$$

$$P = \frac{40000}{(1.08)^{15}} = \frac{40000}{3.1722} = \$12,609.67$$

39. In each situation, we use the formula $I = Prt$, solve for P, and make the appropriate substitution.

   (a) $625,000 $P = \dfrac{I}{rt} = \dfrac{25000}{(0.04)(1)}$

   (b) $416,667 $P = \dfrac{25000}{(0.06)(1)}$

   (c) $250,000 $P = \dfrac{25000}{(0.10)(1)}$

41. We use the given formula, with $P_1 = 4500$ and $P_0 = 3000$

$$r = \left(\frac{4500}{3000}\right)^{1/5} - 1 = 0.084472$$

Approximately 8.45%

47. Using the Rule of 72, we get $t \approx 72 \div 9 = 8$.

Check: $2500 \times (1 + \frac{0.09}{12})^{96} = 5122.30$

$2500 would become $5122.30 in 8 years

49. Using the Rule of 72, we get $t \approx 72 \div 4.5 = 16$.

Check: $2500 \times (1 + \frac{0.045}{12})^{192} = 5129.17$

After 16 years $2500 would become $5129.17

Section 6.2   Simple Interest Loans

Goals
1. Become familiar with various types of loans and how to compute finance charges (interest).

Key Ideas and Questions
1. Why is add-on interest less advantageous to the customer than simple interest?
2. How can you compare different loan arrangements such as add-on interest and rent-to-own at different rates?

Vocabulary

Simple Interest Loan	Average Daily Balance	Annual Percentage Rate (APR)
Finance Charge	Add-on Interest	

· ·

Overview

In this section, we look at types of loans based on simple interest.

Simple Interest Loans

The interest on a **simple interest loan** is simple interest on the amount currently owed. Credit card accounts are a common example. Each month the bank or charge card company charges simple interest, called the **finance charge**, based on the balance owed. Many credit cards also have a grace period during which no interest is charged if full payment is received by the payment due date.

The most common method for calculating finance charges uses the average daily balance. When the **average daily balance** is used, a card holder is only charged for the actual number of days each amount owed was carried on the bill. This method converts the annual percentage rate to a daily interest rate.

To calculate the average daily balance, you determine the outstanding balance for each day and divide the sum of these daily balances by the number of days in the monthly billing period. Any payments or other credits are subtracted from the previous day's balance as they occur. In general, the monthly statement includes the current month's charges, any unpaid balance, and finance charges.

Add-On Interest

Sometimes businesses offer to finance the purchase of furniture, appliances, or automobiles with monthly payments using what is called **add-on interest**. To find the monthly payment for such a purchase, calculate simple interest at the annual interest rate over the length of the loan agreement. Then divide the sum of the purchase price and the interest into equal monthly payments.

Annual Percentage Rate

The **annual percentage rate (APR)** is the simple interest rate that would require the same payments to pay off the debt. Notice that add-on interest loans charge interest on the *entire* principal over the life of the loan even though you don't have use of all of the money during that period. Computing the annual percentage rate for add-on interest loans is difficult to do from formulas. Instead, tables showing the APRs for different combinations of rates and loan periods are used.

Rent-To-Own

Add-on interest loans have become rare. The replacement for add-on interest is the rent-to-own transaction, in which you rent an item you cannot afford to buy outright. After a contracted number of payments, the item becomes yours. Of course, the rental may be for a shorter period of time, and the item is then returned. The effect of a rent-to-own transaction carried to its full term is the same as buying on credit, but technically it is not a credit purchase.

For comparison shopping on rates, you can still treat a rent-to-own transaction as a loan and compute the annual percentage rate. You do this as follows: Find the total of all the payments required to buy the item and subtract the best purchase price available at an ordinary retail store; the difference is essentially the add-on interest. Find the simple interest rate that would have to be charged on the retail purchase price to get the amount of the add-on interest; then multiply by 1.8 to get the approximate annual percentage rate.

The annual percentage rate on rent-to-own transactions will usually turn out to be very high. If possible, you would be well advised to go elsewhere to borrow the money needed to make the ordinary retail purchase.

. . . .

Suggestions and Comments for Odd-numbered problems

1. and 3.
The finance charge is computed with the simple interest formula, but the annual percentage rate must be changed to a daily rate. Since no exact dates are given, assume this is not a leap year.

5. through 9.
The finance charge is computed with the simple interest formula, and the new balance is the sum of the previous ending balance and the finance charge.

11. and 13.
The key to these problems is in correctly counting the number of days that each distinct balance is in effect. Set up a table for the transactions and count the number of days between consecutive entries. Begin with the first day of the billing period and be sure the total number of days is the same as the number in the month when the billing period began. That is, June has 30 days; a billing period from June 15 through July 14 would have 30 days. Errors in counting the number of days are usually in the first period.

15. through 19.
To find the monthly payment when add-on interest is used, you find the finance charge using the simple interest formula. The finance charge is then added to the price of the item(s) or the amount being borrowed, and the total is divided by the number of months covered by the agreement.

21. When simple interest is involved, the final amount is given by the formula $F = P + Prt$ or $F = P(1 + rt)$. Substitute the values that are given and solve for the principal.

23. We will want to use the formula $F = P + Prt$ or $F = P(1 + rt)$.
To do this, we need the total amount that is paid. This is the product of the monthly payment and the number of months. How is the monthly payment calculated?

25. Finding the annual percentage rate from the nominal add-on interest rate with a formula is fairly difficult. If the values cannot be found in a table, they can be approximated with a factor of 1.8.

29. and 31.
The nominal add-on interest rate can be found with the formula $F = P(1 + rt)$, where F is the total amount of the contract.

33. The APR can be approximated using a factor of 1.8.

35. through 39.
First, you need to find the total amount of the contract. The interest rate (as add-on interest) can be found using the simple interest formula. To find the APR, either consult Table 6.2 or approximate the APR with a factor of 1.8.

41. You need to find the APR that corresponds to the nominal add-on interest rate when the difference between the purchase price and the total amount of the contract is treated as a finance charge. First, find the difference. Next, you find the nominal add-on interest rate by using the simple interest formula. Then you can approximate the APR.

Solutions to Odd-numbered Problems

1. The finance charge is the interest that is due. Use $I = Prt$. We will change the annual rate to a daily rate by dividing by 365.

(a) The daily rate is $\frac{0.129}{365}$

Finance charge $= 255.00 \times \frac{0.129}{365} \times 30 = 2.703698$

Finance charge $= \$2.70$ We will generally round to the nearest cent.

(b) The daily rate is $\frac{0.149}{365}$

Finance charge $= 425.80 \times \frac{0.149}{365} \times 31 = 5.388411$

Finance charge $= \$5.39$

3. **(a)** The daily rate is $\frac{0.149}{365}$

   Finance charge $= 315.42 \times \frac{0.149}{365} \times 31 = 3991575$

   Finance charge $= \$3.99$

   (b) The daily rate is $\frac{0.159}{365}$

   Finance charge $= 275.65 \times \frac{0.159}{365} \times 31 = 3.722407$

   Finance charge $= \$3.72$

5. The daily rate is $\frac{0.189}{365}$

   Finance charge $= 275.00 \times \frac{0.189}{365} \times 30 = \$4.27$

   New Balance $= 320.50 + 4.27 = \$324.77$

7. The daily rate is $\frac{0.159}{365}$

   Finance charge $= 155.00 \times \frac{0.159}{365} \times 30 = \$2.03$

   New Balance $= 147.85 + 2.03 = \$149.88$

9. The daily rate is $\frac{0.169}{365}$

   Finance charge $= 105.00 \times \frac{0.169}{365} \times 31 = \$1.51$

   New Balance $= 135.92 + 1.51 = \$137.43$

11.

Date	Transaction	Amount	Daily Bal.	Time Period	Days
10/11	Bal. Fwd	$165.45	165.45	10/11-10/17	7
10/18	Payment	100.00	65.45	10/18-10/24	7
10/25	Restaurant	28.90	94.35	10/25-11/4	11
11/5	Software	85.64	179.99	11/5-11/10	6
					31

   (a) Average daily balance:

   $$\frac{7(165.45) + 7(65.45) + 11(94.35) + 6(179.99)}{31} = \frac{3734.09}{31} = \$120.45$$

   The finance charge is the simple interest on the average daily balance using a daily interest rate.

   Finance charge $= 120.45 \times \frac{0.129}{365} \times 31 = \$1.32$

   (b) The New Balance is the ending balance plus the finance charge.

   New Balance $= \$179.99 + \$1.32 = \$181.31$

13.

Date	Transaction	Amount	Daily Bal.	Time Period	Days
6/11	Bal. Fwd	$225.85	225.85	6/11-6/19	9
6/20	Shoes	79.95	305.80	6/20-6/24	5
6/25	Payment	125.00	180.80	6/25-6/27	3
6/28	Books	34.65	215.45	6/28-7/4	7
7/5	Radio	69.50	284.95	7/5-7/10	6
					30

13. **(a)** Average daily balance = \$244.06

$$\frac{9(225.85) + 5(305.80) + 3(180.80) + 7(215.45) + 6(284.95)}{30} = \frac{7321.90}{30}$$

Finance charge = $244.06 \times \frac{0.149}{365} \times 30 = \$2.99$

(b) The New Balance is the ending balance plus the finance charge.
New Balance = \$284.95 + \$2.99 = \$287.94

15. Finance charge = $675 \times 0.15 \times 2 = \$202.50$
Total contract = 675.00 + 202.50 = \$877.50
Monthly payment = 877.50 ÷ 24 = \$36.56

17. Finance charge = $425 \times 0.15 \times \frac{18}{12} = \$95.63$
Total contract = 425.00 + 95.63 = \$520.63
Monthly payment = 520.63 ÷ 18 = \$28.92

19. Finance charge = $755 \times 0.105 \times \frac{30}{12} = \$198.19$
Total contract = 755.00 + 198.19 = \$953.19
Monthly payment = 953.19 ÷ 30 = \$31.77

21. We use the formula $F = P(1 + rt)$, making the appropriate substitutions:
$4340 = P(1 + (0.08)(3)) = P(1.24)$
Solving for P:
$P = \frac{4340}{1.24} = \$3500$

23. Total amount of the contract = $20 \times 48.03 = \$960.60$
$960.60 = P(1 + (0.125)(\frac{20}{12})) = P(1.20833)$
$P = \frac{960.60}{1.20833} = 794.98$
To the nearest dollar, the original purchase price is \$795

25. All values can be found in Table 6.2

   (a) 12% for 5 years: APR = 20.3%   **(b)** 6% for 2 years: APR = 11.1%

   (c) 10% for 1 year: APR = 18.0%   **(d)** 8% for 2 years: APR = 14.7%

27. To find the amount paid above the purchase price, we first need to know the total amount of the contract. The total contract amount is the product of the monthly payments and the number of months.
   Total contract = 32 × 24 = \$768
The purchase price of the television was \$600, so the difference is \$168.

29. **(a)** The total amount of the contract is $12 \times 46.47 = \$557.64$
   Substituting the known values into the formula $F = P(1 + rt)$,
   we have $557.64 = 500(1 + r(1))$.
   Solving for r: $500(1 + r) = 557.64$
   $1 + r = \frac{557.64}{500} = 1.11528$; or $r = 0.11528$
   The add-on interest rate is approximately 11.5%

29. (b) The total amount of the contract is $24 \times 37.50 = \$900.00$
Substituting, we have $900 = 750(1 + r(2))$.
Solving for r: $750(1 + 2r) = 900$
$$1 + 2r = \frac{900}{750} = 1.20; \text{ or } 2r = 0.20$$
The add-on interest rate is 10%

31. (a) The total amount of the contract is $18 \times 40.58 = \$730.44$
Substituting, we have $730.44 = 600(1 + r(\frac{18}{12}))$.
Solving for r: $600(1 + 1.5r) = 730.44$
$$1 + 1.5r = \frac{730.44}{600} = 1.2174; \; 1.5r = 0.2174 \text{ or } r = 0.14493$$
The add-on interest rate is 14.5%

(b) The total amount of the contract is $20 \times 28.69 = \$573.80$
Substituting, we have $573.80 = 450(1 + r(\frac{20}{12}))$
Solving for r: $450(1 + \frac{5}{3} r) = 573.80$
$$1 + \frac{5}{3} r = \frac{573.80}{450} = 1.27511; \; \frac{5}{3} r = 0.27511 \text{ or } r = 0.16506$$
The add-on interest rate is 16.5%

33. APR $\approx 1.8 \times 12.5\% \approx 22.5\%$
The APR is approximately 22.5%

35. The total amount of the contract is $24 \times 32 = \$768$
Substituting, we have $768 = 600(1 + r(\frac{24}{12}))$
Solving for r: $600(1 + 2r) = 768$
$$1 + 2r = \frac{768}{600} = 1.28; \; 2r = 0.28 \text{ or } r = 0.14$$
The nominal add-on interest rate 14%. Since this isn't found in Table 6.2, we approximate the APR with a factor of 1.8; $1.8 \times 14\% = 25.2\%$.
APR is approximately 25.2%

37. The total amount of the contract is $30 \times 70 = \$2100$
Substituting, we have $2100 = 1500(1 + r(\frac{30}{12}))$
Solving for r: $1500(1 + 2.5r) = 2100$
$$1 + 2.5r = \frac{2100}{1500} = 1.4; \; 2.5r = 0.4 \text{ or } r = 0.16$$
The nominal add-on interest rate 16%. Since this isn't found in Table 6.2, we approximate the APR with a factor of 1.8. $1.8 \times 16\% = 28.8\%$
The APR is approximately 28.8%

39. The total amount of the contract is $24 \times 125 = \$3000$
Substituting, we have $3000 = 2500(1 + r(2))$
Solving for r: $2500(1 + 2r) = 3000$
$$1 + 2r = \frac{3000}{2500} = 1.20; \; 2r = 0.20 \text{ or } r = 0.10$$
The nominal add-on interest rate 10%.
The APR is approximately 18%

41. For the rent-to-own transaction, we can approximate the APR from the nominal add-on interest rate.

The total amount of the contract is $30 \times 32 = \$960$

Substituting, we have $960 = 629(1 + r(\frac{30}{12}))$

Solving for r: $629(1 + 2.5r) = 960$

$1 + 2.5r = \frac{960}{629} = 1.52623$; $2.5r = 0.52623$ or $r = 0.2105$

The nominal add-on interest rate 21.05%

APR $\approx 1.8 \times 21.05\% \approx 38\%$

Section 6.3   Amortized Loans

Goals

1. Learn about amortized loans and the calculation of the loan history.

Key Ideas and Questions

1. Describe the history of an amortized loan.
2. Explain why the amount of principal paid increases each month.

Vocabulary

Amortized Loan	Monthly Payment	Amortization Table
Term of Loan	Net Payment	

· ·

Overview

The majority of loans or debts involve regular payments over a period of several, or many, years. In this section we look at the common time payment loans in which the rates accurately reflect the APR.

Terminology

An **amortized loan** is a simple interest loan with equal monthly payments made over the length of the loan. Each payment includes the interest charged since the previous payment; the remainder of the payment reduces the balance owed. The size of the equal payments is chosen so that once all the payments are made, the balance is zero. At that point, the loan is paid-off. Because of rounding, the last payment may be slightly more or less than the other monthly payments.

The important factors related to an amortized loan are the **principal**, the amount borrowed; the **interest rate**, the annual percentage rate; the **length of the loan** (also called the **term**); and the **monthly payment**. These four factors are interrelated. If you know any three of them, the fourth can be found.

Finding the Monthly Payment

An important problem in dealing with amortized loans is finding the payment when you are given the loan amount, the interest rate, and the length of the loan. There are three ways to do this: (a) by using a financial or business calculator, (b) by looking it up in a table, and (c) by applying a formula and using a scientific calculator.

One simple way to find a monthly payment is to use a financial calculator. These have built-in functions to calculate the payment required for any amortized loan. Another common way to find the payment required for an amortized loan is to consult an **amortization table**. You can look up the payment required for a variety of typical loans. Such amortization tables can be found in most business stationery stores, are relatively inexpensive and last a lifetime.

When you have access to a scientific calculator having an $\boxed{x^y}$ key, the following formula can be used to find a monthly payment.

Monthly Payment Formula

If P is the amount of the loan, r is the annual percentage rate as a decimal, and t is the length of the loan in years, then the monthly payment, PMT, is given by

$$PMT = \frac{P \times \dfrac{r}{12} \times (1 + \dfrac{r}{12})^{12t}}{[(1 + \dfrac{r}{12})^{12t} - 1]}$$

Clearly, you do not use this formula without a calculator, and it must be a calculator with an $\boxed{x^y}$ key or the equivalent for calculating exponentials.

. . . .

Suggestions and Comments for Odd-numbered problems

1. and 3.

When the first payment is made, the interest on the previous balance is calculated and subtracted from the payment. The difference is then used to reduce the outstanding balance. When calculating the interest, be sure to change the time period to $\frac{1}{12}$ of a year.

5. and 7.

As you go from month to month, the following pattern should be used:
 a. Calculate the interest due on the previous balance.
 b. Subtract the interest from the amount of the payment.
 c. Subtract the net payment from the previous balance.
At each successive step, the amount of interest should decrease and the net payment should increase. Use a table to organize the process.

9. and 11.
If the loan amount does not appear in the left-most column, break the total into smaller amounts that *can* be found in the table.

13. and 15.
Table 6.5 gives the payment needed for a given rate and term. The key to using the table is to multiply the payment factor in the table by the number of *thousands* of dollars in the loan principal.

17. and 19.
The formula is complex. If you try to do it with one calculation, make sure you have your parentheses properly placed. You might also consider making the calculations one step at a time, or calculate the numerator and denominator separately.

21. through 25.
Use the formula for monthly payments unless the correct rate and time period can be found in one of the tables.

27. and 29.
First, you need to find the amount that needs to be financed by calculating the down payment and subtracting it from the purchase price.

31. Table 6.5 is usually used to find the monthly payment when the loan conditions are known.
monthly payment = (loan size) × (payment factor)
If we want the loan size (in 1000's) we can solve as follows:
loan size = (monthly payment) ÷ (payment factor)

33. A close look at the table shows that the values change in a fairly consistent way. Any factor in the table is approximately the average of the two factors on either side of it. That is, the factor for 9% for 15 years is 10.142666.
The average for 8% and 10% is $\frac{9.556521 + 10.746051}{2} = 10.151286$.
The factor for 9.5% is approximately the average of the factors for 9% and 10%.

37. and 39.
Table 6.5 is usually used as follows:
monthly payment = (loan size) × (payment factor)
If we have the monthly payment and the size of the loan, we can find the payment for $1000 by dividing the monthly payment by the loan size (in 1000's). To find the interest rate, we locate this payment factor within the column for the correct number of years.

45. The formula is complex. If you try to do it with one calculation, make sure you have your parentheses properly placed. You might also consider making the calculations one step at a time.

Solutions to Odd-numbered Problems

1. First, we calculate the interest due on $2000 since this is the current balance:
$$I = 2000 \times 0.10 \times \tfrac{1}{12} = \$16.67$$
   The net payment is $NP = \$42.50 - \$16.67 = \$25.83$
   The new balance after the payment is $NB = \$2000 - \$25.83 = \$1974.17$

3. When the first payment is made, the interest is;
$$I = 13000 \times 0.049 \times \tfrac{1}{12} = \$53.08$$
   The net payment is $NP = \$298.79 - \$53.08 = \$245.71$
   The new balance after the payment is $NB = \$13,000 - \$245.71 = \$12,754.29$

5. We set up a table that shows all the relevant information:

Month	Payment	Interest	Net Payment	Balance
0				$5000.00

   a. The first month's interest is $I = 5000 \times 0.12 \times \tfrac{1}{12} = \$50$.
   b. The first month's net payment is $NP = \$111.23 - \$50 = \$61.23$
   c. The new balance is $NB = \$5000 - \$61.23 = \$4938.77$

   The three month history of the loan is

Month	Payment	Interest	Net Payment	Balance
0				$5000.00
1	$111.23	$50.00	$61.23	$4938.77
2	$111.23	$49.39	$61.84	$4876.93
3	$111.23	$48.77	$62.46	$4814.47

7. As in problem 5, we use a table to organize the information and show the results of the calculations. The first four month's history of the loan:

Month	Payment	Interest	Net Payment	Balance
0				$600.00
1	$123.78	$6.25	$117.53	$482.47
2	$123.78	$5.03	$118.75	$363.72
3	$123.78	$3.79	$119.99	$243.73
4	$123.78	$2.54	$121.24	$122.49

   The last payment is equal to the last balance plus interest on that balance.
$$LP = \$122.49(1 + 0.125(\tfrac{1}{12})) = \$123.77$$

5	$123.77	$1.28	$122.49	$0.00

9. The amount is found directly in the table: $132.16

11. We don't find $18000 in the left-most column, so we break the total into smaller amounts and record the payments related to each one.

$10,000	$212.48
5,000	$106.24
2,000	42.50
1,000	21.25
$18,000	$382.47

    The required monthly payment is $382.47

13. For a $1000 loan at 12% for 5 years, the payment factor is 22.244448
    The required monthly payment on an $18,000 loan is
    $$PMT = 18 \times 22.244448 = \$400.40$$

15. For a $1000 loan at 6% for 15 years, the payment factor is 8.438568
    The required monthly payment on an $22,500 loan is
    $$PMT = 22.5 \times 8.438568 = \$189.87$$

17. When we make the substitutions into the formula, we have
    $$PMT = \frac{20000 \times (\frac{0.10}{12}) \times (1 + (\frac{0.10}{12}))^{12 \times 10}}{(1 + (\frac{0.10}{12}))^{12 \times 10} - 1}$$

    or $\quad PMT = \dfrac{20000 \times \frac{0.10}{12} \times (1 + \frac{0.10}{12})^{120}}{(1 + \frac{0.10}{12})^{120} - 1} = \dfrac{451.173582}{1.707041} = \$264.30$

    The corresponding value in Table 6.2 is $264.31. The difference is due to rounding.

19. Making appropriate substitutions into the formula, we have:
    $$PMT = \frac{1000 \times \frac{0.09}{12} \times (1 + \frac{0.09}{12})^{60}}{(1 + \frac{0.09}{12})^{60} - 1}$$

    or $\quad PMT = \dfrac{11.742608}{0.565681} = \$20.76$

21. $PMT = \dfrac{6500 \times \frac{0.09}{12} \times (1 + \frac{0.09}{12})^{30}}{(1 + \frac{0.09}{12})^{30} - 1} = \dfrac{60.999498}{0.251272} = \$242.77$

23. $PMT = \dfrac{4850 \times \frac{0.0725}{12} \times (1 + \frac{0.0725}{12})^{20}}{(1 + \frac{0.0725}{12})^{20} - 1} = \dfrac{33.053531}{0.128027} = \$258.18$

25. $PMT = \dfrac{6725 \times \frac{0.128}{12} \times (1 + \frac{0.128}{12})^{48}}{(1 + \frac{0.128}{12})^{48} - 1} = \dfrac{19.371937}{0.664107} = \$179.75$

27. The down payment is $0.2 \times 3250 = \$650$, leaving \$2600 to be financed.
    We need to find the monthly payment on a loan of \$2600 at 10.5% for 2 years.
    Substituting in the formula, we have

$$PMT = \frac{2600 \times \dfrac{0.105}{12} \times (1 + \dfrac{0.105}{12})^{24}}{(1 + \dfrac{0.105}{12})^{24} - 1} = \frac{28.040551}{0.232552} = \$120.58$$

29. The down payment is $0.2 \times 16285 = \$3257$, leaving \$13,028 to be financed. We need
    to find the monthly payment on a loan of \$13.028 at 11% for 5 years.
    In Table 6.5, we find that a loan of \$1000 at 11% for 5 years has a monthly payment
    factor of 21.742423. The required monthly payment will be:
    $$PMT = 13.028 \times 21.742423 = \$283.27$$

31. In Table 6.5, we find that a \$1000 loan at 9% for 20 years has a monthly loan factor
    of 8.997260. Since the monthly payment is to be \$500, the size of the loan (L) that
    can be financed is
    $$L = \frac{500}{8.997260} = \$55,572.47$$

33. To find the size of the loan that can be financed with a payment of \$600 at 9.5% for
    15 years, we will need an approximation to the payment factor. The payment factors
    9% and 10% are 10.142666 and 10.746051. Their average is 10.444359. Since the
    difference in factors actually increase slightly as you go from one entry down to
    another, we will round down. Using a factor of 10.44 we get an approximate answer
    $$L = \frac{600}{10.44} \times 1000 = 57.471 \times 1000 = \$57,471.$$

35. From Table 6.5, we see that the payment on \$1000 at 8% for 5 years is \$20.276394.
    To find the size of the loan (in 1000's) that can be financed with a payment of \$250,
    we divide as follows:
    $$L = \frac{250}{20.276394} \times 1000 = 12.329609 \times 1000 = \$12,330$$

37. Since a monthly payment of \$321.85 is required on a loan of \$40,000, the payment
    per \$1000 is $\dfrac{321.85}{40} = 8.04625$

    Since the loan is for 30 years, we look down the 30 year column and find that the
    payment corresponds to an interest rate of 9%.

39. Given a monthly payment of \$292.68 on a loan of \$25,000, the payment per \$1000 is
    $\dfrac{292.68}{25} = 11.7072$

    Looking in the 15 year column, we find that the payment falls between two of the
    entries in the table;

percentage	payment
11%	11.365969
12%	12.001681

    Since the payment is approximately halfway between these two values, we assume the
    interest rate is halfway between as well: approximately 11.5%.

45. Substitution into the formula gives:

$$P = 420 \times \frac{12}{0.1075} \times \left[1 - \frac{1}{(1 + \frac{0.1075}{12})^{240}}\right]$$

$$P = 420 \times \frac{12}{0.1075} \times [1 - 0.117605] \approx \$41,370$$

Section 6.4 Buying a House

Goals
1. Learn about financing a house.

Key Ideas and Questions
1. Describe the process of buying a house and the affordability guidelines designed to prevent a buyer from becoming overextended.
2. How do "points" affect the decision to buy and finance?

Vocabulary

Mortgage	Term	Discount Charge
Fixed Rate Mortgage	Points	Down Payment
Adjustable Rate Mortgage	Loan Origination Fee	Closing Costs
Cap		

. . . .

Overview

The purchase of a new home represents a large financial commitment, and it is a complicated transaction. Many factors beyond the price of the home have to be considered, beginning with the ability to pay the initial costs as well as the monthly payments that may last for 30 years. This section will cover the basic mechanics of a home mortgage.

Affordability Guidelines

A few general guidelines have been used to estimate how much a buyer could afford to spend on a house. Here are two of the most common ones.

(1) The home you purchase should not cost more than three times your annual family income; this assumes a standard down payment of 20%.
(2) You should limit your monthly housing expenses including mortgage payment, property taxes, and homeowner's insurance to no more than 25% of your monthly gross income (that is, income *before* deductions).

If your planned house purchase fits under both of these guidelines, then you can almost surely afford it.

The most important question for the buyer is whether or not a bank or other financial institution will approve the loan application. Among the things which will be considered is your other debt. Having car payments and credit card balances may affect your ability to buy a house. As of this writing, many banks are allowing up to 38% of the borrower's monthly income to go for mortgage payment, property taxes, and homeowner's insurance. A good credit history is also a valuable asset.

The Mortgage

A **mortgage** is a loan that is guaranteed by real estate. The two main categories of mortgage are *fixed rate* and *adjustable rate*. For a **fixed rate mortgage**, the interest rate is set, once and for all, at the time the loan is made. For an **adjustable rate mortgage** the interest rate can change from year to year. The actual interest rates that are quoted are usually a specified amount higher than some particular financial index. Often there is also a limit (called a **cap**) on how much the interest rate is allowed to rise in a single year.

A common distinction among mortgages is the **term** of the mortgage. Typically the choices are **15 year** or **30 year**. Longer term loans usually carry a higher interest rate because the money is used for a longer period of time and the lender's risk is extended.

Another variable in choosing a mortgage loan is commonly referred to as **points**. One "point" is one percent of the amount of the loan. Points are generally charged in two ways: (1) a **loan origination fee** for making the loan at all and (2) a **discount charge** for offering a lower interest rate. For a fixed term mortgage the combined effect of the interest rate, fee, and discount charge can be summed up in the **annual percentage rate**. For an adjustable rate mortgage the fee and discount charge are typically smaller, and the annual percentage rate cannot be computed because the interest rate will be changing.

The Down Payment

The most typical down payment on a house is 20% of the total value of the house. The published rates in an interest rate table assume such a 20% down payment. If the down payment is 20% of the value of the property, then the value of the property you can buy is the amount you can afford for the down payment divided by 0.20. A lender prefers a large percentage down payment as protection against the borrower defaulting on the loan. If payments are missed and the lender has to take control of the property, the down payment is a significant part of the value that can be recovered.

If a smaller down payment is made, you must expect to pay in some way for the lender's increased risk. The interest rate may be higher, or you may be required to purchase Private Mortgage Insurance - which has a price in points - to insure the lender against a default on the loan. Additionally, there may be other charges when the transaction is finalized. This finalizing of the purchase is called the **closing**, and the additional expenses are called **closing costs**. Down payments smaller than 20% are becoming more common, in particular in state or federal programs aimed at assisting first-time or low-income buyers.

Suggestions and Comments for Odd-numbered problems

5. The key to using the affordability guidelines involve the gross annual household income for the purchase and the gross monthly income for the monthly payments.

9. A "point" refers to one percent of the purchase price or the loan amount, depending on the reference.

13. and 15.
To find a 20% down payment for a given house price, multiply the price by 20% (0.2). If the down payment is known, you can divide by 20% to find the maximum purchase price that can be negotiated.

17. Although some closing costs can be included in the amount of a loan, we will assume that all closing costs must be made before the loan is finalized.

19. Table 6.5 (in Section 6.3) gives the payment for each $1000 that must be financed.

21. through 25.
Approach each of these as a sequence of shorter problems:
How much is the down payment?
What amount must be financed?
What payment is required?

27. and 29.
The monthly payment for principal and interest is based on the loan amount. The monthly payment for taxes and insurance is based on the assessed value.

Solutions to Odd-numbered Problems

1. Maximum house price = $3 \times \$35,000 = \$105,000$.
   Maximum monthly payment = $0.25 \times \dfrac{35000}{12} = \$729.17$

3. Maximum house price = $3 \times \$43,500 = \$130,650$
   Maximum monthly payment = $0.25 \times \dfrac{43500}{12} = \$907.29$

   Note: These are guidelines, and not hard and fast rules. Credit histories are taken into consideration by most lenders. A monthly payment of $950 represents 26.2%, and may be close enough; most banks have been approving payments that represent up to 28%.

5. The gross combined annual income is $53,000, and the gross monthly income is $\dfrac{53000}{12}$, or approximately $4417

   Maximum house price = $3 \times \$53,000 = \$159,000$

   Maximum monthly payment = $0.25 \times \dfrac{53000}{12} = \$1104.17$

7. Low estimate = $0.25 \times 3650 = \$912.50$
   High estimate = $0.38 \times 3650 = \$1387.00$

9. A "point" refers to one percent of the purchase price.
   The two points represent $0.02 \times 95000 = \$1900$

11. The 2.5 points represent 2.5% of the cost of the house.
   $0.025 \times 92000 = \$2300$

13. $0.20 \times$ home price = down payment, so home price = down payment $\div 0.20$
   If the available down payment is \$23,000, then the home price (HP) is
   $$HP = \frac{23000}{0.20} = \$115,000$$

15. Home price = down payment $\div 0.20$
   If the down payment is \$18,000, the home price is
   $$HP = \frac{18000}{0.20} = \$90,000$$

   According to the affordability guidelines given in this section, the home price should be no more than 3 times the gross annual income for the household. Therefore, the annual household income should be at least \$30,000.

17. The Davis family must pay \$2500 in closing costs, and will have only \$15,500 left to use as a down payment. Since they are financing the loan with a 10% down payment, the home price is $HP = \frac{15500}{0.10} = \$155,000$

19. The entry corresponding to a 12% loan for 30 years in Table 6.5 is 10.286125. For a \$72,000 loan, the payment is $72 \infty 10.286125 = 740.601$
   The monthly payment for principal and interest will be \$740.61. By rounding up, the last payment will be less than this amount.

21. If the home price is \$135,000, then a 20% down payment is \$27,000.
   This will leave a balance of \$108,000 to be financed at 8% for 30 years.
   Referring to Table 6.5, the corresponding entry is 7.337646 (this is the payment for each \$1000). For a loan of \$108,000 the monthly payment for principal and interest is: $PMT = 108 \times 7.337647 = 792.465768$ or \$792.47

23. For a home priced at \$127,000 a 10% down payment is \$12,700.
   This will leave a balance of \$114,750 to be financed. We must estimate the payment factor since 9.5% does not appear in Table 6.5. The portion of the table of interest to us has the following entries for 30 year loans.

9%	8.046226
10%	8.775716

   Since 9.5% is the average of 9% and 10%, we will estimate the monthly payment factor for 9.5% as the average of the factors for 9% and 10%, or
   $$\frac{8.046226 + 8.775716}{2} = 8.410971$$

   The required monthly payment for principal and interest on a loan of \$114,750 is approximately $PMT = 114.75 \times 8.410971 \approx \$966$

25. For a home priced at $95,500 a 10% down payment is $9,550. This leaves a balance of $85,950 to be financed at 9.6% for 30 years (or 360 months)
Substituting in the monthly payment formula (Section 6.3), we have

$$PMT = \frac{85950 \times \dfrac{0.096}{12} \times (1 + \dfrac{0.096}{12})^{360}}{(1 + \dfrac{0.096}{12})^{360} - 1} = \frac{12109.5339}{16.61131} \approx \$729$$

27. For the monthly payment of principal and interest, we use Table 6.5.
The payment per $1000 at 9% for 30 years is 8.046226.
For a loan of $115,000 the payment is $115 \times 8.046226 = \$925.32$
The taxes are based on the assessed value; the taxes are 2.5% of $150,000
$$0.025 \times 150000 = \$3750$$
Taxes and insurance are $3750 + $650 = $4400, requiring a monthly payment of
$M = \dfrac{4400}{12} = \$366.67$. The **total** monthly payment is $925.32 + $366.67 = $1291.99
Note: When calculating required monthly payments, we round up in all cases.

29. For the monthly payment of principal and interest, we use Table 6.5.
The payment per $1000 at 10% for 15 years is 10.746051.
For a loan of $75,000 the payment is $75 \times 10.746051 = \$805.96$
The taxes are based on the assessed value; the taxes are 2.25% of $127,700
$$0.0225 \times 127700 = \$2873.25$$
Taxes and insurance are $2873.25 + $740 = $3613.25, requiring a monthly payment of
$M = \dfrac{3613.25}{120} = \$301.11$. The **total** monthly payment is $805.96 + $301.11 = $1107.07
Note: When calculating required monthly payments, we round up in all cases.

Section 6.5   Annuities and Sinking Funds

Goals
1. Learn about sinking funds and annuities and how to calculate the future value of an annuity.

Key Ideas and Questions
1. How does the purchase of an income annuity work?
2. Why is an annuity like an amortized loan in reverse?

Vocabulary

Annuity	Simple Annuity	Annuitant
Income Annuity	Annuity Due	Annuity Certain
Sinking Fund	Ordinary Annuity	Life Annuity
Payment Period	Future Value of Annuity	Annuity Factor

Overview

Although important purchases or financial commitments are often taken care of "after the fact" through loans and mortgages, many people prefer to have their funds set aside before obligating them. Also, some government entities may legally be required to have funds set aside. Such situations require thoughtful planning, and the key to success is making appropriate regular payments to carry out the plan.

Terminology

Suppose you want to pay for something that costs more than you can afford from your paycheck; for example, a vacation. One strategy is to borrow the money; another is to set aside a smaller amount from each paycheck until you have accumulated enough for what you want to do. If you regularly set aside a given amount of money in an account that pays interest, then you have an annuity.

A typical **annuity** is an account into which a sequence of equal, regular payments are made. An annuity is a saving or income strategy. It is a useful strategy for two reasons. First, the regular nature of the deposits helps the saver maintain discipline. Second, if the payments are made over a long time period, the interest received (especially the *compounding* of the interest) contributes significantly to the growth of the annuity.

A **sinking fund** is a type of annuity in which the goal is to have a particular amount of money saved at the end of a given time period.

There is another type of annuity where, if you *purchase* the annuity, you will *receive* a sequence of equal regular payments. We will call this second type of annuity an **income annuity**.

There are two basic questions with annuities. First, how much will accumulate over time when given periodic payments are made? This is called the **future value** of the annuity. Second, what periodic payments will be necessary in order to accumulate a specific amount in a given period of time? This is especially important when establishing sinking funds.

The compounding period is the length of time between payments. The length of time between payments is also called the **payment period**. An annuity for which the interest compounding period is the same as the payment period is called a **simple annuity**. We will restrict our attention to simple annuities. The **term** of an annuity is the length of time from the beginning of the first payment period until the end of the last payment period. At the end of the term of the annuity, the annuity is said to have **expired** and the money may be withdrawn. For an income annuity, the funds will have been depleted.

An annuity for which payments are due at the beginning of each payment period is called an **annuity due.** An annuity for which payments are due at the end of each payment period is called an **ordinary annuity**. Each payment to an ordinary annuity will be on deposit for one less payment period. Therefore, every payment to an ordinary annuity will earn interest for one less compounding period than an annuity due. The difference amount to one payment's interest on the future value.

Finding the Future Value of an Annuity

The following formulas can be used to find the total amount of money available at the end of the term. The amount of money available for withdrawal when the annuity expires is called the **future value** of the annuity and will be represented by the letter F in the various formulas in this section. We will also use i to represent the **periodic interest rate**, which is the annual interest rate, r, divided by the number of periods.

Future Value of an Annuity

If PMT is the payment size, r is the annual rate (expressed as a decimal), m is the number of payment periods per year, and t is the number of years; then the total

number of payment periods is $n = mt$, and the periodic rate is $i = \dfrac{r}{m}$, and the

future value is

$$F = PMT \times \frac{(1+i)^n - 1}{i} \text{ for an ordinary annuity,}$$

$$F = PMT \times (1+i) \times \frac{(1+i)^n - 1}{i} \text{ for an annuity due.}$$

An annuity due (with payment at the beginning of the period) earns one more period of interest than an ordinary annuity with the same rate and term. This difference is reflected in the formula: The annuity due has an additional factor of $(1 + i)$.

Finding Payments for Sinking Funds

The distinguishing feature of a *sinking fund* is that the future value, F, is chosen in advance and the payment size has to be found. We will continue to use PMT for the payment size, i for the periodic interest rate, and n for the total number of periods.

The formulas for finding payments to sinking funds are as follows:

Payments for Sinking Funds

If PMT is the payment size, F is the future value, r is the annual rate (expressed as a decimal), and m is the number of payment periods per year and t is the number of years; then the total number of payment periods is $n = mt$, and the periodic rate

is $i = \dfrac{r}{m}$, and the payment size is

$$PMT = F \times \frac{i}{(1+1)^n - 1} \text{ for an ordinary annuity, and}$$

$$PMT = F \times \frac{i}{(1+i) \times [(1+1)^n - 1]} \text{ for an annuity due.}$$

Income Annuities

An income annuity provides a sequence of regular equal payments to an individual, the **annuitant**, or to several individuals, the **annuitants**. The payments may be made over a fixed period of time, in which case the annuity is called an **annuity certain**, or the payments may be continue indefinitely until some contingency event occurs. The most common contingency that ends an annuity is the death of the annuitant. An annuity that is terminated by a death is called a **life annuity**; it is good for the lifetime of the annuitant. Therefore, the life expectancy of the annuitant is a major factor in determining the cost of the annuity. By contrast, the cost of an annuity certain can be readily computed without reference to the age or health status of the annuitant.

It is expected that an annuity certain will earn interest. The process involved in an annuity certain is the same as putting a specific amount of money into an interest paying account and making periodic equal withdrawals, but no deposits, until the account is empty. The key to success is discipline.

· ·

Suggestions and Comments for Odd-numbered problems

1. and 3.
When payments are made at the beginning of each period, it is referred to as an "annuity due." The value at the end of the annuity is known as the "future value" of the annuity. The rate given is usually an annual rate, but the rate must correspond to the length of the period. $i = \dfrac{r}{m}$, where r is the annual rate and m is the number of periods in a year.

5. and 7.
When payments are due at the end of each period, an annuity is known as an ordinary annuity.

9. through 15.
Be sure that you have correctly identified the type of annuity involved, and that you are using the correct formula and periodic rate.

17. through 23.
You should be sure to correctly identify the type of annuity involved and the relevant formula, as well as the number of payments per year. In using the formula, you may find it easier to calculate the denominator separately.

29. Since we are interested in the future value of the annuity when the last payment is actually made, this is an ordinary annuity.

31. Since there are two sources of money for the fund (the one-time investment with compound interest and an annuity), these should be treated as two separate funds whose future value will be combined.

33. The key to the problem is finding out how much Merrie must have in the sinking fund when she turns 50. One way to do this is to work backwards from the monthly payments she will receive. Think of the sinking fund as money Merrie is loaning to the bank, which in turn will make payments of $2500 a month to her. In Section 6.3, which covered amortized loans, there is a formula that is used to find the monthly payment when the principal is known. In the extended problems of Section 6.3, this formula is modified so that you can find the principal when the amount of the payment is known.

Solutions to Odd-numbered Problems

1. For the future value of an annuity due, we use the formula

$$F = PMT \times (1 + i) \times \frac{(1+i)^n - 1}{i}$$, where PMT is the periodic payment, and i is the

periodic rate. In this case, with monthly payments ($m = 12$), and a rate of 6%,

$i = \frac{0.06}{12} = 0.005$ and $1 + i = 1.005$. Substituting, we have

$$F = 300 \times (1.005) \times \frac{(1.005)^6 - 1}{0.005} = \$1831.76$$

Note: You should always check the reasonableness of your answer. If there are 6 payments of $300, the total has to be greater than $1800. For long term annuities the effect of compounding the interest will have a significant effect and the future value can be two or more times the total of the payments.

3. With monthly payments ($m = 12$), and a rate of 8%, the periodic rate is

$i = \frac{0.08}{12}$. Since the periodic rate is a repeating decimal fraction, we have two choices: use an approximation or carry the fraction in our calculations. First, with the fraction we get

$$F = 150 \times (1 + \frac{0.08}{12}) \times \frac{(1 + \frac{0.08}{12})^{10} - 1}{\frac{0.08}{12}} = \$1556.11$$

Since i = 0.006666..., an approximation to use is 0.00667. This gives us

$$F = 150 \times (1.00667) \times \frac{(1.00667)^{10} - 1}{0.00667} = \$1556.14$$

Note: Our answers will be based on using the fractions, whenever they occur. For long term annuities the differences due to rounding will be greater than in this problem.

5. For the future value of an ordinary annuity, we use the formula

$$F = PMT \times \frac{(1 + i)^n - 1}{i}$$, where PMT is the periodic payment, and i is the periodic

rate. Since $r = 0.06$, $i = 0.005$ and $1 + i = 1.005$. With monthly payments for 25 years, $n = 25 \times 12 = 300$. Substituting, we have

$$F = 50 \times \frac{(1.005)^{300} - 1}{0.005} = \$34,649.70$$

7. Since $r = 0.05$, $i = \frac{0.05}{12}$ and $1 + i = 1 + \frac{0.05}{12}$ (or approximately 1. 00417) With
    monthly payments of \$60 for 2 years, $n = 2 \times 12 = 24$. Substituting, we have
    $$F = 60 \times \frac{(1 + \frac{0.05}{12})^{24} - 1}{\frac{0.05}{12}} = \$1511.16$$

9. This is an annuity due with $P = \$50$, $i = \frac{0.06}{12} = 0.005$, and $n = 300$.
    $$F = 50 \times (1.005) \times \frac{(1.005)^{300} - 1}{0.005} = \$34{,}822.94$$
    Note: Compare this problem to problem 5, which is an ordinary annuity with the
    same terms. The annuity due, with payments at the beginning of the month, earns an
    extra month's income compared to the ordinary annuity.
    \$34,649.70 $\times$ 1.005 = \$34,822.95 (difference due to rounding).

11. This is an ordinary annuity with $PMT = \$80$, $i = \frac{0.06}{12} = 0.005$, and $n = 60$.
    $$F = 80 \times \frac{(1.005)^{60} - 1}{0.005} = \$5581.60$$
    Note: Compare the effects of the length of the annuity on the future value.
    In problem 9, the total payments over 25 years are \$15,000, while the future value is
    nearly \$35,000. In problem 11, the total payments over 5 years are \$4800, while the
    future value is approximately \$5600.

13. This is an annuity due with $PMT = \$100$, $i = \frac{0.06}{4} = 0.015$ and $n = 4 \times 25 = 100$.
    Payments are made quarterly.
    $$F = 100 \times 1.015 \times \frac{(1.015)^{100} - 1}{0.015} = \$23{,}223.51$$

15. This is an ordinary annuity with $PMT = \$200$, $i = \frac{0.06}{2} = 0.03$, and $n = 2 \times 10 = 20$.
    Payments are made semi-annually.
    $$F = 200 \times \frac{(1.03)^{20} - 1}{0.03} = \$5374.07$$

17. This is an annuity due with $F = \$25{,}000$, $i = \frac{0.12}{12} = 0.01$, and $n = 60$.
    $$PMT = 25000 \times \frac{0.01}{(1.01) \times ((1.01)^{60} - 1)} = \$303.08$$

    Note: Is the answer reasonable? Total payments would be \$18,184.80 over the 5 year
    period.

    This calculation can be put directly into most calculators as
    $PMT = 25000 \times .01 \div (1.01 \times (1.01^{60} - 1))$. Make sure parentheses are placed
    correctly.

19. This is an annuity due with $F = \$10,000$, $i = \dfrac{0.12}{12} = 0.01$, and $n = 60$.

$$PMT = 10000 \times \frac{0.01}{(1.01) \times ((1.01)^{60} - 1)} = \$121.23$$

21. This is an ordinary annuity with $F = \$15,000$, $i = 0.008$ and $n = 48$

$$PMT = 15000 \times \frac{0.008}{(1.008)^{48} - 1} = \$257.56$$

23. This is an annuity due with $F = \$8,000,000$, $i = \dfrac{0.06}{12} = 0.005$, and $n = 120$.

$$PMT = 8000000 \times \frac{0.005}{(1.005) \times ((1.005)^{120} - 1)} = \$48,573.53$$

25. We refer to Table 6.7. The entry in the 20 year row (240 months) and the 10% column is 103.6246. This means that $103.6246 is needed to purchase an annuity of $1 a month for 20 years. To purchase an annuity that pays $500 a month would require $103.6246 \times 500 = \$51,812.30$

For a one-time investment of $51,812.30, the annuitant will receive a total return of $120,000 over 20 years.

27. Referring to Table 6.7, the entry in the 30 year row (360 months) and 8% column is 136.2835. This means that $136.2835 is needed to purchase each dollar in the income annuity. To purchase an income annuity of $1,500 per month would require

$$136.2835 \times 1500 = \$204,425.25$$

The investment of $204,425.25 provides a return of $540,000 over 30 years.

29. This is an ordinary annuity with $PMT = \$40$, $i = \dfrac{0.054}{12} = 0.0045$, and $n = 181$.

Note: With the third payment on his third birthday, there are 15 full years with monthly payments before the last payment on the 18th birthday.

$$F = 40 \times \frac{(1.0045)^{181} - 1}{0.0045} = \$11,145.99$$

31. The two sources of money are treated separately.

   (a) $5000 invested at 5.75% compounded monthly for 3 years.

$$F = 5000 \times (1 + \frac{0.0575}{12})^{36} = \$5938.91$$

   (b) An ordinary annuity with $PMT = 200$, $i = \dfrac{0.0575}{12}$, and $n = 36$.

$$F = 200 \times \frac{(1 + \frac{0.0575}{12})^{36} - 1}{\frac{0.0575}{12}} = \$7837.87. \text{ The total fund will equal } \$13,776.78$$

33. **(a)** The first question to answer is: How much must Merrie have on deposit in a sinking fund to withdraw $2500 a month for two years? Referring to Table 6.7, the entry in the 2 year row (24 months) and 6% column is 22.5629. This means that $22.5629 is needed to purchase each dollar in the income annuity. To purchase an income annuity of $2,500 per month would require $22.5629 \times 2500 = \$56,407.25$

33. **(b)** Merrie will have 15 years in which to make monthly payments into the sinking fund. Assume she starts making her payments the next month, ending on her 50th birthday, so this is an ordinary annuity.

$$PMT = \$56,407.25 \times \frac{0.005}{(1.005)^{180} - 1} = \$193.96$$

Merrie should deposit at least $193.96 each month to reach her goal.

Chapter 6 Review Problems

Solutions to Odd-numbered Problems

1. For the simple interest loan, $I = Prt$, where $P = \$650$, $r = 0.08$, and $t = 4$.
   $I = 650 \times 0.08 \times 4 = \$208$

3. We use the formula $A = P \times (1 + r)^n$ where $P = \$650$, $r = 0.08$, and $n = 4$.
   $F = 650 \times (1.08)^4 = \$884.32$
   The amount of interest is $234.32

5. $EAR = (1 + \frac{0.09}{12})^{12} - 1 = (1.0075)^{12} - 1 = 0.0938$ or 9.38%

7. We make a short table to organize the information:

Date	Transaction	Amount	Daily Bal.	Time Period	Days
6/1	Bal. Fwd	$0.00	0.00	6/1-6/19	19
6/20	Television	612.00	612.00	6/20-6/30	11
					30

   Average daily balance $= \dfrac{19 \times 0 + 11 \times 612}{30} = \$224.40$

   Finance charge $= 224.40 \times \dfrac{0.21}{365} \times 30 = \$3.87$

9. The first step is to determine the total that you will pay, since this will be treated as principal and interest. There are 24 payments of $46, so
   Total $= 46 \times 24 = \$1104$
   To estimate the APR, we treat this total as the principal and interest of an **add-on** interest loan in which the principal is $819. Making the appropriate substitutions, we have
   $1104 = P(1 + rt) = 819 \times (1 + 2r)$
   $1 + 2r = 1104 \div 819 = 1.3480$
   $2r = 0.3480;\quad r = 0.174$ or 17.4%
   The simple interest rate is 17.4%. To estimate the APR for the add-on interest loan we multiply by 1.8. $1.8 \times 17.4\% \approx 31\%$
   The APR is approximately 31%

11. Using Table 6.5, the entry in the 10 year column and 11% row is 13.775001.
    The monthly payment for a $16,000 loan is
    $$16 \times 13.775001 = \$220.40$$

13. The affordability guideline is that the price of the home should not exceed three times
    the annual gross income.
    $$\text{Maximum} = 3 \times \$51,000 = \$153,000$$

15. Pat and Chris have $14,000 available, but $3000 is needed for closing costs. That
    leaves $11,000 for the down payment. After making the down payment, the couple
    will have to finance the balance of $99,000 at 8% for 30 years. Using Table 6.5, the
    entry in the 30 year column and the 8% row is 7.337646
    $$\text{Mortgage payment} = 99 \times 7.337646 = \$726.43 \text{ per month}$$
    Taxes for the year are 3.5% of $110,000; or $0.035 \times 110,000 = \$3850$
    Yearly total for taxes and insurance $= 3850 + 450 = \$4300$
    $$\text{Taxes and insurance} = 4300 \div 12 = \$358.33 \text{ per month}$$
    Total monthly payment $= 726.43 + 358.33 = \$1084.76$

    The couple should easily qualify on the basis of being able to make the monthly
    payments. The low estimate of what they can afford is $1062; the high estimate is
    $1615.

17. If the deposits are made at the beginning of each period, then this is an annuity due.
    If interest of 6% is compounded semiannually for 10 years, then $i = 3\%$ and $n = 20$.
    With a future value of $50,000, the monthly payment needed is

    $$PMT = 50000 \times \frac{0.03}{(1.03) \times ((1.03)^{20} - 1)} = \$1806.59$$

19. In Table 6.7, the annuity factor for a 25 year annuity (300 months) based on a rate of
    6% is 155.2069. That means that a one-time payment of $155.2069 provides a return
    of $1 a month for 25 years. With a 6% return on investment, an annuity paying $800
    a month for 25 years would cost
    $$155.2069 \times 800 = \$124,165.52$$

Game Theory

Section 7.1   The Description of a Game

Goals
1. Compute outcomes in two-person games.

Key Ideas and Questions
1. How is a game matrix for a zero-sum game constructed?
2. Why is the most aggressive strategy typically a poor strategy?

Vocabulary

Players	Zero-sum Games	Row/Column
Strategies	Game Matrix	Size of a Matrix
Payoffs	Entry/Element	Most Aggressive Strategy
Preferences		

. . . .

Overview

Game Theory is a branch of mathematics developed to analyze competitive situations and make decisions. These situations need not be games in the usual sense. Game Theory is included in a broader topic called Decision Theory. In this chapter, we introduce the mathematical framework for analyzing such situations and drawing conclusions regarding the best course of action.

Games

The following features are typically present when game theory is used:

1. Conflicting interests of two or more participants. All participants want to win and will behave rationally in their own interests.
2. Incomplete information. Sometimes there is no missing information--as in a chess game where both players see all the board.
3. An interplay of rational choice and chance.

Participants in a game are called **players**. Each player has possible actions to take or choices to make, which are called **strategies**. Once the choices have been made, the players receive a **payoff**. Each player has **preferences** among the payoffs. We expect players to prefer to win and receive the "best" payoff. A game where the amounts won by one player equal the amounts lost by another player is called a **zero-sum** game since the second player's payoff is automatically the opposite of the first player's. When the game being considered involves only two players, it is called a **two-person** game.

Games need not be zero-sum, nor two-person, but we will limit most of our discussion to this type. However, our interpretation will allow such "players" as the weather, and payoffs that amount to a splitting or division of some quantity such as "market share". If one player has 65%, the other will have 35%. In this sense, one player's gain is the other player's loss. An increase for one player means a decrease for the other.

Game Matrices

When working with games, it is customary to put the information describing the game into a rectangular array called a **game matrix**, The term matrix (plural is matrices) refers to any rectangular array of numbers.

A matrix is usually displayed between a large pair of parentheses or brackets as in the matrix M, below:

$$M = \begin{bmatrix} 17 & 2.5 & 4 \\ 8 & -2 & -1 \\ 3 & 1 & 43 \\ -6 & 33 & 8 \end{bmatrix}$$

Each number in the matrix is called an **entry**, or **element**, of the matrix. All the entries in a horizontal line form a **row** of the matrix. All entries in a vertical line form a **column** of the matrix. Rows are numbered from top to bottom and columns are labeled from left to right. The **size of the matrix** is given as the number of rows and the number of columns. You pick an entry in a matrix by specifying its row and column position.

Abstract Game Matrices

An abstract game matrix just uses the row numbers as names for the first player's strategies (example: first player's **second** strategy goes with **second** row of the matrix). Similarly, the column numbers are used as names of the second player's strategies (example: second player's **third** strategy is associated with the **third** column of the matrix.) It is also common practice to eliminate any actual grid and just keep the matrix of payoff numbers in the abstract game matrix. The matrix is constructed from the first player's perspective and includes only the payoffs for the first player. Recall that in a zero-sum game the payoff for the second player is the opposite of the payoff for the first player. It does not mean that any row sum or column sum of the game matrix is zero or, for that matter, that the sum of all the elements is zero.

The strategy that offers the possibility of the maximum payoff, without regard for risk, is called the **most aggressive strategy.** To find the most aggressive strategy for the first player, find the largest element in the game matrix. The most aggressive strategy for the first player will be the strategy represented by the row in which this largest element is located.

To find the most aggressive strategy for the second player, find the smallest element in the game matrix. Remember, since the payoff to the second player is the opposite of the entry in the matrix, smaller is better from the second player's point of view. The most aggressive strategy for the second player corresponds to the column in which this smallest element is located.

A player's most aggressive strategy offers the chance for the greatest possible gain, but it depends on the other player doing just the right thing. This, of course, is not likely to be the kind of cooperation the game generates. Adopting the most aggressive strategy does not guarantee the best payoff, and often runs the greatest risk of loss.

Suggestions and Comments for Odd-numbered problems

13. through **19.**
The game matrix shows the payoffs to the first player for each combination of choices made by the two players. If the description says "the second player wins 3 points", then the entry in the matrix is "-3", because the first player's payoff is the opposite of the second player's payoff in a zero-sum game.

17. Although it is not necessary for constructing the game matrix, this game is not a zero-sum game in the usual sense. If the batter's success rate is .300, then the pitcher's success rate is .700 (the total is 1.000). "Zero-sum" means that one player's success is at the other player's expense.

27. and **29.**
The most aggressive strategy is the "greediest", without regard to risk. The first player goes for maximum payoff for the game. The second player attempts to minimize the payoff to the first player.

31. Make a preliminary sketch. How many possible sequences of moves are there? The first player has 4 choices, the next player has 3, and so on. Allow for all possibilities. When the conditions for winning have been met, the branch of the tree should be ended (x-1, o-4, x-2 is a win for x, and the branch is drawn no farther).

Solutions to Odd-numbered Problems

1. **(a)** This is a 3 by 3 matrix; 3 rows and 3 columns.
(b) The element in the first row and second column is -2.

3. **(a)** This is a 3 by 4 matrix; 3 rows and 4 columns.
(b) The element in the second row and third column is 1.

5. The matrix is constructed using the payoffs for player 1.
$$\begin{bmatrix} 2 & 4 \\ -3 & -1 \end{bmatrix}$$

7.

$$\begin{bmatrix} 3 & 4 & -2 \\ -3 & -1 & 2 \end{bmatrix}$$

9.

$$\begin{bmatrix} 1 & 2 & 3 \\ 4 & 5 & -4 \\ -3 & -2 & -1 \end{bmatrix}$$

11.

$$\begin{bmatrix} 1 & -6 & 2 & -5 \\ 3 & -4 & 4 & -3 \\ 5 & -2 & 6 & -1 \end{bmatrix}$$

13.

		Patrick	
		One Finger	Two Fingers
Patti	One Finger	10	−10
	Two Fingers	−10	10

15.

		Rose	
		Ace	King
Carmen	Ace	1	2
	King	−2	−3

17.

		Pitcher	
		Fastball	Curveball
Batter	Fastball	.450	.200
	Curveball	.240	.400

19.

		IRS	
		Audit	Don't Audit
Taxpayer	Cheat	−3000	1500
	Don't Cheat	−200	0

21.

1stPlayer Strategy	2ndPlayer Strategy	1stPlayer Payoff	2ndPlayer Payoff
I	I	1	-1
I	II	-2	2
II	I	-1	1
II	II	2	-2

23.

1stPlayer Strategy	2ndPlayer Strategy	1stPlayer Payoff	2ndPlayer Payoff
I	I	0	0
I	II	2	-2
I	III	-1	1
II	I	2	-2
II	II	1	-1
II	III	-2	2
III	I	-1	1
III	II	-2	2
III	III	2	-2

25.

1stPlayer Strategy	2ndPlayer Strategy	1stPlayer Payoff	2ndPlayer Payoff
I	I	0	0
I	II	-2	2
I	III	0	0
I	IV	-2	2
II	I	-1	1
II	II	1	-1
II	III	-1	1
II	IV	1	-1
III	I	2	-2
III	II	0	0
III	III	2	-2
III	IV	0	0

27. The game matrix in question is:

$$\begin{bmatrix} 1 & -2 \\ 2 & -1 \end{bmatrix}$$

(a) The greatest payoff for the first player is 2 (in the second row).
The second strategy is the most aggressive one for the first player.

(b) The worst payoff for the first player is -2 (in the second column).
Therefore, this is the greatest payoff for the second player.
The second player's most aggressive strategy is the second strategy.

29. The game matrix in question is

$$\begin{bmatrix} 4 & 1 & -1 \\ -2 & 3 & 0 \\ -4 & -3 & 2 \end{bmatrix}$$

(a) The greatest payoff for the first player is 4 (in the first row).
The first player's most aggressive strategy is the first strategy.

(b) The worst payoff for the first player is -4 (in the first column).
The first strategy for the second player is the most aggressive.

31. (a) The two "outside" branches are "pruned" from the tree.

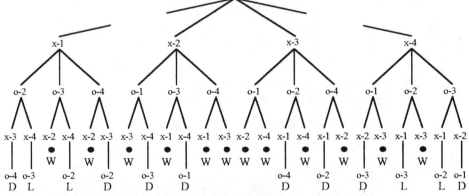

(b) The winning strategy for x is to play either a 2 or a 3 first.
Then, the first player can always win on the third move.
A mistake on the third move will lead to a draw, but never a loss.

(c) Yes, this approach can be employed in tic-tac-toe.

33. Rob will win if he plays a 5. Carlos has no winning strategy. Carlos can only win if Rob plays a 2 or a 4, but that is not a strategy for Carlos (it's only a hope.)

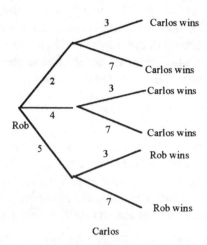

Section 7.2   Determined Games

Goals
1. Determine when an optimal strategy is forced on the players.

Key Ideas and Questions
1. Describe the most conservative strategy.
2. Suppose that both players use their most conservative strategies and in this case the optimal outcome occurs for both. Why is this game determined?

Vocabulary

Most Conservative Strategy	Determined Game	Fair Game
Maximin Procedure	Optimal Strategy	Dominants
Minimax Procedure	Value of the Game	

. . . .

Overview

This section focuses on identifying and selecting the most appropriate strategies to maximize the result of the game for each player.

Most Conservative Strategy

The opposite approach to maximizing possible gain is to minimize possible loss. This is known as the **most conservative strategy**. To find the first player's strategy that minimizes possible loss, find the **minimum** entry in each row of the matrix and then select the **maximum** from among these minima. This procedure is called the **maximin**. The first player is selecting the best outcome from among the worst outcomes from each of his strategies. Whereas the most aggressive strategy is optimistic, the most conservative strategy is cautious or even pessimistic.

You can do a similar analysis for most conservative strategy for the second player. Remember that when considering payoffs for the second player, you have to think of the opposites of the numbers in the matrix. The worst case for the second player is the best case for the first player. So, in looking for the worst cases for the second player, you should look for the largest entries in the columns. These are called **maxima**, plural of maximum. The second player's most conservative strategy is selecting the minimum from among the maxima. This is called the **minimax**.

Determined Games

A two-person zero-sum game for which each player adopting the most conservative strategy makes the other player's most conservative strategy the best choice is called a **determined game**. The resulting choices of strategies are called **optimal** because neither player can reliably improve his or her payoff. However, if one player deviates from the optimal strategy, the other player automatically receives an improved payoff.

In a determined zero-sum game, one player can always have a positive advantage: a "win" every time! Why play the game, then? With some "games", you don't have a choice: life goes on and decisions must be made.

A useful shortcut with matrices is to do "circling" and "boxing" to find the most conservative strategies on the same copy of the matrix. The circles indicate the smallest entry in each row, and the boxes indicate the largest entry in each column. When an entry of the matrix ends up with both a box and a circle around it, then the game is **determined** (these are the choices that would be reached rationally). An entry that is both boxed and circled represents a rational combination of strategies by the players, and is called the **value of the game**. The row and column containing this entry represent the **optimal strategies**. A game with value 0 is called a **fair game**.

Suppose the payoffs related to one strategy for a player are as good as, or better than, the corresponding payoffs related to another strategy for the player. The first of the strategies would always be preferred by a rational player, and the second strategy is dropped or eliminated from consideration. We say that the first strategy **dominates** the other

· ·

Suggestions and Comments for Odd-numbered problems

1. through 9.
First look at the payoffs for each strategy from the first player's (row player's) perspective: "What's the worst that can happen if I use this strategy?" The first player is interested in the lowest payoff in each row, then picks the best of these worst payoffs. The second player is interested in the largest payoff in each column, since these go to the first player.

11. through 19.
Correct analysis of the game depends upon correct construction of the game matrix. Make sure the matrix represents the game correctly.

21. through 27.

Another way to look at dominating strategies is that one strategy dominates another if it is never worse for any choice of the other player's strategies.

Consider the matrix

$$\begin{bmatrix} 1 & 2 & 3 & 4 \\ 1 & 2 & 3 & 5 \\ 10 & 20 & 30 & 3 \end{bmatrix}$$

Row 2 dominates row 1

Row 3 does not dominate either row 1 or 2

The term "dominates" is relative to the current state of analysis of the matrix. Suppose you compare the three rows in a matrix and find that no row dominates any other. Next, you compare the columns and find that one of them (say column 2) dominates another (say column 4). The column that is dominated represents a strategy that a rational person would never use when the other strategy is available, so it is "dropped"; re-write the matrix without the dominated column. Now, go back and compare the rows in the reduced game matrix.

29. Assume that all tickets will be sold unless indicated otherwise. The weather is the second player, and its "strategies" are what it might "decide" to do on the day of the concert. Calculate the proceeds to the promoter for each of the possibilities.

35. The payoffs to General Kennedy are the number of days of bombing that he hopes to attain. What decisions are available to Kennedy? to the Japanese?

Solutions to Odd-numbered Problems

1. (a) Most conservative strategies:

$$\begin{bmatrix} 1 & \boxed{-2} \\ \boxed{-1} & 2 \end{bmatrix} \qquad \begin{bmatrix} \boxed{1} & -2 \\ -1 & \boxed{2} \end{bmatrix}$$

1st player: 2nd strategy;   2nd player: 1st strategy

For the first player, the maximin (maximum of the row minima) is the -1 in the second row (the second strategy's lowest payoff).

For the second player, the minimax (minimum of the column maxima) is the 1 in the first column (the first strategy's largest payoff). The most conservative strategy for a player is choosing the option that places a lower limit to the value of the game.

(b) Not determined. None of the row minima correspond to a column maximum.

$$\begin{bmatrix} \boxed{1} & \boxed{-2} \\ \boxed{-1} & \boxed{2} \end{bmatrix}$$

3. (a) Most conservative strategies: **(b)** This is a determined game

1st player: 1st strategy   2nd player: 1st strategy

For the first player, the maximin (maximum of the row minima) is the 1 in the first row (the first strategy's lowest payoff). For the second player, the minimax (minimum of the column maxima) is the 1 in the first column (the first strategy's largest payoff). For each player, the most conservative strategy is to choose the best of the worst thing that can happen with each strategy.

(c) The value of the game is 1.

5. (a) Most conservative strategies: **(b)** This is a determined game

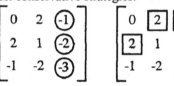

     1st player: 2nd player:
     1st strategy   3rd strategy

(c) The value of the game is -1.

7. (a) Most conservative strategies: **(b)** This is a determined game

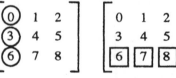

     1st player: 2nd player:
     3rd strategy   1st strategy

(c) The value of the game is 6.

9. (a) Most conservative strategies: **(b)** The game is not determined

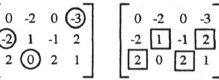 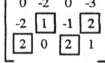 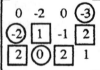

     1st player: 2nd player:
     3rd strategy   2nd strategy

11. (a) The game matrix:

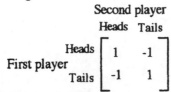

(b) Most conservative strategies

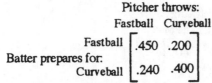

All strategies are "most conservative".

(c) The game is not determined

13. (a) The game matrix:

(b) Most conservative strategies:

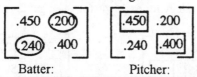

Batter: Pitcher:
prepare for a curve   throw a curve

(c) The game is not determined.

Note: This may seem paradoxical, but isn't. The pitcher is preparing to throw a curve and the batter is preparing for a curve. As a result, the batter will average .400. However, if the pitcher switches to a fastball, the results will be better for the pitcher; the hitter would then switch to preparing for a fastball. The pitcher cannot afford to stick with either of his available strategies..

15. (a) The "game" matrix:

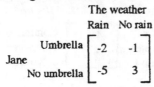

(b) Most conservative "strategies":

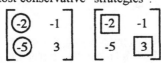

Jane: The weather:
take umbrella   rain

(c) The "game" is determined.

(d) The value of the game is -2. Jane's best guess is to take the umbrella and be prepared for rain.

17. (a) The game matrix:

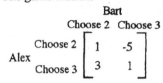

(b) The most conservative strategies: **(c)** The game is determined.

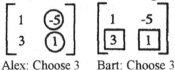

Alex: Choose 3   Bart: Choose 3

(d) The value of the game is 1.

19. (a) The game matrix:

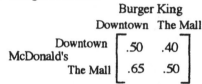

(b) The most conservative strategy for each   **(c)** The game is determined
is locating at the mall.

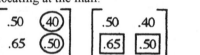

(d) They will split the business.   McDonald's will get 50%, leaving the other 50% for Burger King.

21. For the row player, larger numbers (larger payoffs) are preferred.
For the column player, smaller numbers (smaller payoffs) are preferred.
We compare the payoffs in the 2nd column to each of the other columns.
If the payoffs in the second column are always as good as or better than those in another column, then the second column is preferred. That is, it *dominates.*

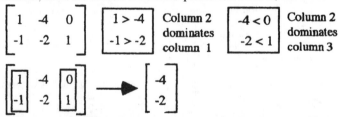

The second player has what is called a "pure" strategy - always use the second strategy.
The row player has to decide which of the remaining payoffs is preferred: -2 or -4.
The value of the game is -2.   The larger number is preferred by the row player.
In this limited sense, the remaining second row "dominates" the remaining first row.

23. We compare the strategies of the first player two rows at a time and see that row 2 dominates row 3.

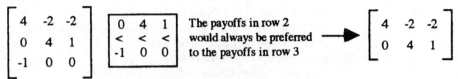

$$\begin{bmatrix} 4 & -2 & -2 \\ 0 & 4 & 1 \\ -1 & 0 & 0 \end{bmatrix} \qquad \boxed{\begin{matrix} 0 & 4 & 1 \\ < & < & < \\ -1 & 0 & 0 \end{matrix}} \quad \begin{matrix} \text{The payoffs in row 2} \\ \text{would always be preferred} \\ \text{to the payoffs in row 3} \end{matrix} \longrightarrow \begin{bmatrix} 4 & -2 & -2 \\ 0 & 4 & 1 \end{bmatrix}$$

Then, column 3 dominates column 2.

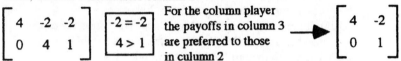

$$\begin{bmatrix} 4 & -2 & -2 \\ 0 & 4 & 1 \end{bmatrix} \qquad \boxed{\begin{matrix} -2 = -2 \\ 4 > 1 \end{matrix}} \quad \begin{matrix} \text{For the column player} \\ \text{the payoffs in column 3} \\ \text{are preferred to those} \\ \text{in culumn 2} \end{matrix} \longrightarrow \begin{bmatrix} 4 & -2 \\ 0 & 1 \end{bmatrix}$$

Alternatively, in the original matrix, column 2 dominates column 3.
Then, row 2 dominates row 3.

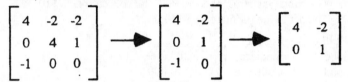

$$\begin{bmatrix} 4 & -2 & -2 \\ 0 & 4 & 1 \\ -1 & 0 & 0 \end{bmatrix} \longrightarrow \begin{bmatrix} 4 & -2 \\ 0 & 1 \\ -1 & 0 \end{bmatrix} \longrightarrow \begin{bmatrix} 4 & -2 \\ 0 & 1 \end{bmatrix}$$

25. Column 1 dominates all other columns.

$$\begin{bmatrix} 0 & 2 & 1 & 3 \\ -4 & -1 & 2 & -1 \\ -5 & 3 & 0 & -2 \end{bmatrix} \qquad \begin{matrix} 0 & 2 \\ -4 & -1 \\ -5 & 3 \end{matrix} \qquad \begin{matrix} 0 & 1 \\ -4 & 2 \\ -5 & 0 \end{matrix} \qquad \begin{matrix} 0 & 3 \\ -4 & -1 \\ -5 & -2 \end{matrix}$$

Column 1 dominates column 2   Column 1 dominates column 3   Column 1 dominates column 4

From the perspective of the column player, the payoffs in column 1 are always as good as, or better than, the payoffs in the other columns.

Since Column 1 dominates all other columns, the matrix reduces:

$$\begin{bmatrix} 0 & 2 & 1 & 3 \\ -4 & -1 & 2 & -1 \\ -5 & 3 & 0 & -2 \end{bmatrix} \longrightarrow \begin{bmatrix} 0 \\ -4 \\ -5 \end{bmatrix}$$

The column player will always pick column 1 (a pure strategy); therefore, the row player should pick the first strategy ; i.e. the first row. The value of the game is 0.

27. No reduction in the matrix is possible.

$$\begin{bmatrix} 3 & -2 & -1 \\ -2 & 3 & -2 \\ 1 & 1 & 2 \end{bmatrix}$$

29. (a) First, we analyze the possible scenarios for the concert.

Using the stadium in good weather:

    Total receipts: $30 \times 20{,}000 = \$600{,}000$.

    The band gets 50%, or \$300,000; the stadium gets \$15,000.

    The net to the promoter is \$285,000.

Using the stadium in bad weather:

    Total receipts are 70% of before, or \$420,000.

    The band gets \$210,000; the stadium gets \$15,000.

    The net to the promoter is \$195,000.

Using the stadium in very bad weather: The concert is canceled.

    The band gets \$50,000; the stadium gets \$15,000.

    The net to the promoter is - \$65,000.

Using the coliseum in any kind of weather:

    Total receipts: $30 \times 10{,}000 = \$300{,}000$.

    The band gets 50%, or \$150,000; the coliseum gets \$10,000.

    The net to the promoter is \$140,000.

Weather	Location	Payoff
Good	Stadium	285,000
Good	Coliseum	140,000
Bad	Stadium	195,000
Bad	Coliseum	140,000
Very bad	Stadium	-65,000
Very bad	Coliseum	140,000

(b)

$$
\begin{array}{cc}
 & \text{Weather} \\
 & \begin{array}{ccc} \text{Good} & \text{Bad} & \text{Very bad} \end{array}
\end{array}
$$

$$
\text{Promoter} \begin{array}{c} \text{Stadium} \\ \text{Coliseum} \end{array}
\begin{bmatrix}
285{,}000 & 195{,}000 & -65{,}000 \\
140{,}000 & 140{,}000 & 140{,}000
\end{bmatrix}
$$

(c) The most aggressive strategy is to rent the stadium.
Renting the stadium has the greatest potential profit.

(d) The most conservative strategy is to rent the coliseum.
Renting the coliseum has the least risk.

37. (a) Strategy I for Japanese: go north.
Strategy II for Japanese: go south.
Strategy I for Kennedy: look north.
Strategy II for Kennedy: look south.   **(b)** This is a determined game

$$
\text{Kennedy}
\begin{array}{c} \text{Northern} \\ \text{Southern} \end{array}
\begin{array}{c}
\begin{array}{cc} \text{Northern} & \text{Southern} \end{array} \\
\begin{bmatrix} 2 & 2 \\ 1 & 3 \end{bmatrix}
\end{array}
\quad
\begin{bmatrix} ② & 2 \\ ① & 3 \end{bmatrix}
\quad
\begin{bmatrix} \boxed{2} & \boxed{2} \\ 1 & 3 \end{bmatrix}
\longrightarrow
\begin{bmatrix} \boxed{②} & \boxed{2} \\ ① & 3 \end{bmatrix}
$$

The value of the game is 2. Kennedy's decision is to concentrate on the northern route which would assure a minimum of two days of bombing if the Japanese moved a force into the region.

Section 7.3   Mixed Strategies

Goals
1. Compute that optimal mixed strategy for a player and the value of a game.
2. Find the average payoff for a player.

Key Ideas and Questions

1. Describe why a mixed strategy may be beneficial over a strategy in which your opponent knows what you will do.

Vocabulary

    Mixed Strategy   Average Payoff of a Game   Optimal Mixed Strategy

· ·

Overview

In this section, we will look at more general games and techniques for finding the optimal (most effective) strategies for the players.

Optimal Mixed Strategy

In a determined game, each player can use just one strategy all the time and be assured as to what is the worst that can happen. If one player switches from the most conservative strategy and the other doesn't, the player who made the switch will suffer a penalty (to the benefit of the one who didn't). Remember, the "value of the game" represents the worst that you will do unless you make a strategic mistake. It exists independently for each of the players, and it's not the best they hope to do. In a determined game, you can eliminate one source of advantage for the other player.

However, not all games are determined games. If a game is not determined, it is not possible to select one strategy for regular use without increasing your risk of lower payoffs or increased risk. The most effective strategy for maintaining this type of control is called the optimal strategy.

To optimize results in a game that is not determined, a player must use a **mixed strategy**. This means that two or more strategies are used at random; there is no pattern to their use. The trick is to determine what percentage of time to use each strategy.

Average Payoff of a Game

The **average payoff** of a game is computed by multiplying each payoff by the fraction of the time the related strategies are used and adding those products.

In a two-person zero-sum game in which each player has two strategies (or the game has been reduced to that basis because of dominating strategies) there is a formula which allows you to find the **optimal mixed strategy** for the players.

The optimal mixed strategy tells you the relative frequency with which each player should use the available strategies in order to maintain control of the lower limit of the payoffs. With the optimal mixed strategy, neither player can do better without the cooperation of the opponent: "cooperation" in the form of not playing the game well enough.

Optimal Mixed Strategy

Given the two-person zero-sum game with matrix:

$$\begin{array}{cc} A & B \\ C & D \end{array}$$

If the game is not a determined game, the optimal mixed strategy is for the first player to choose her first strategy with relative frequency

$$\frac{C - D}{(B + C) - (A + D)}$$

and the second player to choose his first strategy with relative frequency

$$\frac{B - D}{(B + C) - (A + D)} .$$

· ·

Suggestions and Comments for Odd-numbered problems

1. through 5.
Make a second matrix that has the relative frequencies each strategy is used as the labels on the matrix, rather than using the names of the strategies. For the elements of the matrix, compute the products for the combined relative frequencies of the strategies used: if the first player uses the first strategy 1/2 of the time and the second player uses his first strategy 1/3 of the time, then the relative frequency for the combination of the two strategies is 1/6. The payoff in the game matrix for this combination of strategies will be multiplied by 1/6 when calculating the value of the game.

11. through 15.
When one player's mixed strategy is known, it is possible to pick the strategy most beneficial to the other player until there is a change in strategy by the first player. The solutions can be obtained by a formal means using the definition of the average payoff of a game as described in Example 7.13. Rather than construct a table, we can organize the information around the framework of a matrix.

17. through 21.
We say that one strategy dominates the other if the first of the strategies would always be preferred by a rational player. The second strategy is dropped or eliminated from consideration. Remember that a matrix that has been reduced should be re-analyzed for dominant strategies.

23. Is the game determined? If not, how is the average payoff of the game found?

25. Is this a determined game? If the matrix is not a 2 by 2 matrix, can it be reduced to one?

27. and 29.
  When one player's mixed strategy is known, the other player's optimum strategy is a pure strategy.

33. and 35.
  The directions produce a "system of equations". The techniques for solving systems of equations like these are covered in Section 1.8 of the text.

Solutions to Odd-numbered Problems

1. The game matrix   Factors
                                       1/2   1/2

$$\begin{bmatrix} 1 & -2 \\ -1 & 3 \end{bmatrix} \quad \begin{matrix} 1/2 \\ 1/2 \end{matrix} \begin{bmatrix} 1/4 & 1/4 \\ 1/4 & 1/4 \end{bmatrix}$$

Value of game $= \frac{1}{4} \times 1 + \frac{1}{4} \times (-2) + \frac{1}{4} \times (-1) + \frac{1}{4} \times 3 = \frac{1}{4}$

3. The game matrix   Factors
                                       1/3   2/3

$$\begin{bmatrix} 1 & -2 \\ -1 & 2 \end{bmatrix} \quad \begin{matrix} 1/2 \\ 1/2 \end{matrix} \begin{bmatrix} 1/6 & 1/3 \\ 1/6 & 1/3 \end{bmatrix}$$

Value of game $= \frac{1}{6} \times 1 + \frac{1}{3} \times (-2) + \frac{1}{6} \times (-1) + \frac{1}{3} \times 2 = 0$

5. The game matrix   Factors
                                       1/3 1/3 1/3

$$\begin{bmatrix} 0 & 2 & -1 \\ 2 & 1 & -2 \\ -1 & -2 & -3 \end{bmatrix} \quad \begin{matrix} 1/3 \\ 1/3 \\ 1/3 \end{matrix} \begin{bmatrix} 1/9 & 1/9 & 1/9 \\ 1/9 & 1/9 & 1/9 \\ 1/9 & 1/9 & 1/9 \end{bmatrix}$$

Since every entry in the game matrix will be multiplied by $\frac{1}{9}$ and then the products will be added, we use the distributive law to shorten the calculation.

Value of game $= \frac{1}{9} \times (0 + 2 - 1 + 2 + 1 - 2 - 1 - 2 - 3) = -\frac{4}{9}$

7. The game matrix

$$\begin{bmatrix} 1 & -2 \\ -1 & 2 \end{bmatrix} \qquad A = 1, B = -2, C = -1, D = 2$$

Optimal strategies:

First player, first strategy: $\dfrac{(-1) - 2}{((-2) + (-1)) - (1 + 2)} = \dfrac{-3}{(-3) - 3} = \dfrac{1}{2}$

First player: $(\frac{1}{2}, \frac{1}{2})$

Second player, first strategy: $\dfrac{(-2) - 2}{((-2) + (-1)) - (1 + 2)} = \dfrac{-4}{(-3) - 3} = \dfrac{2}{3}$

Second player: $(\frac{2}{3}, \frac{1}{3})$

9. The game matrix

$$\begin{bmatrix} 1 & -3 \\ -1 & 3 \end{bmatrix} \qquad A = 1, B = -3, C = -1, D = 3$$

Optimal strategies:

First player, first strategy: $\dfrac{(-1) - 3}{((-3) + (-1)) - (1 + 3)} = \dfrac{-4}{(-4) - 4} = \dfrac{1}{2}$

First player: $(\frac{1}{2}, \frac{1}{2})$

Second player, first strategy: $\dfrac{(-3) - 3}{((-3) + (-1)) - (1 + 3)} = \dfrac{-6}{(-4) - 4} = \dfrac{3}{4}$

Second player: $(\frac{3}{4}, \frac{1}{4})$

11. Game matrix   Factors

			1/3	2/3
1	-2	p	1/3 x (p)	2/3 x (p)
-1	2	1 - p	1/3 x (1 - p)	2/3 x (1 - p)

The value of the game:

When p = 1 (and 1 - p = 0), the average payoff of the game is;

$$v = \frac{1}{3} \times (1) + \frac{2}{3} \times (-2) = -1$$

When p = 0 (and 1 - p = 1), the average payoff of the game is

$$v = \frac{1}{3} \times (-1) + \frac{2}{3} \times (2) = 1$$

The first player's optimum strategy is to use the second strategy as a pure strategy.
The average payoff of the game is 1.

Alternatively, since the first player has two options and will use one of them as a pure
strategy (relative frequency of 1), we can see what happens with each option and pick
the better of the two.

	1/3	2/3			1/3	2/3
1	1	-2		0	1	-2
0	-1	2		1	-1	2

$v = (1/3)(1) + (2/3)(-2) = -1 \qquad v = (1/3)(-1) + (2/3)(2) = 1$

13. Game matrix   Factors

$$\begin{bmatrix} 1 & -3 & 2 \\ -1 & 3 & -2 \end{bmatrix} \quad \begin{matrix} p \\ 1-p \end{matrix} \begin{bmatrix} 1/4 \times p & 1/4 \times p & 1/2 \times p \\ 1/4 \times (1-p) & 1/4 \times (1-p) & 1/2 \times (1-p) \end{bmatrix}$$

The value of the game:

When p = 1 (and 1 - p = 0), the average payoff of the game is

$$v = \frac{1}{4} \times (1) + \frac{1}{4} \times (-3) + \frac{1}{2} \times (2) = \frac{1}{2}$$

When p = 0 (and 1 - p = 1), the average payoff of the game is

$$v = \frac{1}{4} \times (-1) + \frac{1}{4} \times (3) + \frac{1}{2} \times (-2) = -\frac{1}{2}$$

The first player's optimum strategy is to use the first strategy as a pure strategy. The average payoff of the game is $\frac{1}{2}$.

15. There are three possible situations if the second player uses a known strategy, and the first player chooses a pure strategy.

$$\begin{matrix} & 1/4 & 1/4 & 1/2 \\ 1 \\ 0 \\ 0 \end{matrix} \begin{bmatrix} 1 & -3 & 2 \\ -2 & 4 & -1 \\ 1 & -1 & -1 \end{bmatrix} \qquad \begin{matrix} & 1/4 & 1/4 & 1/2 \\ 0 \\ 1 \\ 0 \end{matrix} \begin{bmatrix} 1 & -3 & 2 \\ -2 & 4 & -1 \\ 1 & -1 & -1 \end{bmatrix} \qquad \begin{matrix} & 1/4 & 1/4 & 1/2 \\ 0 \\ 0 \\ 1 \end{matrix} \begin{bmatrix} 1 & -3 & 2 \\ -2 & 4 & -1 \\ 1 & -1 & -1 \end{bmatrix}$$

  Pure Strategy 1   Pure strategy 2   Pure strategy 3

For the first strategy, $v = \frac{1}{4} \times (1) + \frac{1}{4} \times (-3) + \frac{1}{2} \times (2) = \frac{1}{4} - \frac{3}{4} + 1 = \frac{1}{2}$

For the second strategy, $v = \frac{1}{4} \times (-2) + \frac{1}{4} \times (4) + \frac{1}{2} \times (-1) = \frac{-1}{2} + 1 - \frac{1}{2} = 0$

For the third strategy, $v = \frac{1}{4} \times (1) + \frac{1}{4} \times (-1) + \frac{1}{2} \times (-1) = \frac{1}{4} - \frac{1}{4} - \frac{1}{2} = -\frac{1}{2}$

The optimum strategy for the first player is the first strategy.

The value of the game is $\frac{1}{2}$.

17. The game matrix Column 2 dominates column 3

$$\begin{bmatrix} 3 & -1 & -1 \\ -2 & 2 & 3 \end{bmatrix} \quad \boxed{\begin{matrix} -1 = -1 \\ 2 < 3 \end{matrix}} \longrightarrow \begin{bmatrix} 3 & -1 \\ -2 & 2 \end{bmatrix} \quad \begin{matrix} A = 3, B = -1 \\ C = -2, D = 3 \end{matrix}$$

Optimal strategies for the reduced matrix:

First player, first strategy: $\dfrac{(-2) - 2}{((-2) + (-1)) - (2 + 3)} = \dfrac{-4}{(-3) - 5} = \dfrac{1}{2}$

    First player: $(\frac{1}{2}, \frac{1}{2})$.

Second player, first strategy: $\dfrac{(-1) - 2}{((-2) + (-1)) - (2 + 3)} = \dfrac{-3}{(-3) - 5} = \dfrac{3}{8}$

    Second player: $(\frac{3}{8}, \frac{3}{8}, 0)$.

The third strategy was dominated, and it will never be used.

19. Game matrix　　　Col 2 dominates Col 1　　Col 3 dominates Col 4

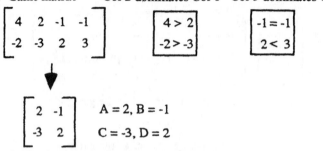

$$\begin{bmatrix} 2 & -1 \\ -3 & 2 \end{bmatrix}$$　　$A = 2, B = -1$

　　　　　　　　　$C = -3, D = 2$

Optimal strategies for the reduced matrix:

First player, first strategy: $\dfrac{(-3)-2}{((-1)+(-3))-(2+2)} = \dfrac{-5}{(-4)-4} = \dfrac{5}{8}$　　First player: $(\frac{5}{8}, \frac{3}{8})$.

Second player, "first" strategy: $\dfrac{(-1)-2}{((-1)+(-3))-(2+2)} = \dfrac{-3}{(-4)-4} = \dfrac{3}{8}$

Second player: $(0, \frac{3}{8}, \frac{5}{8}, 0)$. The first and fourth strategies are dominated.

21. The game matrix

$$\begin{bmatrix} 1 & -2 & -1 \\ -3 & 3 & 2 \\ 2 & -2 & 3 \end{bmatrix}$$
$$\begin{array}{ccc} 1 & -2 & -1 \\ < & = & < \\ 2 & -2 & 3 \end{array}$$
Row 3
dominates　⟶
Row 1
$$\begin{bmatrix} -3 & 3 & 2 \\ 2 & -2 & 3 \end{bmatrix}$$

$$\begin{bmatrix} -3 & 3 & 2 \\ 2 & -2 & 3 \end{bmatrix}$$
$$\begin{array}{c} -3 < 2 \\ 2 < 3 \end{array}$$
Column 1
dominates　⟶
Column 3
$$\begin{bmatrix} -3 & 3 \\ 2 & -2 \end{bmatrix}$$
$A = -3, B = 3$
$C = 2, D = -2$

Now we find the optimal mixed strategies for the reduced matrix:

First player, "first" strategy: $\dfrac{2-(-2)}{(3+2)-((-3)+(-2))} = \dfrac{4}{5-(-5)} = \dfrac{2}{5}$

Note: This is the relative frequency of the first player's first <u>remaining</u> strategies.

Second player, "first" strategy: $\dfrac{3-(-2)}{(3+2)-((-3)+(-2))} = \dfrac{5}{5-(-5)} = \dfrac{1}{2}$

First player: $(0, \frac{2}{5}, \frac{3}{5})$.　　Second player: $(\frac{1}{2}, \frac{1}{2}, 0)$.

23. See problem 13 in Section 7.2 for the derivation of the game matrix, where it was shown that this is not a determined game. Therefore, we find the optimal mixed strategies for the players.

$$\begin{bmatrix} .450 & .200 \\ .240 & .400 \end{bmatrix}$$
$A = .450, B = .200$
$C = .240, D = .400$

Optimal strategies:

Batter, first strategy: $\dfrac{.240 - .400}{(.200 + .240) - (.450 + .400)} = \dfrac{-.160}{.440 - .850} = \dfrac{.160}{.410} \approx 0.39$

The batter should prepare for fastballs 39% of the time, curveballs 61% of the time.

Pitcher, first strategy: $\dfrac{.200 - .400}{(.200 + .240) - (.450 + .400)} = \dfrac{-.200}{.440 - .850} = \dfrac{.200}{.410} \approx 0.49$

The pitcher should throw fastballs 49% of the time, curveballs 51% of the time.

25. The matrix and some analysis:

Interest Rates

Rise   Fall

Charlane
invests in:

		Rise	Fall
Stocks		.10	.12
Bonds		.06	.11
$ Market		.15	.10

Whether interest rates rise or fall, the return on stocks will be better than the return on bonds.

The strategy of investing in stocks dominates the strategy of investing in bonds. The matrix can be reduced, and we will find the optimal mix for Charlane's remaining strategies. We won't find the optimal mix related to interest rates, because they don't have any options: the rates will either rise or fall.

$$\begin{bmatrix} .10 & .12 \\ .15 & .10 \end{bmatrix} \quad \begin{matrix} A = .10, B = .12 \\ C = .15, D = .10 \end{matrix}$$

Charlane, first strategy: $\dfrac{.15 - .10}{(.12 + .15) - (.10 + .10)} = \dfrac{.05}{.27 - .20} = \dfrac{.05}{.07} \approx 0.71$

Charlane should put 71% of her savings into stocks, 29% of her savings into money market funds, and none in bonds.

27. Since the mix of strategies for Mega Sports is known, we will consider the payoff when Action Faction uses a pure strategy.

Mega Sports
0.25   0.75

Action	1	3	-2
Faction	0	-1	2

Newspaper ads only

Mega Sports
0.25   0.75

Action	0	3	-2
Faction	1	-1	2

TV ads only

If Action Faction uses only newspaper ads, the payoff is

$v = 0.25 \times 3 + 0.75 \times (-2) = -0.75$

If Action Faction uses only TV ads, the payoff is

$v = 0.25 \times (-1) + 0.75 \times 2 = 1.25$

Action Faction should put its entire budget into TV advertising. The value of the game is an increase of 1.25 million dollars in sales.

29. Since the mix of strategies for the Republican candidate is known, we will consider the payoff when the Democratic candidate uses a pure strategy.

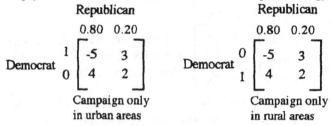

If the Democratic candidate campaigns only in urban areas, the payoff is
$$v = 0.80 \times (-5) + 0.20 \times 2 = -3.6$$
If the Democratic candidate campaigns only in rural areas, the payoff is
$$v = 0.80 \times 4 + 0.20 \times 2 = 3.6$$

The Democrat should only campaign in the rural areas. The payoff is an increase of 3.6 (36,000 votes).

31. Action Faction and Mega Sports

$$\begin{bmatrix} 3 & -2 \\ -1 & 2 \end{bmatrix} \quad A = 3, B = -2 \\ C = -1, D = 2$$

Optimal strategies: Faction Faction: $(\frac{3}{8}, \frac{5}{8})$ Mega Sports: $(\frac{1}{2}, \frac{1}{2})$

Action Faction, first strategy: $\dfrac{(-1) - 2}{((-2) + (-1)) - (3 + 2)} = \dfrac{-3}{(-3) - 5} = \dfrac{3}{8}$;

Mega Sports, first strategy: $\dfrac{(-2) - 2}{((-2) + (-1)) - (3 + 2)} = \dfrac{-4}{(-3) - 5} = \dfrac{1}{2}$

Next, we find the value of the game:

	1/2	1/2		Factors	
3/8	3	-2		3/16	3/16
5/8	-1	2		5/16	5/16

$$v = \frac{3}{16} \times 3 + \frac{5}{16} \times (-1) + \frac{3}{16} \times (-2) + \frac{5}{16} \times 2 = \frac{9}{16} - \frac{5}{16} - \frac{6}{16} + \frac{10}{16} = \frac{8}{16} = \frac{1}{2} = 0.5$$

Action Faction should use 37.5% (3/8) of its advertising for newspaper ads and 62.5% for television ads, with a payoff 0.5 million dollars in increased sales. Mega Sports should evenly divide its advertising, but will still have decreased sales of 0.5 million.

33. The game matrix and the factors associated with each element in the matrix.

$$
\begin{array}{c}
\quad\begin{array}{ccc} r & s & 1\text{-}r\text{-}s \end{array} \\
\begin{array}{c} p \\ q \\ 1\text{-}p\text{-}q \end{array}
\left[\begin{array}{ccc}
1 & -3 & 2 \\
-2 & 4 & -1 \\
1 & -1 & -1
\end{array}\right]
\end{array}
\qquad
\begin{array}{c}
\text{Factors} \\
\left[\begin{array}{ccc}
pr & ps & p(1\text{-}r\text{-}s) \\
qr & qs & q(1\text{-}r\text{-}s) \\
r(1\text{-}p\text{-}q) & s(1\text{-}p\text{-}q) & (1\text{-}p\text{-}q)(1\text{-}r\text{-}s)
\end{array}\right]
\end{array}
$$

The value of the game can be expressed as follows:

$$v = pr(1) + ps(-3) + p(1\text{-}r\text{-}s)(2) + qr(-2) + qs(4) + q(1\text{-}r\text{-}s)(-1) + r(1\text{-}p\text{-}q)(1)$$
$$+ s(1\text{-}p\text{-}q)(-1) + (1\text{-}p\text{-}q)(1\text{-}r\text{-}s)(-1)$$

After carrying out the multiplication and collecting like terms, we have:

$$v = p(3 - 3r - 5s) + q(5s - 3r) + (2r - 1)$$

We set the coefficients of p and q equal to zero: $3 - 3r - 5s = 0$ and $5s - 3r = 0$

The second equation can be written $5s = 3r$, and we can substitute directly for 5s in the first equation:

$$3 - 3r - (5s) = 0 \implies 3 - 3r - (3r) = 0, \text{ or } 3 - 6r = 0$$

From the last equation, we have $r = \frac{1}{2}$. Substituting back into $5s = 3r$, we have $5s = \frac{3}{2}$,

or $s = \frac{3}{10}$. Finally, $1 - r - s = 1 - \frac{1}{2} - \frac{3}{10} = \frac{2}{10} = \frac{1}{5}$.

Putting it all together, the optimum mix of strategies for the second player is:

$$(\frac{1}{2}, \frac{3}{10}, \frac{1}{5}) \text{ or } (\frac{5}{10}, \frac{3}{10}, \frac{2}{10})$$

Now, we need the optimal strategy mix for the first player.

The equation $v = p(3 - 3r - 5s) + q(5s - 3r) + (2r - 1)$ can be re-written as

$$v = r(2 - 3p - 3q) + s(5q - 5p) + (3p - 1)$$

Setting the coefficients of r and s equal to zero, we have:

$$2 - 3p - 3q = 0 \text{ and } 5q - 5p = 0$$

The second equation implies that $p = q$, so the first equation can be written

$$2 - 6p = 0 \text{ or } p = \frac{1}{3}. \text{ This means that } q = \frac{1}{3} \text{ and } 1 - p - q = \frac{1}{3}.$$

The optimum mix of strategies for the first player is: $(\frac{1}{3}, \frac{1}{3}, \frac{1}{3})$

Finally, we can compute the value of the game.

$$
\begin{array}{c}
\quad\begin{array}{ccc} 5/10 & 3/10 & 2/10 \end{array} \\
\begin{array}{c} 1/3 \\ 1/3 \\ 1/3 \end{array}
\left[\begin{array}{ccc}
1 & -3 & 2 \\
-2 & 4 & -1 \\
1 & -1 & -1
\end{array}\right]
\end{array}
\qquad
\begin{array}{c}
\text{Factors} \\
\left[\begin{array}{ccc}
1/6 & 1/10 & 1/15 \\
1/6 & 1/10 & 1/15 \\
1/6 & 1/10 & 1/15
\end{array}\right]
\end{array}
$$

Each element is multiplied by the corresponding factor. We group, making use of the common factors in the columns:

$$v = \frac{1}{6} \times (1 - 2 + 1) + \frac{1}{10} \times (-3 + 4 - 1) + \frac{1}{15} \times (2 - 1 - 1) = 0 + 0 + 0 = 0$$

When the value of the game is zero, we say it is a **"fair"** game.

35. First, the game matrix is as follows:

		Pitcher		
		fastball	**curveball**	**screwball**
Batter	fastball	.400	.200	.100
	curveball	.220	.400	.160
	screwball	.120	.280	.380

The factors are assigned as in the previous problem, with the decimals shortened.

$$\begin{array}{c} \\ p \\ q \\ 1\text{-}p\text{-}q \end{array} \begin{array}{c} r \quad s \quad 1\text{-}r\text{-}s \\ \begin{bmatrix} .40 & .20 & .10 \\ .22 & .40 & .16 \\ .12 & .28 & .38 \end{bmatrix} \end{array} \qquad \begin{array}{c} \text{Factors} \\ \begin{bmatrix} pr & ps & p(1\text{-}r\text{-}s) \\ qr & qs & q(1\text{-}r\text{-}s) \\ r(1\text{-}p\text{-}q) & s(1\text{-}p\text{-}q) & (1\text{-}p\text{-}q)(1\text{-}r\text{-}s) \end{bmatrix} \end{array}$$

The value of the game can be expressed as follows:

$$v = pr(.40) + ps(.20) + p(1\text{-}r\text{-}s)(.10) + qr(.22) + qs(.40) + q(1\text{-}r\text{-}s)(.16)$$
$$+ r(1\text{-}p\text{-}q)(.12) + s(1\text{-}p\text{-}q)(.28) + (1\text{-}p\text{-}q)(1\text{-}r\text{-}s)(.38)$$

After carrying out the multiplication and collecting like terms, we have:

$$v = p(.56r + .20s - .28) + q(.32r + .34s + .22) + (.38 - .26r - .10s)$$

We set the coefficients of p and q equal to zero and multiply by 100 to eliminate the decimals: $56r + 20s - 28 = 0$ and $32r + 34s - 22 = 0$

These are labeled and simplified as follows:

Eq. 1: $14r + 5s = 7$ (dividing by 4)

Eq. 2: $16r + 17s = 11$ (dividing by 2)

We solve this system of equations by the technique of elimination: (See Topic 4 in the text; essentially, we are using multiples of the equations selected so that when "equals are added to equals" only one variable will remain)

$$\begin{array}{ll} 7 \times \text{Eq 2:} & 112r + 119s = 77 \\ -8 \times \text{Eq 1:} & \underline{-112r - 40s = -56} \\ & 79s = 21 \end{array}$$

This gives us $s = \frac{21}{79} \approx 0.266$

Using this value for s and substitution, we get $r \approx 0.405$ and $1 - r - s \approx 0.329$

To solve the game for the batter, we take the original equation and re-write it in terms of r and s (multiply it out and collect terms). We have:

$$v = r(.56p + .32q - .26) + s(.20p + .34q - .10) + (.38 - .28p - .10q)$$

Setting the coefficients of r and s equal to zero, and multiplying to eliminate decimals, we have $56p + 32q - 26 = 0$ and $20p + 34q - 10 = 0$

or Eq. 1: $28p + 16q = 13$

Eq. 2: $10p + 17q = 5$

This system has the solution: $p \approx 0.446$, $q \approx 0.032$, and $1 - p - q \approx 0.522$

Summary: The batter should prepare for fastballs approximately 45% of the time, curveballs 3% of the time, and screwballs 52% of the time.

The pitcher should throw fastballs approximately 41% of the time, curveballs 27% of the time, and screwballs 33% of the time. The batter will average about .250 (verify).

Chapter 7 Review Problems

Solutions to Odd-numbered Problems

1. For the first player, the payoff is 5.
   For the second player, the payoff is -5.

3. The first strategy is the most conservative for both players.

$$\begin{bmatrix} 3 & -1 & \boxed{-2} & 4 \\ 1 & 4 & \boxed{-3} & 1 \\ \boxed{-3} & 2 & 5 & -1 \\ -1 & 3 & 2 & \boxed{-4} \end{bmatrix} \quad \begin{bmatrix} \boxed{3} & -1 & -2 & \boxed{4} \\ 1 & \boxed{4} & -3 & 1 \\ -3 & 2 & \boxed{5} & -1 \\ -1 & 3 & 2 & -4 \end{bmatrix}$$

The first strategy   The first strategy
has the greatest   has the smallest
minimum   maximum

5. The labeled game matrix:

$$\begin{array}{cc} & \textbf{Bob} \\ & \text{tight \quad loose} \end{array}$$

$$\text{Alice} \begin{array}{c} \text{pitch} \\ \text{kick} \end{array} \begin{bmatrix} -1 & 4 \\ 5 & -10 \end{bmatrix}$$

7. This is a determined game; the value of the game is -1.

$$\begin{bmatrix} 0 & 1 & 2 & \boxed{-1} \\ \boxed{-3} & 1 & \boxed{3} & -2 \\ \boxed{1} & 2 & \boxed{-3} & \boxed{-1} \\ -1 & \boxed{3} & 2 & \boxed{-4} \end{bmatrix}$$

9. We analyze the game and choices as follows:

$$\begin{array}{c} \text{Game} \\ \text{matrix} \end{array} \qquad \begin{array}{c} \text{Factors} \\ 1/3 \qquad\qquad 2/3 \end{array}$$

$$\begin{bmatrix} -3 & 3 \\ 4 & -2 \end{bmatrix} \quad \begin{array}{c} p \\ 1-p \end{array} \begin{bmatrix} (1/3)p & (2/3)p \\ (1/3)(1-p) & (2/3)(1-p) \end{bmatrix}$$

The value of the game:
When $p = 1$ and $1 - p = 0$, the average payoff of the game is

$$v = \frac{1}{3} \times (-3) + \frac{2}{3} \times (3) = 1$$

When $p = 0$ and $1 - p = 1$, the average payoff of the game is

$$v = \frac{1}{3} \times (4) + \frac{2}{3} \times (-2) = 0$$

The first player's optimum strategy is to use the first strategy as a pure strategy.
The average payoff of the game is 1. Any mixed strategy will have an average payoff
between 0 and 1.

Management Mathematics

Section 8.1   Linear Constraints

Goals
1. Identify the region of the plane that contains the solutions to a set of linear inequalities and constraints.

Key Ideas and Questions
1. How is a set of linear inequalities graphed in the plane?
2. How do you determine which ordered pairs in the plane are solutions to a set of linear inequalities?

Vocabulary

Linear Inequality	Strict/Inclusive	Linear Constraint
Solution Set	System of Inequalities	Feasible Region

• •

Overview

One of the most fundamental management problems is the allocation of resources among various alternative uses. Solving such problems involves describing them mathematically by using linear inequalities. An inequality of the form $ax + by < c$, where a, b, and c are constants and x and y are variables, is called a linear inequality. The set of points that satisfy a linear inequality is called the **solution set**. The solution set consists of the points on one side or the other of the line determined by replacing the inequality sign with an equal sign. If the inequality is **strict**, the sign is " <" or ">", and the boundary line is not part of the solution set. If the inequality is **inclusive**, the sign is "≤" or "≥", and the boundary line is part of the solution set. Plotting the points satisfying a linear inequality involves:

1. Plotting the line of points satisfying the associated equation.
2. Determining on which side of the line the points satisfy the inequality.

As far as a linear inequality is concerned, either the points on one side of the line all satisfy the inequality or they all do not satisfy it. When you check only one convenient point on a side of the inequality to see if it satisfies the inequality, you know that all points on that side will yield the same result.

Whenever there is more than one inequality that must be satisfied in a problem, we call it a **system of inequalities.**

Linear Constraints

Conditions that must be satisfied in an allocation problem are called **constraints**.
When the constraints are linear inequalities, they are called **linear constraints**. The
region consisting of the points that satisfy all the constraints is called the **feasible
region** determined by the constraints.

The following points summarize how you find the feasible region of a system of linear
inequalities.

The Feasible Region of a System of Linear Inequalities

1. Graph the lines determined by the given "equations".
2. Determine if the boundary lines are included based on the inequalities. The
   lines are solid if boundaries is included, or dashed if they are not.
3. Determine the feasible region for each inequality.
4. Find the intersection (the overlap) for all feasible regions in step 3.

Suggestions and Comments for Odd-numbered problems

1. through 7.
The points in question satisfy the inequality if they make it true. Be careful as to
whether the inequality is strict or inclusive.

9. through 15.
Replace the inequality by "=" and graph the boundary line: solid for "inclusive" and
dashed for "strict". Use the origin $(0,0)$ to test which side of the boundary line satisfies
the condition. If the boundary line passes through the origin, use a point on one of the
axes such as $(1,0)$ or $(0,1)$.

17. through 23.
Test each of the constraints separately, shading in portion of the plane on the side of the
boundary line that satisfies the constraint. What does it mean when they overlap?
For problems 21 and 23, there are the additional conditions that the variables be
nonnegative. How does that affect the solution?

25. through 37.
These problems require you to put together the "skills" from the previous sections. All
of the problems include the nonnegativity constraints. At least on a conceptual level,
these are the most important. Linear programming deals with the allocation of
resources, and assigning negative amounts is not a viable choice to make.

39. through 43.
How much of each resource should be allocated to a given product and what limitations
on the resources exist? Inequalities are appropriate because (1) you can't exceed the
amounts available, and (2) you don't have to use all of a given resource (the limitations
on another resource may limit production sooner). The point where two boundary lines
cross indicates that two resources are being used to the maximum.

45. through 49.
Here you will be describing the entire linear programming problem. In particular, make sure that your set of inequalities is a correct translation of the problem into symbolic form, since everything else is dependent upon the system of inequalities.

47. One of the conditions is that the capacity of the planes must be <u>at least</u> 1200, since we are assuming all tickets will be sold. Equality is not used since additional seats can be used for other purposes. An additional restriction is that the number of planes of each type has to be a whole number.

49. As in problem 47, the solutions must be whole numbers. Also, you can use your common sense on this one; mathematical solutions are a guide to decision-making, and other considerations must often be taken into account.

Solutions to Odd-numbered Problems

1. The points satisfy the statement $x + y \geq 5$ if they make the statement true.
  (a)   No $0 + 0 = 0$; $0 \geq 5$ is not true.
  (b)   No $6 + (-2) = 4$; $4 \geq 5$ is not true.
  (c)   Yes $3 + 4 = 7$; $7 \geq 5$ is true.

3. Do the points satisfy the inequality $2x - 3y \leq 5$?
  (a)   Yes $2(0) - 3(0) = 0$; $0 \leq 5$ is true.
  (b)   Yes $2(4) - 3(1) = 5$; $5 \leq 5$ is true.
  (c)   No $2(6) - 3(2) = 6$; $6 \leq 5$ is not true.

5. Do the points satisfy the inequality $x + y > 7$?
  (a)   No $0 + 0 = 0$; $0 > 7$ is not true.
  (b)   Yes $2 + 6 = 8$; $8 > 7$ is true.
  (c)   No $9 + (-2) = 7$; $7 > 7$ is not true.

7. Do the points satisfy the inequality $x + 3y < 8$?
  (a)   Yes $0 + 3(0) = 0$; $0 < 8$ is true.
  (b)   No $2 + 3(2) = 8$; $8 < 8$ is not true.
  (c)   Yes $10 + 3(-1) = 7$; $7 < 8$ is true.

9.

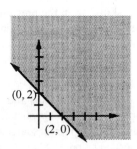

Graph the line $x + y = 2$
Intercepts: when $x = 0$, $y = 2$
                   when $y = 0$, $x = 2$
The point $(0,0)$ does not satisfy the inequality, so the feasible region is on the other side of the line from the origin.

11.

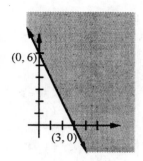

Graph the line $2x + y = 6$
Intercepts: when $x = 0$, $y = 6$
                   when $y = 0$, $x = 3$
Since the point $(0,0)$ does not satisfy the
inequality, the feasible region is on the
other side of the line from the origin.

13.

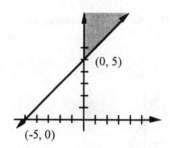

Graph the line $-x + y = 5$
Intercepts: when $x = 0$, $y = 5$
                   when $y = 0$, $x = -5$
Since the point $(0,0)$ does not satisfy the
inequality, the feasible region is on the
other side of the line from the origin.
The restrictions $x \geq 0$ and $y \geq 0$ limit the
feasible region to the first quadrant only.

15.

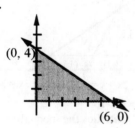

Graph the line $2x + 3y = 12$
Intercepts: when $x = 0$, $y = 4$
                   when $y = 0$, $x = 6$
Since the point $(0,0)$ satisfies the
inequality, the feasible region is on the
same side of the line as the origin. The
restrictions $x \geq 0$ and $y \geq 0$ limit the
feasible region to the first quadrant only.

17. Region III satisfies both inequalities.

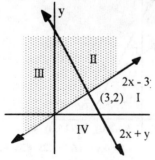

$2x - 3y \leq 0$
Test with a point on an
axis away from the origin

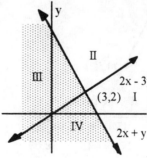

$2x + y \leq 8$
$(0,0)$ satisfies the inequality

19. Region IV satisfies both inequalities.

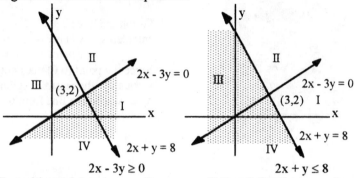

$2x - 3y \geq 0$
Test with a point on an
axis away from the origin

$2x + y \leq 8$
$(0,0)$ satisfies the inequality

21. Region II is the solution region

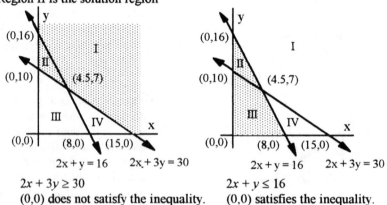

$2x + 3y \geq 30$
$(0,0)$ does not satisfy the inequality.

$2x + y \leq 16$
$(0,0)$ satisfies the inequality.

23. Region IV is the solution region.

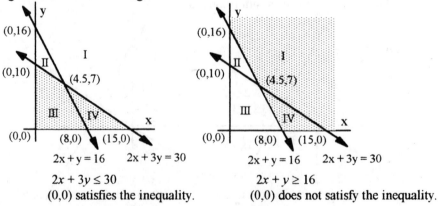

$2x + 3y \leq 30$
$(0,0)$ satisfies the inequality.

$2x + y \geq 16$
$(0,0)$ does not satisfy the inequality.

25. To find the feasible region, we first need an accurate graph, including the points of intersection for boundary lines and axes.

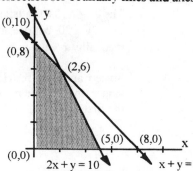

The origin (0,0) satisfies both inequalities, so the feasible region is on the same side of both boundary lines as the origin.

Since x and y are restricted to nonnegative values, the feasible region is restricted to the first quadrant.

27.

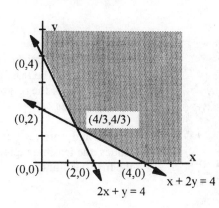

The origin (0,0) satisfies neither of the inequalities, so the feasible region is on the opposite side of the origin from both boundary lines.

Since x and y are restricted to non-negative values, the feasible region is restricted to the first quadrant.

29.

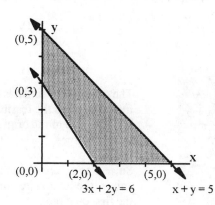

The origin (0,0) satisfies the inequality $x + y < 5$, but doesn't satisfy the inequality $3x + 2y > 6$.

Since x and y are restricted to non-negative values, the feasible region is restricted to the first quadrant.

31.

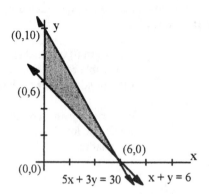

The origin (0,0) satisfies the inequality $5x + 3y < 30$, but doesn't satisfy the inequality $x + y > 6$.

Since x and y are restricted to non-negative values, the feasible region is restricted to the first quadrant.

33.

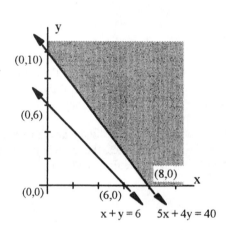

The origin (0,0) satisfies neither of the inequalities, so the feasible region is on the opposite side of the origin from both boundary lines.

Since x and y are restricted to non-negative values, the feasible region is restricted to the first quadrant.

35.

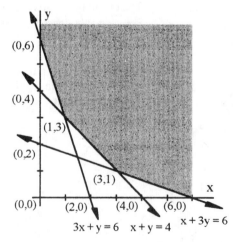

The origin (0,0) satisfies none of the inequalities, so the feasible region is on the opposite side of the origin from all boundary lines.

Since x and y are restricted to non-negative values, the feasible region is restricted to the first quadrant.

37.

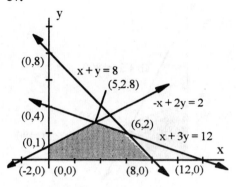

The origin (0,0) satisfies all of the inequalities, so the feasible region is on the same side of the origin as all the boundary lines.

Since x and y are restricted to non-negative values, the feasible region is restricted to the first quadrant.

39. b = number of bookcases
t = number of tables

$40b + 68t \le 800$
$b \ge 0, t \ge 0$
b and t are whole numbers

This is the amount of lumber needed for b bookcases and t tables. He can't use more than 800 feet.

Note: Since we are looking at the total number of pieces of furniture for a one-time only effort, the values for b and t must be whole numbers. This is often a restriction in applied problems.

41. c = number of cord type drills
d = number of cordless drills

$2c + 3d \le 600$
$c \ge 0, d \ge 0$

To produce c cord-type drills takes $2c$ labor hours; d cordless drills take a total of $3d$ hours. Total hours can't exceed 600.

Note: Since there is no restriction that the units be completed by the end of the day, we can assume that these are daily averages, and the values of c and d can be any nonnegative numbers.

43. f = number of 4-person tents
t = number of two-person tents

$100f + 60t \le 9000$
$f \ge 0, t \ge 0$

This is the total cost for producing f four-person tents at \$100 each and t two-person tents at \$60 each.

Fractional parts of tents are worthless, so whole numbers are needed.

45. b = number of bass
  t = number of trout
      $2b + 5t \leq 800$ ("A" food)
      $4b + 2t \leq 800$ ("B" food)
      $b \geq 0, t \geq 0$
      b and t are whole
  numbers

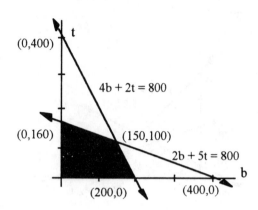

47. x = number of Type 1
  y = number of Type 2
      $0 < x + y \leq 10$
      $100x + 150y \geq 1200$
      $x \geq 0, y \geq 0$
      x and y are whole numbers

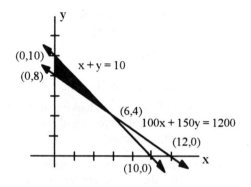

49. m = number of motel rooms
  h = number of hotel rooms

      $3m + 2h \geq 400$
      $20(3m) + 40(2h) \leq 12{,}000$
      $m \geq 0, h \geq 0$
      m and h are whole numbers

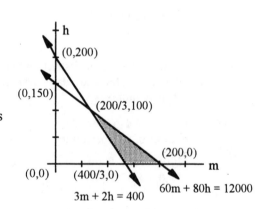

Section 8.2   Linear Programming

Goals
1. Solve linear inequalities to find optimal mixtures of resources under several constraints.

Key Ideas and Questions
1. How do you use regions in the plane as feasible regions for a problem involving linear constraints?
2. What method is used to find the maximum or minimum of a linear objective function over a feasible region?

Vocabulary

Objective Function	Coefficient	Level Lines
Linear Function	Linear Programming	Simplex Algorithm

. . . .

Overview

In this section we learn to define the objective of our problem and find the best solution to it from among the feasible alternatives.

Objective Functions

Linear programming gives a rational process for deciding what to do in a situation involving a number of restrictions that are expressed as linear constraints. Possible solutions are those in the feasible region. The **objective function** provides a numerical assessment of how good any particular course of action is. The objective function has the form $F = ax + by + c$, where a, b, and c are constants and x and y are variables. Such a function is called a **linear function**, and a, b, and c are called **coefficients**. A **linear programming** problem has an objective to maximize or minimize a **linear** function over the set of all points in the feasible region determined by **linear** constraints.

Level Lines

With an objective function that you wish to maximize, the problem is to choose values for x and y in the feasible region that make the objective function as large as possible. Similarly, with an objective function you wish to minimize, the problem is to choose values for x and y in the feasible region that make the objective function as small as possible. Two choices of (x,y) that give the same value for the objective function are equally good. To find the best pairs, plot lines for which all the points on the lines have the same objective function value. Such lines are called **level lines**. All level lines are parallel. Whenever the objective function is linear, the level lines will be parallel lines with the objective value increasing in one direction or the other. Where the level lines first contact the feasible region (usually a corner point) is where you will find minimum or maximum values.

Fundamental Principle of Linear Programming

If the feasible region is bounded, then the maximum or minimum of a linear objective function is attained at a corner point of the feasible region.

If the feasible region is unbounded, there may or may not be a maximum or minimum for any particular objective function. As long as a maximum or minimum exists, it is still going to occur at a corner of the feasible region.

To summarize, in solving linear programming problems expressed in two variables, you list all the corners of the feasible region and the value of the objective function on those corners. From that list, the smallest or largest value of the objective function can be selected. Algebraically, each corner can be found as the simultaneous solution of two of the equations associated with the constraint inequalities.

The Simplex Algorithm

Where the number of variables is larger than two, graphical methods become difficult to apply. A systematic procedure to solve large linear programming problems algebraically was developed in the 1940's and is called the **simplex algorithm**. This procedure moves from corner to corner along the boundary of the feasible region and recognizes when the minimum or maximum has been found. The simplex method is widely used in industry, and computerized models involve thousands of constraints and millions of variables.

. . . .

Suggestions and Comments for Odd-numbered problems

5. through 11.
Substitute each of the values for P (or C or R, etc) and sketch the line corresponding to that value. Use the intercept method: assign $x = 0$ and solve for y to see where the line crosses the y-axis; then let $y = 0$ to see where the line crosses the x-axis. All lines should be parallel.

13. and 15.
The maximum or minimum value occurs where a level line first contacts the feasible region. Since we only have a "sample" of the level lines, the first corner will likely occur between level lines. The closer the corner point is to a level line, the closer the value at the corner point is to the level line's value.

17. and 19.
To find the corner points (other than those on an axis where one of the values is zero) you need to solve two simultaneous equations (a topic covered in Section 1.8). We want to find a point whose coordinates satisfy both equations.

21. through 29.
To find maximum and/or minimum values you have to evaluate the objective function at each of the corner points.

25. through 29.

What happens if you graph the feasible region incorrectly or solve incorrectly for the corner points? Wrong answers! Use the intercept method for graphing boundary lines and make sure the algebraic solution for the corner points agrees visually with the graph of the feasible region.

31. through 37.

These problems are extensions of those in the previous section where you were asked to write the linear constraints. Review those if necessary. The new ingredient is to write the equation for the objective function. The key to the correct solution will be to graph the feasible region correctly and identify the corner points.

39. If the value of P at A is represented by $P(A)$, then $P(A) = a(0) + b(8) = 8b$ Similarly, $P(B) = a(4) + b(6) = 4a + 6b$. If the maximum occurs only at A, then $P(A) > P(B)$, or $8b > 4a + 6b$. Likewise, $P(A) > P(C)$. Inequalities are solved much like equations; however, here there will be two conditions between a and b that must be satisfied. If you have difficulty, review the solution for part (a) before doing parts (b) through (e). Finally, if the maximum occurs at two points, then the values must be equal.

43. The feasible region contains infinitely many points, but the solution must be a pair of whole numbers. After describing and graphing the feasible region, use trial and error to find pairs of whole numbers that satisfy the constraints.

Solutions to Odd-numbered Problems

1. **(a)** Linear; the coefficients are 3 for x and -7 for y.
   (b) Not linear: x has an exponent of 2.
   (c) Not linear; both x and y appear in the denominator.
   (d) Linear; the coefficients are 2 for x and 1 for y.

3. **(a)** Not linear; the xy term is second-degree.
   (b) Linear; the coefficients are -1 for x and 2 for y.
   (c) Linear; the coefficients are 10 for x and 6 for y.
   (d) Not linear; x has an exponent of 3.

5.

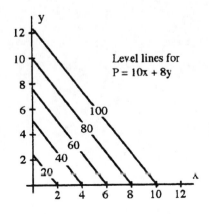

7.

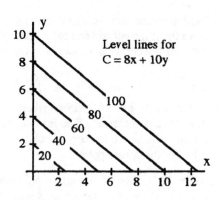

9.

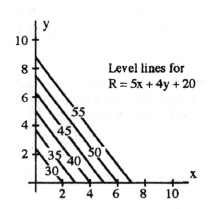

11.

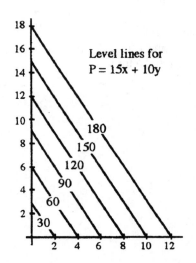

13. The farthest extent of the feasible region is between the level lines for $P = 18$ and $P = 24$. $P = 20$ is the estimated maximum.

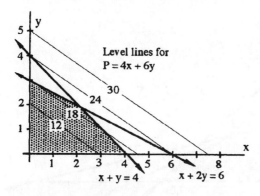

15. The estimated minimum value is $C = 16$.

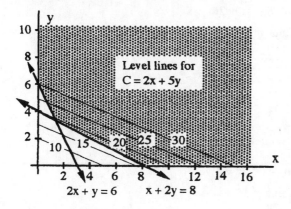

17. The corner points are $(0, 0)$, $(0, 4)$, $(6, 1)$, and $(7, 0)$.

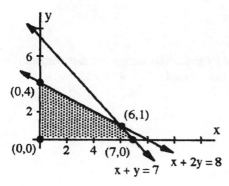

19. The corner points are $(0, 0)$, $(0, 3)$, $(4, 2)$, $(6, 1)$, and $(7, 0)$.

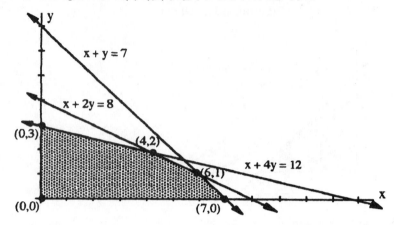

21. Objective Function: $P = 3x + 2y$

Corner Points	Value of P
$(0,0)$	0
$(0,6)$	12
$(3,5)$	19
$(5,4)$	23
$(8,0)$	24

Maximum: $P = 24$
Minimum: $P = 0$

23. Objective Function: $M = 3x + 4y$

Corner Points	Value of M
$(0,7)$	28
$(1,4)$	19
$(3,2)$	17
$(7,0)$	21

Minimum: $M = 17$
There is no maximum value. The feasible region extends without limit in both the x and y directions. For $y > 7$, for example, $3x + 4y > 28$.

25.

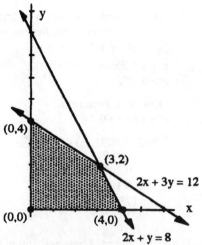

Objective Function:
$F = 5x + 2y$

Corner Points	Value of F
(0,0)	0
(0,4)	8
(3,2)	19
(4,0)	20

Maximum: $F = 20$

27.

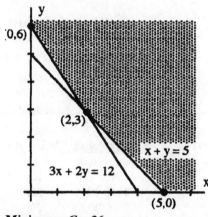

Objective Function:
$C = 10x + 6y$

Corner Points	Value of C
(0,6)	**36**
(2,3)	38
(5,0)	50

Minimum: $C = 36$

29.

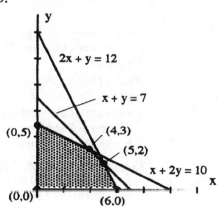

Objective Function:
$P = 7x + 5y$

Corner Points	Value of P
(0,0)	0
(0,5)	25
(4,3)	43
(5,2)	**45**
(6,0)	42

Maximum $P = 45$

31.

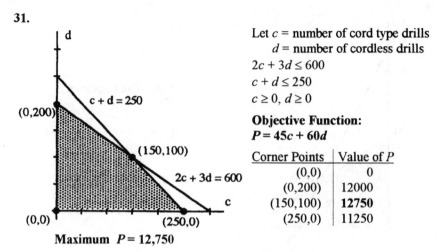

Let c = number of cord type drills
d = number of cordless drills

$2c + 3d \le 600$
$c + d \le 250$
$c \ge 0, d \ge 0$

Objective Function:
$P = 45c + 60d$

Corner Points	Value of P
(0,0)	0
(0,200)	12000
(150,100)	**12750**
(250,0)	11250

Maximum $P = 12,750$

The maximum revenue is $12,750 when 150 of the cord type and 100 of the cordless drills are produced.

33.

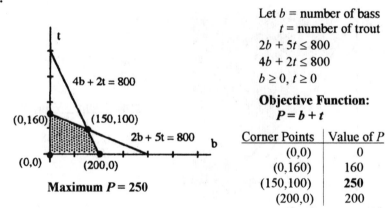

Let b = number of bass
t = number of trout

$2b + 5t \le 800$
$4b + 2t \le 800$
$b \ge 0, t \ge 0$

Objective Function:
$P = b + t$

Corner Points	Value of P
(0,0)	0
(0,160)	160
(150,100)	**250**
(200,0)	200

The maximum number of fish that can be supported is 250: 150 bass and 100 trout.

35.

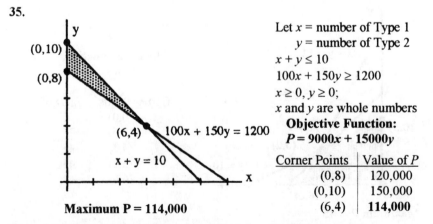

Let x = number of Type 1
y = number of Type 2

$x + y \le 10$
$100x + 150y \ge 1200$
$x \ge 0, y \ge 0$;
x and y are whole numbers

Objective Function:
$P = 9000x + 15000y$

Corner Points	Value of P
(0,8)	120,000
(0,10)	150,000
(6,4)	**114,000**

The minimum cost is $114,000 using 6 Type 1 and 4 Type 2 aircraft.

37.

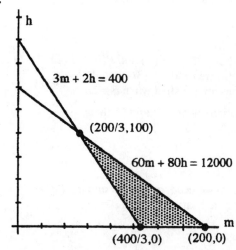

Let m = number of motel rooms
h = number of hotel rooms
$3m + 2h \geq 400$
$20(3m) + 40(2h)$ or $60m + 80h \leq 12,000$
$m \geq 0, h \geq 0$; m and h are whole numbers
Objective Function:
$P = 135m + 120h$

Corner Points	Value of P
(200/3,100)	21,000
(400/3,0)	**18,000**
(200,0)	27,000

A whole number of rooms must be rented. The two obvious choices are:
      134 motel rooms or 133 motel rooms and 1 hotel room.
Both of these choices fall in the feasible region.
For 134 motel rooms, the cost is $18,090.
For 133 motel rooms and 1 hotel room, the cost is $18,075.
Most conference organizers would choose to use 134 motel rooms rather than have 2
participants located elsewhere.

39. Given: $P = ax + by$, $a > 0$ and $b > 0$.
      $A = (0,8)$ $B = (4,6)$ $C = (6,0)$
      $P(A) = 8b$ $P(B) = 4a + 6b$ $P(C) = 6a$

(a) If the maximum value of P occurs only at A, then:
      $P(A) > P(B)$ and $P(A) > P(C)$
      $8b > 4a + 6b$ $8b > 6a$
      $2b > 4a$ $b > (3/4)a$
      $b > 2a$
   If $b > 2a$, then $b > (3/4)a$ since $a > 0$ and $2 > 3/4$.
(b) If the maximum value of P occurs only at B, then:
      $P(B) > P(A)$ and $P(B) > P(C)$
      $4a + 6b > 8b$ $4a + 6b > 6a$
      $a > 2b$ $6b > 2a$
      $a > (1/2)b$ $3b > a$
   $3b > a > (1/2)b$
(c) If the maximum of P occurs only at C, then:
      $P(C) > P(A)$ and $P(C) > P(B)$
      $6a > 8b$ $6a > 4a + 6b$
      $a > (4/3)b$ $2a > 6b$
                           $a > 3b$
   If $a > 3b$, then $a > (3/4)b$ since $b > 0$.

39. (d) If the maximum of P occurs at both A and B, then:

$$P(A) = P(B)$$
$$8b = 4a + 6b$$
$$2b = 4a \quad \text{or} \quad b = 2a$$

However, we must also have $P(A) > P(C)$ and $P(B) > P(C)$.
From parts (a) and (b) we need $b > (3/4)a$ and $3b > a$.
Both these conditions are satisfied when $b = 2a$.

(e) If the maximum value occurs at both B and C, then:

$$P(B) = P(C)$$
$$4a + 6b = 6a$$
$$6b = 2a \quad \text{or} \quad a = 3b$$

We also need $P(B) > P(A)$ and $P(C) > P(A)$.
From parts (b) and (c) we need $a > (1/2)b$ and $a > (3/4)b$.
Both these conditions are satisfied when $a = 3b$.

41. Let d = number of day-time ads
   p = number of prime-time ads
   n = number of late-night ads

Constraints: $1200d + 2250p + 1600n \le 30000$

  $d + p + n \le 15$
  $d + p + n \ge 10$

Minimum value constraints:

  $d \ge 0, p \ge 0, n \ge 0$

Objective function:

  Maximize $A = 12000d + 21000p + 16000n$

43. (a)

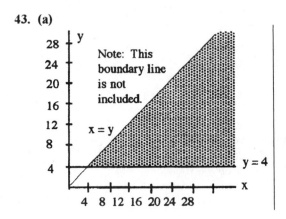

(b) $x = 13, y = 12$ and $x = 20, y = 6$ are the only pairs of whole numbers satisfying the constraints.

(c) Choosing 20 games per division rival and 6 for inter-divisional play gives the divisions more meaning.

Section 8.3   Routing Problems

Goals
1.   Solve routing problems using graph theory.

Key Ideas and Questions
1.   What is the test to determine whether a graph has an Euler path?
2.   What is the test for an Euler circuit?
3.   How does one use Fleury's algorithm to construct Euler circuits?

Vocabulary

Graph	Degree of a Vertex	Connected/Disconnected
Vertex, Vertices	Path	Component of a Graph
Edge	Euler Path	Bridge
Loop	Euler Circuit	Fleury's algorithm

. . . .

Overview

Another important management problem involves finding efficient ways to route goods or services to different destinations. The destination can be thought of as part of a network, together with the connecting links. The goal is to use the network to efficiently accomplish a task. There are different types of routing problems, but here we look at the type where we seek a route, also called a circuit, that uses each connection in a network once and only once.

Problems of this kind are called **Euler circuit problems** after Leonhard Euler who originally solved the **Königsberg bridge problem**. **Graph theory** is a tool used to deal with Euler circuit problems and other routing problems. Here, a graph refers to a collection of points called **vertices**, and the paths connecting them, called **edges**. (The singular form of vertices is vertex.) Edges can be straight or curved. The graph is usually sketched with the vertices represented by black dots, and the edges represented by line segments or arcs.

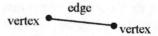

Each edge has either two ends with different vertices, or it is **a loop** connecting a vertex to itself.

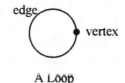

A Loop

When it appears that two arcs representing edges cross, it does not mean that there is a vertex at that place. Two graphs are considered to be the same if they have the same number of vertices connected to each other in the same way. This is the case even if the edges look different.

The two drawings below are considered to be drawings of the same graph. The vertices are labeled to show that the connections are the same.

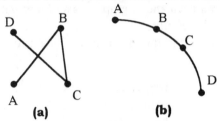

Two vertices are said to be **adjacent** in a graph if there is an edge connecting them.

The **degree** of a vertex in a graph is the total number of edges at that vertex. When a loop connects a vertex to itself, it is agreed that the loop contributes 2 to the degree of the vertex. Visually, the degree of a vertex can be found by counting the number of segments or arcs attached at the vertex. The degree of the vertices will be important in solving a number of the problems. It was the key to the solution of the Königsberg bridge problem.

Since each edge of a graph has two ends with a vertex at each, and since the degree of a vertex is the number of ends of edges attached to the vertex, the sum of the degrees of a graph is always twice the number of edges. A loop has two ends which attach to the same vertex.

> The sum of the degrees of all the vertices in a graph is twice the number of edges in the graph.

A **path** in a graph is a list of vertices in the graph such that each vertex in the list is adjacent to the next vertex in the list and each edge that connects adjacent vertices is used at most once. Think of traveling along a path from vertex to vertex via the edges, with each edge used at most once. A vertex may appear more than once. A path that ends at the same vertex where it started is called a **circuit** (sometimes referred to as a cycle). A path that uses every edge once is called an **Euler path**. A circuit that uses every edge once is called an **Euler circuit**.

It is helpful in drawing paths to indicate a starting vertex and an arrow to show which way to go first. By numbering all the edges, you can simply take the next numbered edge when you arrive at a vertex.

Another way to describe a path is to name the vertices and then list them in the order they are visited on the path. You should number the edges as you go from vertex to vertex. If there are two or more edges between the two vertices, you can only use each edge once. A vertex can be visited more than once.

A graph is **connected** if, for every pair of vertices, there is a path that contains them. By convention, it is agreed that a graph with only one vertex is connected, whether or not there are any edges. If a graph is not connected, then it is called **disconnected**. A disconnected graph can always be separated into connected pieces of the largest possible size (maximal connected pieces) called **components**.

To find the components of a graph, use the following steps:

1. Pick any vertex and highlight it.
2. Highlight all the edges connecting to the highlighted vertex and all the vertices at the ends of those edges.
3. Repeat step 2 for all edges connected to any highlighted vertex.
4. When no new vertices get highlighted, you have a connected part of the graph that is maximal while still being connected. You have a component of the graph.

For simple graphs you may not need to go through the steps in constructing components, but in actual applications such as analyzing computer connections or a highway system, the step-by-step approach is required.

An edge in a connected graph is called a **bridge** if its removal from the graph would leave behind a graph that is not connected.

Remember that the **degree** of a vertex in a graph is the total number of edges at that vertex. Also, it is defined that a loop contributes 2 to the degree of the vertex. The sum of the degrees of all the vertices in a graph is twice the number of edges in the graph. Thus, the sum of degrees is always even, and a graph must have an even number of vertices with odd degree.

Any graph must have an even number of vertices with odd degree.

To answer the question of whether or not there is an Euler circuit in a given graph, Euler arrived at the following theorem.

Theorem (L. Euler)

For a connected graph

1. If the graph has no vertices of odd degree, then it has at least one Euler circuit (which is also an Euler path), and if a graph has an Euler circuit, then it has no vertices of odd degree.

2. If the graph has exactly two vertices of odd degree, then it has at least one Euler path, but does not have an Euler circuit. Any Euler path in the graph must start at one of the two vertices with odd degree and end at the other.

3. If the graph has four or more vertices of odd degree, then it does not have an Euler path.

Fleury's Algorithm

For small graphs, such as those in the examples and exercises, Euler paths and circuits can usually be found through trial and error. However, help is usually needed in finding Euler circuits in very large graphs. **Fleury's Algorithm** is a procedure by which you can find at least one Euler circuit if Euler's Theorem guarantees the existence of one.

· ·

Suggestions and Comments for Odd-numbered problems

19. Valid graphs that satisfy a variety of conditions can be drawn in a great many ways, and the "look" of the graph is often a matter of personal taste. If your graph doesn't look like one of the possible answers provided, review your answer by seeing if it satisfies the problem statements. That's all that really matters, except if someone else is going to use a graph to gain information, it should be as easy to understand as possible.

21. through 27.
Look at each vertex individually (as though it were under a microscope). How many edges are connected to it?

31. and 33.
Since an Euler path uses every edge (only once), a good practice is to make a duplicate drawing of the graph and number each of the edges as you follow a path. An Euler circuit begins and ends at the same vertex.

37. Remember that FE as an edge begins at F. When do you encounter the first decision about the next vertex you can go to on the path?

39. through 43.
If there are no vertices of odd degree, then an Euler circuit exists. If there are two vertices of odd degree, then an Euler path exists beginning at one of the vertices of odd degree and ending at the other.

45. If an Euler path exists, try using a tree diagram to follow the possible choices. Since there are two edges between C and D, label them "1" and "2" and put the number on the tree branch if you are following that edge.

47. through 51.
Draw a simplified graph representing the map. What are the degrees of the vertices on the graph? What does that tell you about possible Euler paths or circuits?

53. Consider the inclusion of either of the additional streets. What is the degree of each vertex in the new network?

Solutions to Odd-numbered Problems

1.

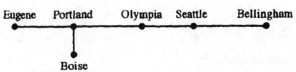

Note: Compare this to the answer in the text. Two graphs are equivalent if they have the same vertices and the same connecting edges.

3.

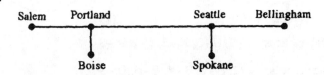

5. **(a)** Vertices: A, B, C, D

   (b) Adjacent vertices: (A, B), (A, D), (B, C), (B, D)

7. **(a)** Vertices: R, S, T, U

   (b) Adjacent vertices: (R, S), (R, T), (R, U), (S, U), (T, U)

9. The complete graph with all possible interstate highway connections

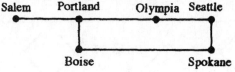

11. **(a)** Vertices: T, U, K, J, R

    (b) Adjacent vertices: (T, U) - 2 edges, (T, R), (R, K), (R, J), (J, K), (U, K)

13. **(a)** Vertices: A, B, C, D, E

    (b) Adjacent vertices: (A, E), (A, C), (C, D) - 2 edges, (B, D), (B, B) (loop), (B, E)

15. **(a)** Vertices: A, B, C, D
    Edges: AB, AC, AD, BC, BD

    (b) Vertices: A, B, C, D
    Edges: AB, AC, AD, BC, BD

17. They have the same vertices and the same edges; thus they are essentially the same graph. They are different only in the arrangement of the vertices.

19. (a)   Three possible solutions are:

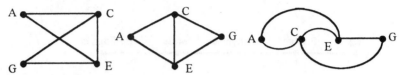

(b) Three possible solutions are:

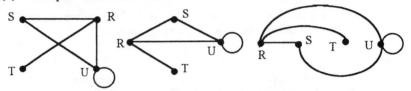

21. Degrees of the vertices: T - 3, U - 3, K - 3, J - 2, R - 3

23. Degree of the vertices: A - 2, E - 2, B - 4, C - 3, D - 3

25. Degree of the vertices: A - 3, B - 4, C - 1, D - 2, E - 4
Sum of the degrees = 14; Number of edges = 7
The sum of the degrees at the vertices is twice the number of edges.

27. Degree of vertices: A - 5, B - 2, C - 3, D - 2, E - 3, F - 2, G - 5, H - 2
Sum of the degrees = 24; Number of edges = 12
The number of edges is one-half the sum of the degrees at the vertices.

29. CD, DF, DE are bridges

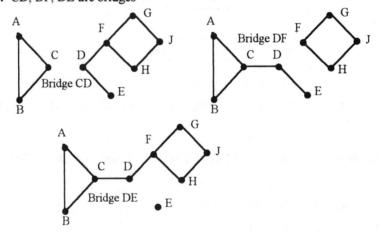

31. (a) Path

(b) Not a path; there is no edge from T to Y

(c) Circuit

(d) Not a path; there is only one edge connecting Y and R, and it has already been used. Each edge in a path can be used only once.

(e) Circuit (but not an Euler circuit); one of the edges connecting U and R is not used. An Euler circuit uses every edge once.

33. **(a)** Path

    (b) Not a path; there is no edge from M to K.

    (c) Euler circuit; the path begins and ends at the same vertex and uses every edge just once.

    (d) Euler circuit

35. U and R are the only two vertices of odd degree, thus this graph has an Euler path, but not an Euler circuit.

37. The vertices of the circuit: F, E, B, A, C, D, A, B, D, F

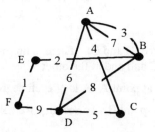

39. Vertex : R S T U Y
    Degree: 5 4 2 3 4
    Since there are exactly two vertices of odd degree, there is at least one Euler path but no Euler circuit. Two such paths follow:

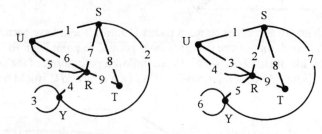

41. Vertex: H J K L M N
    Degree: 5 3 4 4 4 4

    Since there are exactly two vertices of odd degree, there is at least one Euler path but no Euler circuit. Two such paths follow:

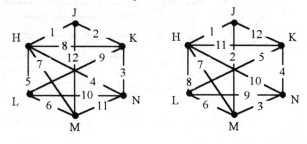

43. Vertex: H K L M N O P R S
Degree: 2 3 2 1 2 4 2 2 4

Since there are exactly two vertices of odd degree, there is at least one Euler path but no Euler circuit. Since there are two bridges which can only be crossed once, all Euler paths must begin at either K or M and end at the other. Two such paths follow. Notice that one path is not simply the reversal of the other.

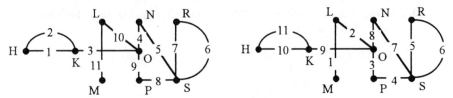

45. There are exactly two vertices of odd degree (B and C). There is no Euler circuit, and any Euler path must begin at either B or C and end at the other.

(a) The are no Euler paths that originate at point A (A is of even degree).

(b) There are 16 Euler paths that begin at point B. To see this, we will draw a tree diagram for the paths.

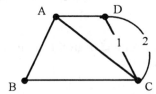

Since there are two edges between points C and D, these have been labeled "1" and "2". A branch on the tree that goes from point C to point D on the second edge will look like this: C -- 2 -- D. As you follow a path, think of the available options. Note: After going from B to C to D, a return to C would be a dead-end..

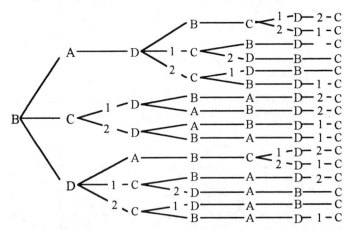

(c) There are 16 Euler paths that originate at C. A tree diagram similar to that in part (b) would show all paths.

(d) The are no Euler paths that originate at point D. Point D is of even degree.

47. Let P = Portland, A = Albany, N = Pendleton, B = Burns, and O = Ontario.

(a) The route could begin in Pendleton and end in Burns, having visited each of these cities twice: N, O, B, A, P, N, B

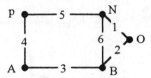

A suggested route the driver could take the second day in order to end up at the home office at the end of the second day:

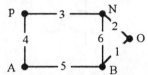

Everyone on the route has pickup and delivery service each day, and the head office has two pickups and deliveries each day.

(b) Pendleton and Burns have an odd degree (3), thus there is an Euler path from Pendleton to Burns or from Burns to Pendleton. An Euler path begins at one odd vertex and ends at the other.

49. There is no Euler path starting at Portland since it has an even degree, thus the courier will have to travel some road twice. The most logical choice would appear to be the Pendleton to Burns road.

51. Let H = I-84 intersection
D = Bend

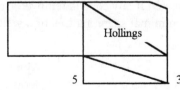

Since D, H, N, and B are of odd degree, there is no Euler path for this network.

53. Hollings Ave. would be the best street to add since this would provide exactly two vertices with odd degree, so there would be an Euler path for the cruisers. If Barton Blvd. is added, there will be four vertices of odd degree, and therefore no Euler path.

55. After we first leave a vertex of odd degree there are an even number of edges remaining, thus every time we come back there is at least one edge leading out. Finally, there would be no way back to the vertex. The circuit cannot be completed.

57 Yes, we can add a bridge from the left bank to the lower island and a bridge from the right bank to the upper island.

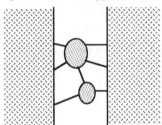 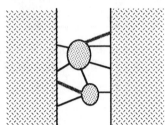

59. Yes, a single bridge can be added to create an Euler path. One possibility:

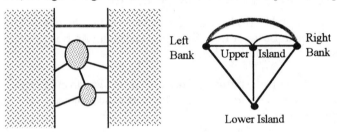

61. No, if a bridge to the upper island is washed out there will still be two vertices of odd degree. There will be an Euler path, but no Euler circuit.

Section 8.4   Network Problems

Goals
1.   Solve network problems using graph theory.

Key Ideas and Questions
1.   Explain why a circuit is created if an edge is added to a spanning tree.
2.   Explain why a disconnected graph is created if any edge is removed from a spanning tree.
3.   Describe the method for constructing a minimal spanning tree of a weighted graph.

Vocabulary
Weighted Graph	Tree	Minimal Spanning Tree
Subgraph	Spanning Tree	Kruskal's Algorithm

Overview

Common business and engineering problems require designing efficient networks. To work with these types of problems, we need to consider graphs that have characteristics to measure efficiency.

Weighted Graphs

We need an enhanced graph whose edges have numbers associated with them. Such a graph is called a **weighted graph**. The number associated with the edge is called the **weight** of the edge. The weights are usually assigned in some way relevant to the problem being solved. The lengths of the edge do not, however, have to be proportional to the weights.

Subgraphs

The efficiency of a network can be increased by removing redundant connections. For weighted graphs, this means selecting a smaller set of edges from the graph. We are thus interested in a **subgraph**. A **subgraph** is a set of vertices and edges chosen from among those of the original graph.

Trees

If a path in a network starts somewhere and returns without using any connection twice, then there must be a redundant connection. When you want efficiency, you want to avoid such redundancy. A graph that is connected (you can go from any vertex to any other vertex) and has no circuits is called a **tree**. Trees have no redundant connections. A subgraph that contains all the original vertices, is connected, and contains no circuits is called a **spanning tree**.

Minimal Spanning Tree

When you want to construct the most efficient network, begin with a connected, weighted graph representing the physical network. Try to find a connected subgraph of smallest total weight that contains all of the vertices. You want to construct a spanning tree with the smallest possible total weight. This tree is called a **minimal spanning tree**. The following method for finding the minimal spanning tree for a graph is called **Kruskal's Algorithm.**

1. Start with only the vertices.
2. Add edges until the resulting graph is connected.
3. At each stage, look at the list of edges that have not been used and add the acceptable edge of smallest weight. Use the following rules:

Acceptable Edges
1. An edge that does not share a vertex with any of the edges already chosen.
2. An edge that connects together two components of the subgraph.
3. An edge that connects to a component of the subgraph and brings a new vertex into the subgraph.

Unacceptable Edges
1. Adding an edge to a component of the subgraph without also adding a vertex.

Suggestions and Comments for Odd-numbered problems

5. and 7.

A good graph does not have to show the correct geographic relationships. Spread the vertices around the circumference of a circle or give yourself ample room to work in some other manner.

11. through 25.

The number of edges in a tree is 1 less than the number of vertices. The number of edges to be removed is a clue to the number of trees, but you have to identify edges that are bridges as well.

11. There are 11 spanning trees possible.

15. There are 9 spanning trees possible. The problem doesn't ask you to find them, but see if you can.

19. There are 8 spanning trees possible.

21. There are 8 spanning trees possible.

23. There are 11 spanning trees possible.

31. and 33.

The process as described works just like the algorithm for finding minimum cost spanning trees, only you start with the largest weights on the graph.

43. Treat this as a six part problem. For small graphs, the Nearest Neighbor Algorithm often gives the same answer regardless of where you start. As the complexity of a graph grows, the possibility of different answers increases.

Solutions to Odd-numbered Problems

1. (a)

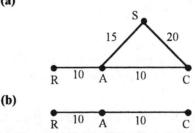

(b)

Note: We've deliberately drawn the two 10-mile segments different lengths to emphasize that it is the relationships between vertices and edges that is most important in graph.

3.

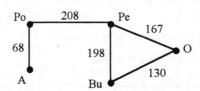

5. Notice that this is a valid graph but is not realistic geographically.
Compare this graph to the one in the answer key of the text. Why are they equal?

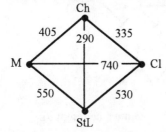

7. Notice that this is a valid graph but is not realistic geographically.
Compare this graph to the one in the answer key of the text. Why are they equal?

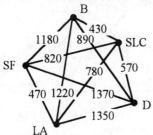

9. (a) This graph is a tree.

(b) This graph has a circuit, and it is not a tree.

(c) This graph has two components, and is it not a tree.

11. There are 11 possible spanning trees for the graph.

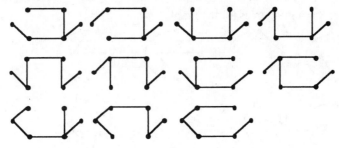

13. There are 3 possible spanning trees for the graph.

15. (i) There are currently 6 edges and 7 vertices. 2 edges must be removed. The number of edges in a tree is 1 less than the number of vertices.

(ii) The top edge (vertical) cannot be removed. It is a bridge.

(iii) 9 spanning trees can be produced from the graph.

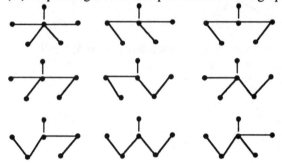

17. (i) There are 6 edges and 6 vertices. 1 edge must be removed.

(ii) The two edges sloping downward from the left to the right cannot be removed since they are bridges connecting single points to the graph.

(iii) 4 spanning trees can be produced from the graph.

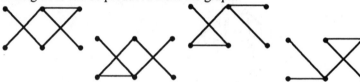

19. There are 8 spanning trees for the graph.

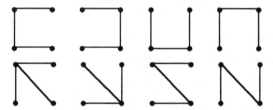

21. There are 8 spanning trees for the graph.

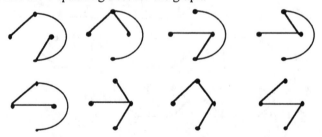

23. There are 11 spanning trees for the graph.

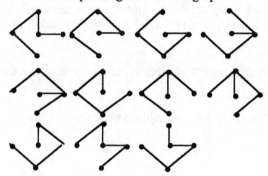

25. There are 5 spanning trees for the graph.

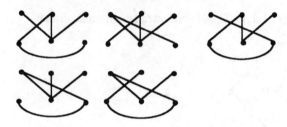

27. (a) 30, 35, 40   **(b)** 7, 8, 9, 10

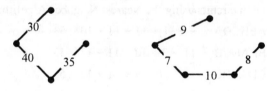

29. 5, 5, 5, 5, 6, 6, 8, 8, 10

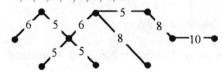

31. 16, 15, 13, 12

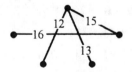

33. 10, 10, 10, 8, 8, 8, 8, 6, 6

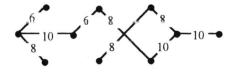

35. Yes, it is possible to have an Euler path that is not a Hamiltonian path.
In the example preceding this problem, the highway inspector's route is an Euler path (each edge used once, and only once) but not a Hamiltonian path (three cities are visited twice each).

37. C must be visited more than once if all other vertices are visited and the path is a circuit.

39. (a) A, B, D, E, C, F, A is one such Hamiltonian circuit.

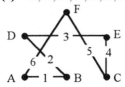

(b) A, B, C, D, E, F, J, I, H, G, A is one such Hamiltonian circuit.

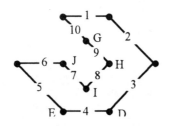

41. Length of Hamiltonian circuits using the Nearest Neighbor Algorithm.

From vertex A: AEBCDA: $9 + 11 + 12 + 11 + 10 = 53$

From vertex B: BEADCB: $11 + 9 + 10 + 11 + 12 = 53$

From vertex C: CDAEBC: $11 + 10 + 9 + 11 + 12 = 53$

From vertex D: DAEBCD: $10 + 9 + 11 + 12 + 11 = 53$

From vertex E: EADCBE : $9 + 10 + 11 + 12 + 11 = 53$

The solution to the Traveling Salesman problem is 53.

43. Length of Hamiltonian circuits using the Nearest Neighbor Algorithm.

From vertex A: ABFCDEA: $7 + 10 + 13 + 5 + 21 + 14 = 70$

From vertex B: BAEFCDB: $7 + 14 + 16 + 13 + 5 + 23 = 78$

From vertex C: CDABFEC: $5 + 18 + 7 + 10 + 16 + 26 = 82$

From vertex D: DCFBAED: $5 + 13 + 10 + 7 + 14 + 21 = 70$

From vertex E: EABFCDE: $14 + 7 + 10 + 13 + 5 + 21 = 70$

From vertex F: FBAEDCF: $10 + 7 + 14 + 21 + 5 + 13 = 70$

The solution to the Traveling Salesman problem is 70.

Chapter 8 Review Problems

Solutions to Odd-numbered Problems

1. The graph of the inequalities
$$x + y \geq 6$$
$$2x + y \geq 8$$
$$x \geq 0;\ y \geq 0$$

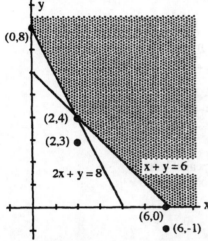

Although the graph indicates that the points (2,3) and (6,-1) are not in the feasible region, we will check them algebraically to be sure. There are cases where this is necessary. For a point to be in the feasible region, the coordinates must satisfy <u>all</u> the constraints.

For (2,3), we have $2 + 3 = 5$, which is not greater than 6 as required.
For (6,-1), the condition y ε 0 is not satisfied. The non-negative restrictions on the variables are the most important constraints.

3. The feasible region for the system of inequalities
$$x + 2y \leq 4$$
$$4x + y \leq 4$$
$$x \geq 0;\ y \geq 0$$

Objective Function:
$$P = 2x - y$$

Corner Points	Value of P
(0,0)	0
(0,2)	-2
$(\frac{4}{7}, \frac{12}{7})$	$-\frac{4}{7}$
(1,0)	2

Maximum of $P = 2$ occurs at (1,0)
Minimum of $P = -2$ occurs at (0,2).

5. Let x = pounds of creamy peanut butter made.
    y = pounds of chunky peanut butter made.

   $0.40x + 0.60y \leq 2400$ (pounds of grade A peanuts)
   $0.60x + 0.40y \leq 2400$ (pounds of grade B peanuts)
   $x \geq 0; y \geq 0$

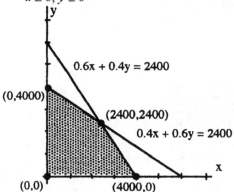

7. $P = \$0.60x + \$0.80y$
   where x = pounds of creamy peanut butter
          y = pounds of chunky peanut butter

Corner points	Value of P
(0,0)	0
(0,4000)	3200
(2400,2400)	**3360**
(4000,0)	2400

   The company should make 2400 pounds of each type of peanut butter for a maximum
   profit is $3360. This requires the company to buy 2400 pounds of each grade of peanut.

9. The graph is a connected graph, and each vertex has degree 3. There is no Euler path
   since there are more than two vertices of odd degree. Since there is no Euler path,
   there can be no Euler circuit.

11. A graph of the network of bridges can be drawn as follows:

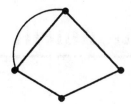

We can represent the path on the graph in which every bridge has to be traveled twice as equivalent to an Euler path on the graph below. Each edge on the original graph considered to have two "lanes", making it the equivalent of two edges. Since every vertex has even degree, there is an Euler circuit. In this case, there are several Euler circuits originating at each vertex. Several examples are shown below, and the edges are numbered to show the order in which they are used.

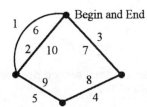
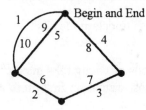
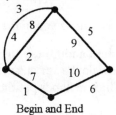

13. There are 8 spanning trees for the graph

15. The edges in the minimum cost spanning tree are AB, BC, and BD.

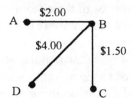

The total cost for using the telephone tree is $7.50

Voting and Apportionment

Section 9.1   Voting Systems

Goals
1. Use various voting methods to determine election results.

Key Ideas and Questions
 1. How do the various methods of voting take into account voters' second preferences?
 2. Which methods put the most weight on first preference?

Vocabulary

Plurality Method	Plurality with Elimination	Preference Table
Borda Count Method	Run-off Election	Pairwise Comparison

Overview

In this section we will describe the most popular voting systems and how they are implemented. A democratic society must determine how officials are elected and how decisions are made. If there are three or more choices in an election, different methods can lead to very different results. The consequences of three viable alternatives is easily seen in several recent elections. Some political analysts point to the conservative candidacy of Ross Perot as the determining factor of the 1992 presidential election won by Bill Clinton with 43% of the popular vote. One western state recently changed from the plurality method to the plurality with elimination method to ensure that the governor was elected by a majority of voters.

Plurality Method

The plurality method is one way of settling who is the winner if there are three or more candidates in an election and no candidate has a majority.

The Plurality Method

Vote for one candidate.
The candidate receiving the most votes is selected.

Borda Count Method

The Borda Count Method is named for the 18th century Frenchman who proposed it--Jean Charles de Borda. In this method, each voter must rank all the candidates. The Borda Method lets the voters provide more information than the plurality method.

The Borda Count Method

Voters rank the m candidates.
A voter's m^{th} choice gets one point, $(m-1)^{st}$ choice gets 2 points,..., second choice gets $m - 1$ points, and first choice gets m points.
The candidate receiving the most points is selected.

Plurality with Elimination Method

In the plurality with elimination method, a series of votes may be required.

The Plurality with Elimination Method

Vote for one candidate.
If a candidate receives a majority of votes, that candidate is selected.
If no candidate receives a majority, eliminate the candidate(s) receiving the fewest votes, and do another round of voting.

Including all its variations, this method is probably the most widely used voting method. Plurality with elimination methods sometimes use other rules to decide which candidates are in the second or later round of voting. Often, the top two may be the only candidates left for the second round. The second round is usually called a **run-off election** in this case.

An assumption used to simplify how people might vote in later rounds of voting is that each voter has a ranking or preference of all the candidates, and that in each election, the voter casts the vote for the highest ranking candidate still in the election. These rankings are displayed in a **preference table**. In the first vote, each voter casts the vote for the first choice. If a candidate must be eliminated, then the first place votes of the eliminated candidate go instead to the second place choices of those voters. This simulates running another ballot.

Pairwise Comparison Method

When a choice between several candidates is made by the pairwise comparison method, each voter must make a choice between every possible pair of candidates. Instead of separate ballots, the voters will be required to rank all the candidates. Using the rankings, we go through every possible pairing of candidates and determine which of the two is preferred based on the rankings. Each candidate is assigned points based on how well they do with respect to the other candidates.

> **The Pairwise Comparison Method**
>
> Voters rank all the candidates.
> For each pair of candidates X and Y determine how many voters prefer X to Y and vice versa.
> If X is preferred to Y, then X receives 1 point.
> If Y is preferred to X, then Y receives 1 point.
> If the candidates tie, then each receives $\frac{1}{2}$ point.
> The candidate receiving the most points is selected.

Tie Breaking

The four voting methods looked at here can produce different winners even when the voter preferences are the same. Any one of them can also produce a tie between two or more of the alternatives. Ties can be broken by making an arbitrary choice or bringing in another voter, or by other means. With the Borda count method, a tie could be broken by basing the winner on who obtained the most first place rankings. The best thing to do is to choose the tie breaking method in advance.

· ·

Suggestions and Comments for Odd-numbered problems

5. and 7.

When using the Borda count method, first make a table showing the number of votes each candidate has for each place.

	1	2	3
G	1	2	4
F	2	4	1
Y	4	1	2

Make sure that all votes are accounted for. Each row and each column should total the same as the number of votes that are cast.

11. If no alternative has a majority, the one with the fewest first place votes is eliminated and the votes are reassigned. Since 9th street has the fewest votes, it is eliminated; councilor D, for example, will rank Beca first and Davis second.

13. The table is a summary of the voters who filled out their ballots a certain way. The first column indicates that 12 voters had Ann as first choice, Eno second, and Pat third. If Ann were eliminated, Eno would be assigned the first place votes and Pat the second place votes.

15. and 17.
In the pairwise comparison method, you have to compare relative placement for each possible pair. If there are n alternatives, there will be n(n - 1)/2 possible pairs; for example, if there are 4 alternatives, there are 6 possible pairs. Look at the pairings systematically. If the alternatives are A, B, C, and D, first consider all the pairings that involve A; pair A with each of the other alternatives. Then consider all the pairings that involve B other than those with A. As you go from one choice to the next, there will be fewer pairings that haven't yet been considered.

19. and 21.
The table in 19 summarizes the preference schedules for 135 voters. Since the total points assigned for each voter is 6 (3-2-1), the total of the Borda counts for the three candidates must be $6 \times 135 = 810$. Problem 21 should be double-checked in the same manner.

25. In some competitions where teams are involved, each alternative (team) may have more than one member who can earn points for the team.

33. Determine which alternative has the most first place votes and do a pairwise comparison between the other two.

35. Remember that the number at the top of each column indicates the number of voters who have a particular set of preferences. There are 21 total votes.

39. First, total the first place votes. This determines the order of the run-offs, and the order of the run-offs may help determine the winner (this is shown in problem 45.)

41. and 43.
Voting is essential to decision-making, but not having significant opposition is as important as having significant support. If the winner in a plurality based election is to be in a leadership role (such as president) and is the type of candidate that you either "love 'em or hate 'em", it may well turn out that in all subsequent activities that require votes, he will be opposed by a majority. In these situations, everyone may eventually end up a loser.

Solutions to Odd-numbered Problems

1. Morrita is elected under the plurality method.

3. Davis Ave. is selected using plurality.
   Davis - 4, Beca - 3, 9th - 2

5. Yamada is selected with the Borda count method.
   Totals: Yamada: $4 \times 3 + 1 \times 2 + 2 \times 1 = 16$
   Finster: $2 \times 3 + 4 \times 2 + 1 \times 1 = 15$
   Gorman: $1 \times 3 + 2 \times 2 + 4 \times 1 = 11$

7. 9th Street is selected using the Borda count method.
   Totals: 9th Street: $2 \times 3 + 6 \times 2 + 1 \times 1 = 19$
   Davis Ave.: $4 \times 3 + 1 \times 2 + 4 \times 1 = 18$
   Beca Blvd.: $3 \times 3 + 2 \times 2 + 4 \times 1 = 17$

9. Yamada is selected using plurality.

      Yamada - 4, Finster - 2, Gorman - 1

11. Beca Blvd. is selected using plurality with elimination.

Since 9th Street has the fewest first place votes, it is eliminated.
The preferences of the councilors are:

Councilor	A	B	C	D	E	F	G	H	I
Davis Ave	1	2	2	2	1	2	1	2	1
Beca Blvd.	2	1	1	1	2	1	2	1	2

Beca is preferred to Davis by a 5 to 4 margin.

13. Ann is selected using the plurality with elimination method

In the original balloting, Ann had 20 first place votes, Eno had 16, and Pat had 12.
Since no one has a majority, the person with the fewest first place votes (Pat) is
eliminated, and the positions on the ballots adjusted as follows:

12	8	6	10	8	4
Ann	Ann	Eno	Eno	Ann	Eno
Eno	Eno	Ann	Ann	Eno	Ann

Ann is preferred to Eno by a 28 to 20 margin.

15. Yellowstone is selected using the pairwise comparison method.

      Yellowstone vs Grand Canyon: Y - 4, G - 3   Y

      Yellowstone vs Mt. St. Helens: Y - 4, M - 3   Y

   Grand Canyon vs Mt. St. Helens: G - 3, M - 4   M

Yellowstone has 2 points, and Mount St. Helens has 1.

17. Albuquerque is selected using the pairwise comparison method.

With 4 choices, there are 6 pairs to be considered.

      Albuquerque vs Phoenix: A - 15, P - 10   A

      Albuquerque vs Santa Fe: A - 17, S - 8   A

      Albuquerque vs Tucson: A - 13, T - 12   A

         Phoenix vs Santa Fe: P - 12, S - 13   S

          Phoenix vs Tucson: P - 14, T - 11   P

         Santa Fe vs Tucson: S - 11, T - 14   T

Albuquerque has 3 points and was preferred to all other choices.

19. Shawna is selected using the Borda count method.

Totals:	Peter:	$33 \times 3 + 68 \times 2 + 34 \times 1 = 269$
	Carmen:	$53 \times 3 + 28 \times 2 + 54 \times 1 = 269$
	Shawna:	$49 \times 3 + 39 \times 2 + 47 \times 1 = 272$

21. Charming is selected using the Borda count method.

Totals:	Able:	$33 \times 3 + 33 \times 2 + 34 \times 1 = 199$
	Boastful:	$39 \times 3 + 19 \times 2 + 42 \times 1 = 197$
	Charming:	$28 \times 3 + 48 \times 2 + 24 \times 1 = 204$

23. Carmen is selected using the Borda count method if 4 points are given for a first place vote instead of 3.

    Totals: Peter: $33 \times 4 + 68 \times 2 + 34 \times 1 = 302$
               Carmen: $53 \times 4 + 28 \times 2 + 54 \times 1 = 322$
               Shawna: $49 \times 4 + 39 \times 2 + 47 \times 1 = 321$

25. Using the Borda count method and scoring on a 4 - 3 - 2 - 1 basis, the teams finish in the following order: Vikings, Spartans, Raiders, Titans.

Summary:	First	Second	Third	Fourth	Total Points
Raiders	1×4	0×3	2×2	1×1	9
Spartans	2×4	1×3	0×2	0×1	11
Titans	0×4	2×3	1×2	0×1	8
Vikings	1×4	1×3	1×2	3×1	12

27. Site A is selected using plurality.

    Site A - 6, site B - 5, site C - 4

29. B and C tie using the Borda count method with 3 - 2 - 1 scoring.

    Totals: Site A: $6 \times 3 + 1 \times 2 + 8 \times 1 = 28$
              Site B: $5 \times 3 + 6 \times 2 + 4 \times 1 = 31$
              Site C: $4 \times 3 + 8 \times 2 + 3 \times 1 = 31$

31. Site C is selected using the pairwise comparison method.

    Site A vs Site B: A - 6, B - 9   B
    Site A vs Site C: A - 7, C - 8   C
    Site B vs Site C: B - 7, C - 8   C
    Site C has 2 points, and Site B has 1.

33. C is the winner using plurality with a run-off.

    A had the most first place votes, so there is a run-off between B and C.
    C is preferred to B by a margin of 8 to 7, so B is eliminated.
    C is then preferred to A by a margin of 8 to 7, and C is the winner.

35. Mike Griffey is selected using plurality with elimination.

    Total first place votes were A - 8, B - 4, G - 6, and R - 3. R is eliminated.
    The votes are adjusted as follows:
    ```
    4 3 2 2 2 2 2 1 1 1 1
    A B B G G A A B G G G
    B G A A B B G G A B B
    G A G B A G B A B A A
    ```
    The first place votes are now: A - 8, B - 6, and G - 7. B is eliminated.
    The votes are adjusted as follows:
    ```
    4 3 2 2 2 2 2 1 1 1 1
    A G A G G A A G G G G
    G A G A A G G A A A A
    ```
    The first place votes are now: A - 10 and G - 11. Griffey is the winner.

37. Billy Bonds is the winner with a modified Borda count method (5-3-1-0).

Totals:
Aaron: $8 \times 5 + 2 \times 3 + 3 \times 1 + 8 \times 0 = 49$
Bonds: $4 \times 5 + 10 \times 3 + 2 \times 1 + 5 \times 0 = 52$
Griffey: $6 \times 5 + 4 \times 3 + 7 \times 1 + 4 \times 0 = 49$
Ruth: $3 \times 5 + 5 \times 3 + 9 \times 1 + 4 \times 0 = 39$

39. Joe Aaron is selected using plurality with elimination using a run-off.

Aaron has the most first place votes on the initial ballot. Griffey and Bonds rank second and third respectively, and Ruth is eliminated. In the run-off election, Bonds is preferred to Griffey by a 12 to 9 margin.

In the run-off between Aaron and Bonds, Aaron is preferred 11 to 10.

41. Billy Bonds is selected using run-off with elimination of the choice with the most last place votes.

Aaron has the most last place votes, and is eliminated in the first round.
The votes are adjusted as follows:

4	3	2	2	2	2	2	1	1	1	1
B	B	R	G	G	B	R	B	G	G	R
G	G	B	R	B	R	G	R	R	R	G
R	R	G	B	R	G	B	G	B	B	B

Ruth has 9 last place votes; Bonds has 7, Griffey has 5. Ruth is eliminated.

In the run-off between Bonds and Griffey, Bonds is preferred 12 to 9.

43. Malcolm Adams and Angela Darden are selected.

Approvals: Adams - 13, Barron - 11, Calderone - 10, Darden - 13

45. The ballots are summarized in the following table:

Number of ballots:	4	3	3
	A	B	C
	B	C	A
	C	A	B

(a) Between A and B, A is preferred 7 to 3. B is eliminated.
Between A and C, C is preferred 6 to 4. C wins.

(b) Between A and C, C is preferred 6 to 4. A is eliminated.
Between B and C, B is preferred 7 to 3. B wins.

(c) Between B and C, B is preferred 7 to 3. C is eliminated.
Between A and B, A is preferred 7 to 3. A wins.

Section 9.2   Flaws of Voting Systems

Goals
1. Show how different voting methods satisfy, or fail to satisfy, the fairness properties.

Key Ideas and Questions
1. What are the desired properties of a voting method?
2. Which voting methods have each of these properties?
3. Explain what the Arrow Impossibility Theorem tells us.

Vocabulary

Majority Criterion	Monotonicity Criterion	Arrow Impossibility Theorem
Head-to-head Criterion	Irrelevant Alternatives	

. . . .

Overview

In this section we look at the properties of a voting system that is believed to be "rational and reasonable." These properties will be called criteria (singular is criterion) since they will be used as standards in judging whether a voting system is always fair and sensible.

The Majority Criterion

The Majority Criterion only tells us who should be elected if there is one candidate who is the first choice of a majority of voters.

Majority Criterion
If a candidate is the first choice of a majority of voters, then that candidate should be selected.

The Head-to-Head Criterion

If there is one candidate the voters favor, in turn, to each of the other candidates, then that candidate ought to win the election.

Head-to-Head Criterion
If a candidate is favored when compared separately with every other candidate, then the favored candidate should be elected.

Notice that if a method fails to satisfy the majority criterion, then it also fails to satisfy the head-to-head criterion, since a majority winner is also the winner of every head-to-head contest in which it is involved.

The Monotonicity Criterion

If a candidate gains support at the expense of the other candidates, then the chances of this candidate winning the election should be increased. This idea leads to the next criterion:

Monotonicity Criterion

Suppose a particular candidate, X, is selected in an election. If there is a reelection and all the voters who change their preferences make X their first choice, X should be selected..

The Irrelevant Alternatives Criterion

This criterion concerns the effect of removing or introducing a candidate that has no chance of winning.

Irrelevant Alternatives Criterion

If an alternative is selected in an election and, in a re-election, one or more of the alternatives is removed, then the previous winner should be selected in the re-election.

You can make each of the voting methods looked at in section 9.1 violate some reasonable criterion. It is natural to want to have a better voting method, one that satisfies all of the criteria all of the time. Not one such method exists, unfortunately! The **Arrow Impossibility Theorem** states that even if all the voters assign preferences to all the alternatives, there is no voting method that will always satisfy the Majority, Head-to-Head, Monotonicity and Irrelevant Alternatives criteria.

The following tables indicate which criteria are satisfied by the various voting methods.

Plurality Method	
Majority Criterion	Always Satisfied
Head-to-Head Criterion	Sometimes Not Satisfied
Monotonicity Criterion	Always Satisfied
Irrelevant Alternatives Criterion	Sometimes Not Satisfied

Borda Count Method	
Majority Criterion	Sometimes Not Satisfied
Head-to-Head Criterion	Sometimes Not Satisfied
Monotonicity Criterion	Sometimes Not Satisfied
Irrelevant Alternatives Criterion	Sometimes Not Satisfied

Plurality with Elimination Method	
Majority Criterion	Always Satisfied
Head-to-Head Criterion	Sometimes Not Satisfied
Monotonicity Criterion	Sometimes Not Satisfied
Irrelevant Alternatives Criterion	Sometimes Not Satisfied

Pairwise Comparison Method	
Majority Criterion	Always Satisfied
Head-to-Head Criterion	Always Satisfied
Monotonicity Criterion	Sometimes Not Satisfied
Irrelevant Alternatives Criterion	Sometimes Not Satisfied

. . . .

Suggestions and Comments for Odd-numbered problems

1. Does any alternative have a majority of first place votes? Is it the winner using the given method?

3. Set up a basic table in which one alternative has 3 first place votes and all other positions are to be filled in. Give one alternative as much strength in the remaining votes as possible.

5. Who is the winner using plurality? Who is the winner using pairwise comparison? Are they the same?

7. Use the suggestion in the problem statement and refer to problem 5 as an example of a set of preferences where this happens.

11. Using the pairwise comparison method with four choices requires six comparisons. Look at the pairings systematically.

13. If B is the winner when C is "removed" from the election, then the inclusion of C must cause B to have fewer first place votes.

17. Work backwards from the result on this problem. If B is to be the winner with C eliminated from the contest, then what would the preference schedule have to look like between just A and B? Then include C so that relative strength is taken away from B and A becomes the winner.

19. Use the hint given with the problem to set up the initial preference schedule. No matter how the table is filled in from there, D will be preferred to all other choices in a pairwise comparison. However, D will be the first to be eliminated when the plurality with elimination method is used. Fill in the rest of the table so that A maintains a plurality at each step.

21. Although there are six ways to rank preferences for three alternatives, the fewer you can use, the easier it will be; at least four types of preference schedules are needed. Find numbers for w, x, y, and z totaling 17, where

         w voters prefer A to B to C
         x voters prefer B to A to C
         y voters prefer C to B to A
and   z voters prefer C to A to B

When we set up the preference schedules, we make sure that no alternative has a majority of first place votes and that A is not eliminated in the first round. In the second table make changes so that A is the only one who benefits from the change but eventually loses because the order of elimination is changed.

23. Begin with the table as suggested in the problem, then give B enough strength so that C is eliminated. C must be stronger than A on a pairwise comparison.

25. When using the pairwise comparison method with 5 choices, there will be a total of 10 pairs. For A to win with this method, A will have to have at least 3 points; that is, A must be preferred over at least 3 of the other 4. Since E is preferred over A, A must be preferred over each of B, C, and D on at least 3 ballots. Moreover, E can't be given any first place votes, since that would result in E winning in too many comparisons.

29. There are only three different ways the run-offs with three choices can be made:

         A and B first, winner against C
         A and C first, winner against B
         B and C first, winner against A

Solutions to Odd-numbered Problems

1. A majority (5) prefer A, however in the Borda count we have the totals:
A - 19, B - 15, C - 20, so C wins the Borda count.

3. One example of a preference table where the Borda count method violates the majority criterion:

A	1	1	1	3	3
B	2	2	2	1	1
C	3	3	3	2	2

Candidate A has a majority of 3. The Borda count method with 3 - 2 - 1:

Totals: A: $3 \times 3 + 0 \times 2 + 2 \times 1 = 11$
             B: $2 \times 3 + 3 \times 2 + 0 \times 1 = 12$
             C: $0 \times 3 + 2 \times 2 + 3 \times 1 = 7$

B is the winner if the Borda count method is used.

Note: If the Borda count uses 4 -2 -1 scoring, the totals are:
         A - 14, B - 14, C - 7

5. A wins using the plurality method: A - 4, B - 2, C - 3.
   However, in pairwise comparison, the results are:
       A vs B: B is preferred to A 5 to 4
       A vs C: C is preferred to A 5 to 4
       B vs C: C is preferred to B 6 to 3
   C is preferred to all other candidates in a pairwise comparison, but is not the winner using the plurality method.

7. Following the guidelines in the problem statement, our table begins:
   ```
   A 1 1 1 1 x x x x x x
   B x x x x 1 1 1 x x x
   C x x x x x x x 1 1 1
   ```
   Then, A's votes must be made as weak as possible in terms of using the pairwise comparison method. This is done by placing A last on the remaining ballots. A is first or last on all ballots ("Love 'em or hate 'em").
   ```
   A 1 1 1 1 3 3 3 3 3 3
   B x x x x 1 1 1 x x x
   C x x x x x x x 1 1 1
   ```
   The table is then filled in so that both B and C are preferred to A on more ballots. One Example:
   ```
   A 1 1 1 1 3 3 3 3 3 3
   B 2 2 2 2 1 1 1 2 2 2
   C 3 3 3 3 2 2 2 1 1 1
   ```
   Both B and C are preferred to A by a 6 to 4 margin in pairwise comparison. B is preferred to C by a 7 to 3 margin.

9. **(a)** Using the pairwise comparison method, the results are:
       A is preferred to B; 5 to 4
       A is preferred to C; 5 to 4
       C is preferred to B; 7 to 2
   A wins.

   (b) Using the Borda count method, the scores are:
       A: $4 \times 3 + 2 \times 2 + 3 \times 1 = 19$
       B: $2 \times 3 + 2 \times 2 + 5 \times 1 = 15$
       C: $3 \times 3 + 5 \times 2 + 1 \times 1 = 20$
   C wins.

11. **(a)** Using pairwise comparison with four choices requires six comparisons.
        A to B: B is preferred to A; 30 to 21   B wins
        A to C: C is preferred to A; 29 to 22   C wins
        A to D: D is preferred to A; 33 to 18   D wins
        B to C: C is preferred to B; 40 to 11   C wins
        B to D: D is preferred to B; 34 to 17   D wins
        C to D: D is preferred to C; 32 to 19   D wins
    D is the winner using the pairwise comparison method.

    (b) In the plurality with elimination method, B and D both have 7 votes and are eliminated. When the votes are adjusted, C has a majority of 29 out of 51 votes.
    C was preferred to A by 29 to 22 in part (a).

13. First table: Alternative A wins by plurality. The table could begin:

```
A  1  1  1  x  x  x  x
B  x  x  x  1  1  x  x
C  x  x  x  x  x  1  1
```

First place votes: A - 3, B - 2, C - 2.

We fill in the table so that B has an advantage over A in pairwise comparison.

```
A  1  1  1  3  3  3  3
B  2  2  2  1  1  2  2
C  3  3  3  2  2  1  1
```

When alternative C is eliminated, the table becomes:

```
A  1  1  1  2  2  2  2
B  2  2  2  1  1  1  1
```

Alternative B has a majority of four to three.

15. If alternative B is to have a majority, then B must begin with 3 first place votes.

```
A  x  x  x  x  x
B  1  1  1  x  x
C  x  x  x  x  x
```

If A is to be favored using the Borda count method, then the table should be filled in giving A as much strength as possible and not giving much to B. One possibility:

```
A  2  2  2  1  1
B  1  1  1  3  3
C  3  3  3  2  2
```

Alternative A receives 12 points under Borda Count rules, B receives 11 points, and C receives 7 points. Alternative A wins under the Borda count method, even though B has a majority of first place votes.

17. If there are 5 voters and the elimination of C makes B the Borda count winner, then the elimination of C must leave us with a table like the following:

```
A  2  2  2  1  1
B  1  1  1  2  2
```

Now, we add the alternative C in such a way that C takes some strength away from B, but not from A. The easiest choice:

```
A  2  2  2  1  1
B  1  1  1  3  3
C  3  3  3  2  2
```

Alternative A receives 12 points under Borda count rules, B receives 11 points, and C receives 7 points. A is the winner under the Borda count method.

When alternative C is eliminated, we have the original table with only A and B. Alternative A receives 7 points under Borda Count rules, but B receives 8 points, so alternative B is now the winner.

19. Using the hint given with the problem, the preference schedule would begin:

```
A  1  1  x  x  x
B  x  x  1  1  x
C  x  x  x  x  1
D  2  2  2  2  2
```

No matter how the table is filled in from here, D will be preferred to all other choices in a pairwise comparison. However, D will be the first to be eliminated when the plurality with elimination method is used. If A is to win when the plurality with elimination method is used, then A must be ranked higher than B on the last ballot. Both A and B must be ranked higher than C on at least one more ballot. Two possibilities:

```
A  1  1  4  4  3      1  1  3  4  3
B  3  3  1  1  4      4  3  1  1  4
C  4  4  3  3  1      3  4  4  3  1
D  2  2  2  2  2      2  2  2  2  2
```

In plurality with elimination, D is eliminated first, then C. A wins, 3 to 2.
In pairwise comparisons, however, D is the winner
  D beats A 3 to 2
  D beats B 3 to 2
  D beats C 4 to 1

21. When we set up the preference schedules, we make sure that no alternative has a majority of first place votes and that A is not eliminated in the first round. Alternative A does not need to have a plurality at this point, but A must benefit from the elimination of another alternative. Consider the following set of alternatives:

  6 voters prefer A to C to B
  5 voters prefer B to A to C
  4 voters prefer C to B to A
  2 voters prefer C to A to B

Using the plurality with elimination method, B has the fewest first place votes and is eliminated. A is preferred to C by an 11 to 6 margin. A is the winner.

Suppose that the two voters who preferred C to A to B change their ballots to show that they prefer A to C to B. Clearly, this benefits A and no one else. The preference tables can now be grouped as follows:

  8 voters prefer A to C to B
  5 voters prefer B to A to C
  4 voters prefer C to B to A

Using the plurality with elimination method, C has the fewest first place votes and is eliminated. B is preferred to A by an 9 to 8 margin. B is the winner.

23. We begin with the table as suggested in the problem:

```
A  1  1  1  1  x  x  x  x  x
B  x  x  x  x  1  1  1  x  x
C  x  x  x  x  x  x  x  1  1
```

Original table — alternative A wins using plurality with elimination method.
C is eliminated, giving A a majority.

```
A  1  1  1  1  3  3  3  2  2
B  2  2  2  2  1  1  1  3  3
C  3  3  3  3  2  2  2  1  1
```

23. *continued*

After alternative B is removed as a choice in the original table and the votes are adjusted, we have

```
A  1  1  1  1  2  2  2  2  2
C  2  2  2  2  1  1  1  1  1
```

C has a majority, and is the winner.

25. See the hints for conditions that must be satisfied. The following preference schedules satisfy these conditions, but many more can be constructed along similar lines of reasoning.

	Ballot	Ballot	Ballot	Ballot	Ballot
first choice	A	A	B	C	D
second choice	B	C	E	E	E
third choice	C	D	A	A	A
fourth choice	D	B	C	D	B
fifth choice	E	E	D	B	C

Pairwise comparison:

A beats B, C, D for 3 points and wins using this method.
B beats C and E for 2 points
C beats D and E for 2 points
D beats B and E for 2 points
E beats only A for 1 point

27. (a) In the run-off between B and C, B receives 7 votes and C receives 2 votes, so B wins the run-off.

In the final election between A and B, A receives 4 votes and B receives 5 votes, so B is the over all winner.

(b) If a minimum of 3 of the voters who prefer A over B over C were to vote (insincerely) for C over B in the run-off between B and C, then the preferences would be

A over B over C: 1 vote
A over C over B: 3 votes
B over A over C: 3 votes
C over B over A; 2 votes

Now, C is preferred over B on 5 of the 9 ballots. C wins the run-off. In the final election, A receives 7 votes and C receives 2 votes, making A the over all winner.

Note: If only two of the voters make the switch, B will still win the run-off over C, only to lose the election to A.

29. There are only three different ways the run-offs with three choices can be made:

A and B first, winner against C
    B beats A, 8 to 7; C beats B, 8 to 7 C wins

B and C first, winner against A
    C beats B, 8 to 7; A beats C, 9 to 6 A wins

A and C first, winner against B
    A beats C, 9 to 6; B beats A, 8 to 7 B wins

Section 9.3 Weighted Voting Systems

Goals
1. Use the Banzhaf Power Index to determine relative voting power..

Key Ideas and Questions
1. How do you measure the relative strength of each voter in a weighted voting system where the voters have different numbers of votes, or "weight"?
2. Explain how coalitions, dictators, dummies, voters with veto power, and critical voters affect the outcomes in voting.
3. How is the Banzhaf Power Index Used?

Vocabulary

Weight	Winning/losing Coalition	Critical Voter
Motion	Dummy	Banzhaf Power
Quota	Dictator	Total Banzhaf Power
Coalition	Voter with Veto Power	Banzhaf Power Index

. . . .

Overview

The voting methods in Section 9.1 dealt with the most common type of decision-making: each voter has a single vote, and all have an equal say in the outcome. In this section, we consider situations where voters may have any *unequal* number of votes. One example is a stockholders' meeting, when stockholders in a corporation have votes that are proportional to the number of shares they hold. Another is the United States electoral college that officially elects the President; each state has a number of votes equal to the number of members it has in Congress - the House of Representatives and the Senate combined. The major question we want to answer in this section is: How do we measure the relative strength of each voter in such a system?

Weighted Voting Systems

In a weighted voting system, the number of votes each voter has is referred to as the voter's **weight**, and the number of votes required for passage of a **motion** is called the **quota**. A weighted voting system is fully described by indicating the quota and the number of votes held by each of the voters. If there are four voters with 10, 9, 6, and 2 votes respectively, and a simple majority is needed to pass a motion, the system is described using the notation [14 | 10, 9, 6, 2]. The quota is 14 since there is a total of 27 votes. We refer to the kth voter as P_k, and the weight of that voter as W_k.

Coalitions

A **coalition** is a combination of voters who vote on the same side of a motion, If the combined number of votes in a coalition is greater than the quota, the motion passes, and the coalition is called a **winning coalition**. If not, it's called a **losing coalition**. A **critical voter** in a coalition is one whose votes are essential to the passage of a motion.

Dictators, Dummies, and Vetoes

If an individual voter has more weight than the required quorum, then the motions passes if the voter is for it and fails if the voter is against it. This kind of voter is referred to as a **dictator**. At the other end of the voting spectrum is the voter whose votes never really make any difference. In the weighted system [14 | 10, 9, 6, 2], the fourth voter is never needed to make a winning coalition. This voter is called a **dummy**, since it has *no voice* in the outcome. Finally, if a voter has enough weight, individually, to stop the passage of a motion by voting against it, the voter is said to have **veto power**. In the weighted system [18 | 10, 9, 6, 2] the first voter has veto power. Note that in this system the fourth voter is not a dummy.

The Banzhaf Power Index

The most common way to measure the relative power of voters in a weighted voting system is the **Banzhaf Power Index**, in which we initially count all the situations in which a voter's change from "yes" to "no" changes a winning coalition to a losing coalition. In order to compute the Banzhaf Power Index, it is necessary to list all the winning coalitions, and determine which voters are critical in each of these. This can be time-consuming. In a weighted voting system with n voters, there are $2^n - 1$ coalitions that have to be considered (if there are 6 voters, there are $2^6 - 1 = 63$ coalitions).

The Banzhaf Power of a voter is the number of times a voter is critical in a winning coalition. An individual voter's Banzhaf Power index is the ratio of the voter's Banzhaf Power to the total of the Banzhaf Power for all the voters in the system. Suppose there is a weighted voting system with four voters and that the first voter is a critical voter six times, while the second and third voters are critical voters twice, and the third voter is a dummy. Then, the total Banzhaf power in the system is 10, and the Banzhaf Power index for the system may be expressed as $\{ \frac{6}{10}, \frac{2}{10}, \frac{2}{10}, 0 \}$ or $\{ \frac{3}{5}, \frac{1}{5}, \frac{1}{5}, 0 \}$.

• •

Suggestions and Comments for Odd-numbered problems

1. - 4.
Find the total number of votes in the system and apply the appropriate percentage. If this results in a decimal fraction, round up to the next whole number.

7. and 8.
Each member has one vote. How many votes are needed for passage?

9. - 12.
How many votes are held by members of the coalition? Does the total exceed the quota?

15. - 18.

The first task is to list ALL coalitions. This can be done systematically by a process similar to the following example. Suppose you have a system with four voters. First, list all the coalitions four voters (of course, there's only one), then all the coalitions with three voters (how many ways can you leave one out?), then all the ones with two voters (watch out for duplicates), and all the single voter "coalitions". There should be fifteen coalitions in all. The pattern for four voters is as follows:

P_1	P_2	P_3	P_4		
x	x	x	x		1 coalition with 4 voters
x	x	x		*	
x	x		x	*	
x		x	x	*	
	x	x	x	*	4 coalitions with 3 voters
x	x			**	
x		x		**	
x			x	**	
	x	x		**	
	x		x	**	
		x	x	**	6 coalitions with 2 voters
x				*	
	x			*	
		x		*	
			x	*	4 coalitions with 1 voter

Which ones have more than the quota? Which ones have less?
To see if a voter is a critical voter in a coalition, "drop" the voter from the coalition and total up the remaining votes. Note: systematic counting is covered in Section 5.4 of the text. What we are looking for is technically called a *combination*.

19. - 24.

Consider all the winning coalitions and find the total number of times each voter is a critical voter; this is the Banzhaf Power for each individual voter. Then, you find the total Banzhaf Power for the system and the relative share for each voter. This can be expressed as a decimal, common fraction, or percentage. Only one voter can have veto power (why?). Exclude the greatest weight; does the total of the rest exceed the quota?

25. and 26.

If two voters *always* vote the same, they can be considered as a single voter.

29. The coalitions can be listed and analyzed as though the chair always voted. The only difference in this case is that the chair doesn't have to disclose it's vote if a majority is already established.

31. You can treat this system as one in which the chair has two votes.

31. and beyond In determining the Banzhaf power index, the key step is determining the number of winning coalitions of each type and then whether or not a given voter will be a critical voter. Suppose there is a committee with a chairman and, say, eight other members who have equal votes to each other, and that a winning coalition can consist of the chairman and at least two other members. We encounter a significant counting problem: In how many ways can we select two of the eight other members to vote with the chair? In how many ways can we select three of the eight? This kind of counting problem is covered in section 5.4 under the topic of *unordered subsets*, or *combinations*.

Solutions for Odd-numbered problems

1. **(a)** [14 | 6, 5, 5, 3, 3, 2, 1, 1]
   (b) [15 | 8, 5, 5, 4, 3, 2, 2]
   (c) [18 | 10, 5, 5, 5, 3, 3, 3]
   (d) [12 | 7, 5, 5, 2, 2 , 1]

3. **(a)** [16 | 8, 5, 5, 3, 3, 2]
   (b) [22 | 9, 6, 5, 5, 3, 2, 2]
   (c) [21 | 7, 5, 5, 5, 3, 2]
   (d) [15 | 5, 5, 4, 4, 3, 3]

5. There are 23 "votes"; 10 is less than a majority.

7. [6 | 1, 1, 1, 1, 1, 1]; "unanimous" means that all 6 votes are needed.

9. **(a)** Winning   Total votes = 10 + 7 + 4 = 21; the quota is matched
   (b) Losing   Total votes = 8 + 7 + 4 = 19; the quota is not met
   (c) Winning   Total votes = 8 + 7 + 7 = 22; the quota is exceeded
   (d) Winning   Total votes = 7 + 7 + 4 + 4 = 22; the quota is exceeded

11. **(a)** Winning   Total votes = 8 + 4 + 3 = 15; the quota is matched
    (b) Losing   Total votes = 4 + 3 + 2 = 9; the quota is not met
    (c) Losing   Total votes = 4 + 3 + 3 + 2 = 12; the quota is not met
    (d) Losing   Total votes = 4 + 3 + 3 + 2 + 2 = 14; the quota is not met
    Note : P_1 has veto power

13. **(a)** $2^8 - 1 = 255$
    (b) $2^{10} - 1 = 1023$

15. **(a)** Losing coalitions: $\{P_1\}$, $\{P_2\}$, $\{P_3\}$, $\{P_2, P_3\}$
        Winning coalitions: $\{P_1, P_2\}$, $\{P_1, P_3\}$, $\{P_1, P_2, P_3\}$
    (b) Losing coalitions: $\{P_1\}$, $\{P_2\}$, $\{P_3\}$, $\{P_4\}$, $\{P_1, P_4\}$, $\{P_2, P_3\}$, $\{P_2, P_4\}$, $\{P_3, P_4\}$
        Winning coalitions: $\{P_1, P_2\}$, $\{P_1, P_3\}$, $\{P_1, P_2, P_3\}$, $\{P_1, P_3, P_4\}$, $\{P_1, P_2, P_4\}$,
        $\{P_2, P_3, P_4\}$, $\{P_1, P_2, P_3, P_4\}$

17. **(a)** $\{P_1, P_2\}$, $\{P_1, P_3\}$
    (b) $\{P_1, P_2\}$, $\{P_1, P_3\}$, $\{P_2, P_3, P_4\}$

19. **(a)** $\{\frac{3}{5}, \frac{1}{5}, \frac{1}{5}\}$
        Winning coalitions:
            $\{P_1, P_2, P_3\}$ P_1 is a critical voter
            $\{P_1, P_2\}$ P_1 and P_2 are critical voters
            $\{P_1, P_3\}$ P_1 and P_3 are critical voters
        P_1 has veto power; no dictator, no dummy

Continued on next page

19. (b) $\{\frac{1}{3}, \frac{1}{3}, \frac{1}{3}, 0\}$

Winning coalitions:	Critical voters
$\{P_1, P_2, P_3, P_4\}$	None
$\{P_1, P_2, P_3\}$	None
$\{P_1, P_2, P_4\}$	P_1 and P_2
$\{P_1, P_3, P_4\}$	P_1 and P_3
$\{P_2, P_3, P_4\}$	P_2 and P_3
$\{P_1, P_2\}$	P_1 and P_2
$\{P_1, P_3\}$	P_1 and P_3
$\{P_2, P_3\}$	P_2 and P_3

The Banzhaf Power for each of the first three voters is 4 and the fourth voter is never a critical voter. The total Banzhaf power in the system is 12.

The Banzhaf Power Index may be expressed as $\{\frac{4}{12}, \frac{4}{12}, \frac{4}{12}, 0\}$ or $\{\frac{1}{3}, \frac{1}{3}, \frac{1}{3}, 0\}$

P_4 is a dummy; no dictator, no voter with veto power

21. (a) $\{1, 0, 0\}$

P_1 has votes enough to match the quota
P_1 is a dictator, the others are dummies

(b) $\{\frac{1}{4}, \frac{1}{4}, \frac{1}{4}, \frac{1}{4}\}$

Winning coalitions:	Critical voters	
$\{P_1, P_2, P_3, P_4\}$	None	Every three voter coalition is a winning
$\{P_1, P_2, P_3\}$	P_1, P_2, P_3	coalition; no two voter coalition has enough
$\{P_1, P_2, P_4\}$	P_1, P_2, P_4	votes. The Banzhaf Power for each voter is
$\{P_1, P_3, P_4\}$	P_1, P_3, P_4	3, and the Total Banzhaf Power is 12.
$\{P_2, P_3, P_4\}$	P_2, P_3, P_4	No dummy, dictator, voter with veto power

23. (a) $\{1, 0, 0\}$

(b) $\{\frac{3}{5}, \frac{1}{5}, \frac{1}{5}\}$

(c) $\{\frac{1}{2}, \frac{1}{2}, 0\}$

(d) $\{\frac{1}{2}, \frac{1}{2}, 0\}$

(e) $\{\frac{1}{3}, \frac{1}{3}, \frac{1}{3}\}$

25. (a) For the original system we have

Winning coalitions:	Critical voters
$\{P_1, P_2, P_3, P_4\}$	None
$\{P_1, P_2, P_3\}$	P_1, P_2, P_3
$\{P_1, P_2, P_4\}$	P_1, P_2, P_4
$\{P_1, P_3, P_4\}$	P_1, P_3, P_4
$\{P_2, P_3, P_4\}$	P_2, P_3, P_4

The Banzhaf power index is $\{\frac{1}{4}, \frac{1}{4}, \frac{1}{4}, \frac{1}{4}\}$

(b) Since P_1 and P_2 always vote the same, they can be considered as a single voter with weight of 5. We'll call the new set of voters $\{R_1, R_2, R_3\}$.
The revised system can be represented as [6 | 5, 2, 2,].

Winning coalitions:	Critical voters
$\{R_1, R_2, R_3\}$	R_1
$\{R_1, R_2\}$	R_1 and R_2
$\{R_1, R_3\}$	R_1 and R_3

The Banzhaf Power Index (using decimals) is $\{0.6, 0.2, 0.2\}$

27. (a) For the system $\{11 \mid 8, 6, 4, 3\}$ the Banzhaf Power Index is $\{\frac{1}{2}, \frac{1}{6}, \frac{1}{6}, \frac{1}{6}\}$

(b) For the system $\{11 \mid 8, 6, 5, 2\}$ the Banzhaf Power Index is $\{\frac{1}{3}, \frac{1}{3}, \frac{1}{3}, 0\}$

29. $\{\frac{1}{5}, \frac{1}{5}, \frac{1}{5}, \frac{1}{5}, \frac{1}{5}\}$

Winning coalitions:	Critical voters	
$\{P_1, P_2, P_3, P_4\}$	None	
$\{P_1, P_2, P_3\}$	P_1, P_2, P_3	Note: C is only in coalitions in
$\{P_1, P_2, P_4\}$	P_1, P_2, P_4	which there are exactly two of
$\{P_1, P_3, P_4\}$	P_1, P_3, P_4	the other voters.
$\{P_2, P_3, P_4\}$	P_2, P_3, P_4	
$\{C, P_1, P_2\}$	C, P_1, P_2	The Banzhaf power of each voter is 6.
$\{C, P_1, P_3\}$	C, P_1, P_3	Power is equally shared.
$\{C, P_1, P_4\}$	C, P_1, P_4	
$\{C, P_2, P_3\}$	C, P_2, P_3	
$\{C, P_2, P_4\}$	C, P_2, P_4	
$\{C, P_3, P_4\}$	C, P_3, P_4	

31. $\{\frac{1}{2}, \frac{1}{6}, \frac{1}{6}, \frac{1}{6}\}$

In order to have a strict majority, three votes are needed. Since the three regular members have equal voice, they can each be assigned 1 vote. Since the chair plus one member is a winning coalition, the chair can be assigned 2 votes, and we can describe the system as $[3 \mid 2, 1, 1, 1]$. The chair is a critical voter in 6 winning coalitions - three with one of each of the regular members, and three with each combination of 2 regular members. Each of the regular members is a critical voter 2 times: once when voting with the two other members against the chair, and once when voting with the chair against the other two members. The Banzhaf power in the system is $\{6, 2, 2, 2\}$.

33. (a) $[6 \mid 4, 1, 1, 1, 1, 1, 1]$

Since all six members can form a winning coalition without the chairman, the quota can be set at 6, and each of the regular members assigned one vote. Since the chair plus two members is also a winning coalition, the chair can be assigned 4 votes. In this situation, it would be 6 votes for, and 4 votes against. The chair is a critical voter in any coalition with at least two, but less than six, regular members. The types of winning coalitions for the chair are:

Chair plus two	15 of this type	$_6C_2 = 15$ (See section 5.4)
Chair plus three	20 of this type	$_6C_3 = 20$
Chair plus four	15 of this type	$_6C_4 = 15$
Chair plus five	6 of this type	$_6C_5 = 6$

The chair is a critical voter in 56 coalitions

Each regular member is a critical voter when all six vote without the chair - 1 way Each regular member is a critical voter when voting with the chair and one other member - there are 5 coalitions of this type. Each regular member is a critical voter in only 6 coalitions. The Banzhaf power in the system is $\{56, 6, 6, 6, 6, 6, 6\}$

(b) $\{\frac{28}{46}, \frac{3}{46}, \frac{3}{46}, \frac{3}{46}, \frac{3}{46}, \frac{3}{46}, \frac{3}{46}\}$

35. (a) $\{\frac{1}{3}, \frac{1}{3}, \frac{1}{3}\}$

   (b) $\{\frac{1}{2}, \frac{1}{6}, \frac{1}{6}, \frac{1}{6}\}$

37. For P_1, there are $10 + 5 + 1$ ($_5C_3 + {}_5C_4 + {}_5C_5$) winning coalitions where P_1 is critical. For P_2, there are $5 + 1$ ($_5C_4 + {}_5C_5$) winning coalitions where P_2 is critical. On issues where P_1 and P_2 are on opposite sides, P_1 can be part of 16 winning coalitions, while P_2 can be part of 6 winning coalitions.

   Therefore, the relative power is $\frac{8}{11}$ to $\frac{3}{11}$.

Section 9.4   Apportionment Methods

Goals
1. Use different apportionment methods to determine fair shares.

Key Ideas and Questions
1. How do the different apportionment methods handle the fractional parts of the standard or modified quotas?
2. Which apportionment methods favor larger states the most?
3. How do quota methods differ from divisor methods?

Vocabulary
Apportionment Problem	Standard Quota	Modified Quota
Hamilton's Method	Quota Rule	Webster's Method
Standard Divisor	Jefferson's Method	

· ·

Overview

Apportionment can be defined simply as the "fair division" of some asset or valued resource. For the states in a national legislature, the counties in a state, the colleges in a university, or the heirs to an estate, the basic problem is the same: How do you divide, on an equitable basis, those things that are not individually divisible?

The apportionment problem arises when one or more of the due portions has a fractional part, but what is being apportioned cannot be divided with fractions. Mathematically speaking, the **apportionment problem** is to determine a method for rounding a collection of numbers, some of which may be fractions, so that the sum of the numbers is unchanged.

The various methods of apportionment we look at were proposed by famous U.S. statesmen and were used for the purpose of apportioning the seats in the U.S. House of Representatives. These methods are named for their authors: Alexander Hamilton, Thomas Jefferson, and Daniel Webster.

Hamilton's Method

There are three steps to Hamilton's method:

1. Total population $= P$ Number of seats to be apportioned $= M$

   Number of persons per seat overall $= \frac{P}{M} = D$

   D is called the **standard divisor**.

   If the seats could be divided into fractions, then we would want to assign

   $$\frac{\text{State's population}}{\text{standard divisor}} = Q.$$

   Q is called the state's **standard quota**.

2. Round each Q down to a whole number. Each state will get at least that many seats, but must have at least one seat.

3. If there are seats left over (and there usually are unless all the standard quotas were whole numbers) then allocate those seats one at a time to the states ordered by the **size of the fractional part** of the standard quota, beginning with the state with the largest fractional part.

When Hamilton's method is applied, every apportionment is either the whole number just below any fractional standard quota, or is the whole number just above any fractional standard quota. Any method having this property of always assigning the whole number either just above, or just below, the standard quota is said to satisfy the **quota rule**.

Jefferson's Method

Jefferson's method changes the standard divisor to obtain modified quotas that can all be rounded down to the desired total number of representatives. There are two steps to Jefferson's method:

1. The number of seats to be apportioned $= M$.

   Find a number d, the **divisor**, such that if all the numbers

   $$mQ = \frac{\text{state's population}}{d},$$

   called **modified quotas**, are rounded down, then the sum is M.

2. Assign each state the rounded down value of its modified quota.

Webster's Method

Webster's method was suggested as a compromise between Hamilton's and Jefferson's methods. Webster's Method has two steps:

1. The number of seats to be apportioned $= M$.

   Find a number d, called the **divisor**, such that if all the numbers

   $$mQ = \frac{\text{state's population}}{d}$$

   called **modified quotas**, are rounded off to the nearest integer, then the sum is M.

2. Assign each state the integer nearest its modified quota.

All three apportionment methods carry the stipulation that when the quotas are rounded each state will get at least that many seats, but must have at least one seat. This is a protection for small states in a country with a small national legislature. If the U. S. House of representatives had only 100 seats, the standard divisor would be 2,487,098.73 (using 1990 census data). Any state with a population less than that would have a standard quota less than 1.0. Worse, any state with a population less than 1,243,550 would have a standard quota less than 0.5. According to the 1990 census, there were 13 states that small! When rounding produces an apportionment of zero seats to states, these seats are assigned first, and then the remaining seats are reapportioned among the remaining states.

Webster's method was used in the U.S. during the 1840's and again from 1900 - 1941. In 1941, Webster's method was modified to give us the Huntington-Hill method which is still in use as of 1996. Instead of using the usual average where 0.5 is the "in-between" mark for two consecutive whole numbers, the Huntington-Hill method uses the geometric mean, which is slightly lower. See extended problem #27 and #29 after section 9.4 and the Hints sections for more information.

Summary

The apportionment methods just described have several major features in common, but handle the decisions regarding fractional parts in different ways. These differences lead to different apportionments, and, in the case of the United States House of Representatives, several legal disputes that have gone all the way to the Supreme Court for resolution. The latest challenge was as recently as 1990.

The following example reviews the methods and highlights the differences.

Example   A country with four states and a national assembly of 40 seats reapportions the legislature based on the following population figures (in thousands):

State	A	B	C	D	Total
Population (1000's)	125	318	418	450	1311

(a) Use Hamilton's method to reapportion the assembly.

Solution   With 40 seats, the standard divisor is $\frac{1311}{40} = 32.75$. We use the standard divisor to get the standard quotas for each state. With Hamilton's method, seats in the assembly are first assigned according the integer part of the standard quotas. The remaining seats are assigned based on the size of the fractional part of the quotas.

State	A	B	C	D	Total	
Population (1000's)	125	318	418	450	1311	
Standard divisor					32.775	
Standard quotas	3.81	9.70	12.75	13.73		
Integer part	3	9	12	13	37	(3 to assign)
fractional rank	(1)		(2)	(3)		
Final Apportionment	**4**	**9**	**13**	**14**	**40**	

(b) Apportion the national assembly using Jefferson's method.

Solution   Jefferson's method rounds all quotas down to the integer part of the quota. For the standard quotas, this only gives a total of 37 seats. A modified divisor will have to be found; one that is smaller. When a smaller divisor is used, all quotas will increase, with the larger ones increasing faster. The question to ask is "Which ones get to the next integers faster?" Although state A has the largest fractional part, its quota is less than half of the others, so it will increase less than half as fast. Since there are three seats to assign, it appears that state B will probably get one of the seats. As our first choice, we try a divisor that would raise that state's standard quota to the next integer (from 9 to 10); divide the state's population by 10 ($318 \div 10 \approx 31.8$).

State	A	B	C	D	Total
Population (1000's)	125	318	418	450	1311
Modified divisor					31.8
Modified quotas	3.93	10.00	13.14	14.15	
Final Apportionment	3	10	13	14	40

Note: If the modified quota for state A had increased beyond 4.00, then it would be the third state getting another seat. A slightly larger divisor would be needed.

(c) Apportion the national assembly using Webster's method.

Solution   Webster's method rounds the quotas to the nearest integer. For the standard quotas, this gives:

State	A	B	C	D	Total
Population (1000's)	125	318	418	450	1311
Standard divisor					32.75
Standard quotas	3.81	9.70	12.75	13.73	
Rounded quotas	4	10	13	14	41

Since the total is too high, this means a modified divisor will have to be found; one that is larger. All the standard quotas will decrease. In contrast to Jefferson's method, which makes adjustments to an appropriate integer or beyond, Webster's method must take the quotas past the fractional value of 0.5 It would appear that the quota for state D can be decreased to below 13.5 the fastest, even if ever so slightly. As our first choice, we try a divisor that would decrease state D's quota to 13.49; divide the state's population by 13.49 ($450 \div 13.49 \approx 33.36$).

State	A	B	C	D	Total
Population (1000's)	125	318	418	450	1311
Modified divisor					33.36
Modified quotas	3.75	9.53	12.53	13.49	
Final Apportionment	4	10	13	13	40

Important Note: The apportionments by different methods are often the same, but as in our example, they can be completely different. It depends on the numbers involved.

State	A	B	C	D	Total
Hamilton Apportionment:	4	9	13	14	40
Jefferson Apportionment:	3	10	13	14	40
Webster Apportionment:	4	10	13	13	40

Suggestions and Comments for Odd-numbered problems

General Considerations for Apportionment Methods

First you compute the standard divisor. In the case of a legislature, this represents how many in the population are represented by a single seat in the legislature. The next step is to compute the standard quota for each state. This is each state's "fair share" of the seats. The fair shares will hardly ever be whole numbers. Every representative has an equal vote, and fractions of seats aren't practical, anyway, so something has to be done with regard to how the fractions are handled.

Using Hamilton's Method

Hamilton's method is very straight forward, but requires accuracy (as do all the methods). With Hamilton's method, the integer part of the standard quota is assigned as the minimum number of seats the state will receive. However, every state is also guaranteed at least one seat. If any state is assigned zero seats on the basis the integer part of its standard quota, it will be assigned 1 seat and the remaining seats will be reapportioned among the remaining states. After seats have been assigned on the basis of the integer parts and guarantees, any remaining seats are assigned on the basis of the size of the fractional part of the standard quotas. Since the quotas are based on a population census, a state that just misses getting an additional seat might well decide to challenge the accuracy of the census. Always double check the final apportionment total. Every state should always have at least as many seats as any state that has a smaller population. Bigger states get more representatives.

Note: In a two-house legislature such as the U. S. Congress, the second chamber (in this case, the Senate) is generally not apportioned.

Using Jefferson's Method

Jefferson's method uses the integer parts of the quotas as the final quota; this is the effect of *rounding down*. Only in rare cases will the standard quotas be correct, because that would require that all the standard quotas were whole numbers. As a result, the total assigned seats will be too low. A smaller divisor is needed to <u>raise</u> the quotas. This is a "dynamic" process. As you decrease the divisor, the quotas will increase. You have to decrease the quota at least enough so that the sum of the integer parts is the right total, but not so much that it exceeds the total.

Finding a workable divisor is a matter of problem solving using the Guess and Test strategy. Take some time to think about your initial guess, and you'll save time later on. How close to the next integers are the fractional parts? Which ones might grow faster than others? Think of it as a race where each quota is headed to the next integer.

Using Webster's Method

After the standard quotas have been calculated, Webster's method rounds each quota to the nearest whole number; we call this *rounding off*. The total may or may not be correct, but usually isn't. A new divisor is selected, and the process is done over until the total is correct. If the total is too high, a larger divisor (therefore smaller quotas) is needed. If the total is too low, a smaller divisor is needed. Whereas using Jefferson's method requires you to raise or lower quotas past the whole numbers, Webster's method requires you to raise or lower them past the halfway mark. How close are the fractional parts to 0.5?

Using Adam's Method   (Extended problems)

Adam's method is similar in its mechanics to Jefferson's except all standard or modifie[c] quotas are rounded up. Philosophically, however, the results of the methods are quite different.

Consider two quotas of 1.5 and 4.5.
Jefferson's method rounds them to 1 and 4; Adam's rounds them to 2 and 5.
The relative power between the two states changes significantly.

Solutions to Odd-numbered Problems

1. Hamilton's method uses the standard divisor. Each individual is assigned the integer part of the standard quota, and the items (seats or bottles or whatever) are assigned in the order of the fractional parts of the standard quotas.

	Jaron	Mikkel	Robert	Total
	295	205	390	890
Bottles				20
Standard divisor				44.5
Standard quotas	6.63	4.61	8.76	
Integer part	6	4	8	18 (2 to be assigned)
Fraction rank	(2)		(1)	
Final Apportion	7	4	9	20

    Jaron -   7 bottles
    Mikkel -   4 bottles
    Robert -   9 bottles

3. Webster's method starts with the standard divisor and standard quota and rounds off to the nearest integer. If the total is not correct, a new divisor is chosen so that the rounded-off quotas have the correct total.

	Jaron	Mikkel	Robert	Total
	295	205	390	890
Bottles				20
Standard divisor				44.5
Standard quota	6.63	4.61	8.76	
Rounded quota	7	5	9	21 (too large)

    A new divisor is needed that will lower the fractional part of one of the quotas just below 0.5. We try $295 \div 6.49 \approx 45.45$.

	Jaron	Mikkel	Robert	Total
Modified divisor				45.45
Modified quota	6.49	4.51	8.58	
Rounded quota	6	5	9	20 (just right)

    Jaron -   6 bottles
    Mikkel -   5 bottles
    Robert -   9 bottles

5. Hamilton's method is being used.

	Gorge	Mane	Organ	Taxes	Total	
	275	767	465	383	1890	
Seats					50	
(a) Standard divisor					37.8	
(b) Standard quotas	7.28	20.29	12.30	10.13		
Integer part	7	20	12	10	49	(1 to assign)
Fraction rank			(1)			
Final Apportion	7	20	13	10	50	

(c) Gorge -   7 seats
      Mane -   20 seats
      Organ -   13 seats
      Taxes -   10 seats

7. Jefferson's method begins with the standard divisor and quota and rounds down to the integer part. The total will generally be too low, so a smaller divisor is chosen until the total is correct.

	Gorge	Mane	Organ	Taxes	Total	
	275	767	465	383	1890	
Seats					60	
Standard divisor					31.5	
Standard quota	8.73	24.35	14.76	12.16		
Rounded quota	8	24	14	12	58	(too low)

Mane will probably get one

(a) Modified divisor   30.68   (767 ÷ 25)

Note: the modified divisor could be as low as 3.60

	Gorge	Mane	Organ	Taxes	Total
(b) Modified quotas	8.96	25.00	15.16	12.48	
Rounded quota	8	25	15	12	60

(c) Gorge -   8 seats
      Mane -   25 seats
      Organ -   15 seats
      Taxes -   12 seats

9. Jefferson's method is being used.

	A&L	Science	Engin.	Soc.Sc.	H. P.	Total	
	2540	3580	1410	1830	1050	10410	
Seats						30	
Standard divisor						**347**	
Standard quota	7.32	10.32	4.06	5.27	3.02		
Rounded quota	7	10	4	5	3	29	
						Science may get one	
Modified divisor						**325.45** (3580÷11)	
Modified quota	7.80	11.00	4.33	5.62	3.23		
Rounded quota	7	11	4	5	3	30	

Apportionment   Arts and Letters -   7 seats
               Sciences -   11 seats
               Engineering -   4 seats
               Social Sciences -   5 seats
               Human Performance -   3 seats

11. Hamilton's method is being used.

	A&L	Science	Engin.	Soc.Sc.	H. P.	Total	
	2930	3320	1290	2140	1180	10860	
Seats						30	
Standard divisor						**362**	
Standard quota	8.09	9.17	3.56	5.91	3.26		
Integer part	8	9	3	5	3	28	(2 needed)
Fractional rank			(2)	(1)			
Final Apportion	8	9	4	6	3	30	

Apportionment   Arts and Letters -   8 seats
     Sciences -   9 seats
     Engineering -   4 seats
     Social Sciences -   6 seats
     Human Performance -   3 seats

13. Hamilton's method is being used.

	Algebrion	Geometria	Analystia	Stochastica	Total	
(1000's)	892	424	664	1162	3142	
Seats					314	
Standard divisor					**10.0064**	
Standard quota	89.14	42.37	66.36	116.13		
Integer part	89	42	66	116	313	(1 needed)
Fractional rank		(1)				
Final Apportion	89	43	66	116	314	

Algebrion -   89 seats   Analystia -   66 seats
Geometria -   43 seats   Stochastica -   116 seats

15. Webster's method is being used.

	Algebrion	Geometria	Analystia	Stochastica	Total	
(1000's)	892	424	664	1162	3142	
Seats					314	
Standard divisor					**10.0064**	
Standard quota	89.14	42.37	66.36	116.13		
Rounded quota	89	42	66	116	313	too low
						Analystia may get one
Modified divisor					**9.98**	(664÷66.51)
Modified quota	89.38	42.48	66.53	116.43		
Rounded quota	89	42	67	116	314	

Algebrion -   89 seats   Analystia -   67 seats
Geometria -   42 seats   Stochastica -   116 seats

17. Jefferson's method is being used.

	Algebrion	Geometria	Analystia	Stochastica	Total
(1000's)	892	424	664	1162	3142
Seats					400
Standard divisor					**7.855**
Standard quota	113.56	53.98	84.53	147.93	
Rounded quota	113	53	84	147	397 too low

Three seats will get assigned; G and S are sure to increase enough;
Algebrion will probably get the third seat

	Algebrion	Geometria	Analystia	Stochastica	Total
Modified divisor					**7.82** (892÷114)
Modified quota	114.07	54.22	84.91	148.59	
Rounded quota	114	54	84	148	400

Algebrion -   114 seats
Geometria -   54 seats
Analystia -   84 seats
Stochastica - 148 seats

19. Hamilton's method is being used.

Algebra	Beginning	Intermediate	Advanced	Total
	130	282	188	600
Sections				20
Standard divisor				**30**
Standard quota	4.33	9.40	6.27	
Integer part	4	9	6	19 (1 needed)
Fraction rank		(1)		
Final Apportion	4	10	6	20

Beginning Algebra -   4 sections
Intermediate Algebra - 10 sections
Advanced Algebra -   6 sections

21. Webster's method is being used.

Algebra	Beginning	Intermediate	Advanced	Total
	130	282	188	600
Sections				20
Standard divisor				**30**
Standard quota	4.33	9.40	6.27	
Rounded quota	4	9	6	19 too low

Algebra	Beginning	Intermediate	Advanced	Total
Modified divisor				**29.65** (282÷9.51)
Modified quota	4.38	9.51	6.34	
Rounded quota	4	10	6	20

Beginning Algebra -   4 sections
Intermediate Algebra - 10 sections
Advanced Algebra -   6 sections

23. Hamilton's method is being used.

	A	B	C	D	E	Total
(1000's)	1592	1596	5462	1323	1087	11060
Seats						200
Standard divisor						**55.3**
Standard quota	28.79	28.86	98.77	23.92	19.66	
Integer part	28	28	98	23	19	196 (4 needed)
Fractional rank	(3)	(2)	(4)	(1)		
Final Apportion	29	29	99	24	19	200

A - 29 seats
B - 29 seats
C - 99 seats
D - 24 seats
E - 19 seats

25. Webster's method is being used.

	A	B	C	D	E	Total
(1000's)	1592	1596	5462	1323	1087	11060
Seats						200
Standard divisor						**55.3**
Standard quota	28.79	28.86	98.77	23.92	19.66	
Rounded quota	29	29	99	24	20	201 too high

C might lose one; we would need a quota of 98.49 for that to happen.

	A	B	C	D	E	Total
Modified divisor						**55.46**
Modified quota	28.71	28.78	98.49	23.86	19.60	
Rounded quota	29	29	98	24	20	200

A - 29 seats
B - 29 seats
C - 98 seats
D - 24 seats
E - 20 seats

27. Jefferson's method is being used.

Precinct	A	B	C	D	E	F	Total
	456	835	227	526	338	446	2828
Officers							180
Standard divisor							**15.71**
Standard quota	29.03	53.15	14.45	33.48	21.51	28.39	
Rounded quota	29	53	14	33	21	28	178 too low

D should increase faster than C and E

Modified divisor							**15.47** (526÷34)
Modified quota	29.48	53.98	14.67	34.00	21.85	28.83	
Rounded quota	29	53	14	34	21	28	179 too low

It should be obvious that a slightly smaller divisor will raise B's quota to 54.
You may wish to verify this.

A -   29 officers   B - 54 officers
C -   14 officers   D - 34 officers
E -   21 officers   F - 28 officers

29. Webster's method is being used.

Precinct	A	B	C	D	E	F	Total	
	456	835	227	526	338	446	2828	
Officers							180	
Standard divisor							**15.71**	
Standard quota	29.03	53.15	14.45	33.48	21.51	28.39		
Rounded quota	29	53	14	33	22	28	179	too low
Modified divisor							**15.69**	(526÷33.51)
Modified quota	29.06	53.22	14.47	33.52	21.54	28.43		
Rounded quota	29	53	14	34	22	28	180	

A - 29 officers   B - 53 officers
C - 14 officers   D - 34 officers
E - 22 officers   F - 28 officers

31. Adams' method is being used.

	A	B	C	D	E	Total	
(1000's)	1320	1515	4935	1118	1112	10000	
Seats						200	
Standard divisor						**50**	
Standard quota	26.4	30.3	98.7	22.36	22.24		
Rounded quota	27	31	99	23	23	203	too large

B will probably lose 1
Remember that Adams' Method rounds up, so B should drop below 30

Modified divisor						**50.52**	(1515÷29.99)
Modified quota	26.13	29.99	97.68	22.13	22.01		
Rounded quota	27	30	98	23	23	201	too large

    It should be obvious that a slightly larger divisor will drop E's quota below 22.
A - 27 seats
B - 30 seats
C - 98 seats
D - 23 seats
E - 22 seats

33. Adams' method is being used.

Precinct	A	B	C	D	E	F	Total	
	456	835	227	526	338	446	2828	
Officers							180	
Standard divisor							**15.71**	
Standard quota	29.03	53.15	14.45	33.48	21.51	28.39		
Rounded quota	30	54	15	34	22	29	184	too large

    Since 4 quotas must be lowered, we will arbitrarily try 16 as a larger divisor, and
make adjustments from there, if needed. It turns out we don't need to.

Modified divisor							**16**	
Modified quota	28.50	52.19	14.19	32.88	21.13	27.88		
Rounded quota	29	53	15	33	22	28	180	

A - 29 officers   B - 53 officers   C - 15 officers
D - 33 officers   E - 22 officers   F - 28 officers

35. Jefferson's method is being used.
Divisor = 33,000   Note: Standard divisor would be 34,437

State	Population	Mod. Quota	Integer
CT	236,841	7.18	7
DE	55,540	1.68	1
GA	70,835	2.15	2
KY	68,705	2.08	2
MD	278,514	8.44	8
MA	475,327	14.40	14
NH	141,822	4.30	4
NJ	179,570	5.44	5
NY	331,589	10.05	10
NC	353,523	10.71	10
PA	432,879	13.12	13
RI	68,446	2.07	2
SC	206,236	6.25	6
VT	85,533	2.59	2
VA	630,560	19.11	<u>19</u>

Total   105

37. Adams' method is being used. Because all quotas will be rounded up, a divisor larger than the standard divisor is needed to produce smaller quotas. After some trial and error, we find that d = 36,100 is a suitable divisor.

State	Population	Mod. Quota	Integer
CT	236,841	6.56	7
DE	55,540	1.54	2 *
GA	70,835	1.96	2
KY	68,705	1.90	2
MD	278,514	7.72	8
MA	475,327	13.17	14
NH	141,822	3.93	4
NJ	179,570	4.97	5
NY	331,589	9.19	10
NC	353,523	9.79	10
PA	432,879	11.99	12 *
RI	68,446	1.90	2
SC	206,236	5.71	6
VT	85,533	2.37	3 *
VA	630,560	17.47	<u>18 *</u>

Total   105

Final note: (refer to problems 34 and 35) Notice the difference when we switch from the apportionment using Jefferson's method to that using Adams method. The two larger states, Pennsylvania and Virginia lose a seat each, while two of the smaller states, Delaware and Vermont, each gain a seat. This is more evidence that Jefferson's method favors larger states, and Adams method favors smaller states. Jefferson was from Virginia; Adams was from Massachusetts.

Section 9.5   Flaws of Apportionment Methods

Goals
1. Identify the various flaws and paradoxes of apportionment methods.

Key Ideas and Questions
1. What are desirable properties of an apportionment method?
2. How does Hamilton's method fail to satisfy each of these desirable properties?
3. Do you think that Webster's method is more fair than Jefferson's? Explain.

Vocabulary
Quota Method	Alabama Paradox	New States Paradox
Divisor Method	Population Paradox	Balinski and Young's Theorem
Quota Rule		

. . . .

Overview

We will consider several fair and reasonable properties that an apportionment method should have, and then see where these properties are not satisfied. Problems can arise in several different situations:

- a reapportionment based on population changes
- a change in the total number of seats
- the addition of one or more new states.

Much of the difficulty concerns the use of quotas.

The Quota Rule

A **quota method** is any apportionment method for which each state's apportionment is either the rounded up or the rounded down standard quota. A **divisor method** is any apportionment method that requires the use of a divisor other than the standard divisor. The **quota rule** requires each state's quota to be the whole number just below or just above the state's standard quota. A quota method automatically satisfies this rule. No divisor method can always satisfy the quota rule.

The Alabama Paradox

In apportionment of the U.S. House of Representatives in the 1880's, it was discovered that adding one more seat to the House of Representatives would actually decrease the number of seats in Alabama. It was the first time this paradoxical behavior was observed in Congressional apportionment, and the paradox became known as the **Alabama Paradox**.

When the number of seats is increased, the standard quota must also increase. For a state to lose a seat under a quota method, the decrease must be because of changing from rounding up to rounding down. The seat in question must go to some other state, so *that* increase must be because of changing from rounding down to rounding up.

Population Paradox

It is possible to construct a quota method that avoids the Alabama Paradox, but there are other paradoxical situations that can arise instead. The **Population Paradox** involves two states with growing populations. When the legislature is reapportioned based on the new census, there is a transfer of a seat between the two states, but paradoxically, the faster growing state is the one that loses the seat.

New States Paradox

The **New State's Paradox** occurs when a re-calculation of the apportionment results in the change of the apportionments of some of the other states, other than the new state. It is also possible that the new state is included in the changes, but that would mean it gained a seat at the expense of another state. For example, if a state is added to a country with a legislature with 100 seats and its standard quota is calculated at 7.92, it would be assigned 7 seats and the legislature increased to 107. When the legislature is reapportioned, the new state would most likely receive 8 seats.
It gains a seat, so another state loses a seat.

Summary

There is no perfect apportionment method. There is no method that satisfies the quota rule and always avoids a paradox. You can have some of the good features such as obeying the quota rule and avoiding the paradoxes, but you cannot have all the good features all of the time, no matter what method is used. The choice of an apportionment method is ultimately a political decision.

. . . .

Suggestions and Comments for Odd-numbered problems

3. A divisor between 142 and 149 will raise one quota past the next integer.
If the divisor is less than 142, two of the quotas will increase past the next integer.

11. The quota rule is that each state's quota is a whole number just below or just above the standard quota.

13. through 19.
Make three separate apportionments and compare the results. The Alabama Paradox occurs if an increase in the size of the legislature leads to a decrease in a state's apportionment (fair share). It doesn't seem fair, does it?

21. and 23.
The Population Paradox is also a question of fairness: the state that has a faster rate of increase in population should do as well, or better, than a slower growing state. This is usually true, but if the fractional parts are very close, the opposite might happen. That's the paradox!

25. Apportion the legislature with 338 seats and the five states (including the new state) and compare to the original apportionment with only four states. Do any of the states lose (or gain) seats in the legislature?

Geometric Mean (Extended Problems)

The *mean* for two numbers usually refers to the algebraic mean, or the algebraic *average*, of the two numbers. The mean of 4 and 16 is 10. This idea of the mean treats the difference between numbers as the most important way to compare them; $10 - 4 = 6$, and $16 - 10 = 6$. In terms of differences, the mean is the average for the two numbers.

By contrast, the geometric mean treats the relative sizes of the two numbers as the most important way to compare them. For the numbers 4 and 16, the geometric mean is 8. Here, relative size is maintained: $8 \div 4 = 2$; $16 \div 8 = 2$.

Consider the numbers 2 and 102. The mean is 52, and the geometric mean is $\sqrt{2 \times 102} \approx 14.3$. For 1002 and 1102, the mean is 1052 and the geometric mean is $\sqrt{1002 \times 1102} \approx 1050.8$. For consecutive whole numbers, there is also a special feature that the fractional part of the mean is 0.5, and the fractional part of the geometric mean is greater than 0.41 and less than 0.5.

As the numbers get larger, the fractional part gets closer to 0.5. This is proved in problem 29.

29. Mathematical proof often resorts to tricks. In this case, assume that the statement is true and see what the consequences are. If you arrive at a statement that is obviously true and all the steps are reversible, then begin with this obvious statement and retrace your steps for the proof. In part (a), begin with $\sqrt{n(n+1)} < n + \frac{1}{2}$ and square both sides. Then you just keep simplifying the expressions.

In part (b), begin with $n + 0.41 < \sqrt{n(n+1)}$.

Solutions to Odd-numbered Problems

1. We are using Hamilton's method where seats are assigned based on the size of the fractional part of the standard quotas.
   With **24 seats**, the apportionment is

	Medina	Alvare	Loranne	Total	
(1000's)	530	990	2240	3760	
Seats				24	
Standard divisor				**156.67**	(1000)
Standard quota	3.38	6.32	14.30		
Integer part	3	6	14	23	(1 to be assigned)
Fraction rank	(1)				
Final Apportion	**4**	**6**	**14**	**24**	

   With **25 seats**, the apportionment is

	Medina	Alvare	Loranne	Total	
(1000's)	530	990	2240	3760	
Seats				25	
Standard divisor				**150.4**	(1000)
Standard quota	3.52	6.58	14.89		
Integer part	3	6	14	23	(2 to be assigned)
Fraction rank		(2)	(1)		
Final Apportion	**3**	**7**	**15**	**25**	

   Medina loses a seat when the legislature is increased from 24 to 25.
   The Alabama Paradox occurs.

3. We are using Jefferson's method. See Hints if review is needed.
   (a) With **24 seats**, the apportionment is

	Medina	Alvare	Loranne	Total	
(1000's)	530	990	2240	3760	
Seats				24	
Standard divisor				**156.67**	(1000)
Standard quota	3.38	6.32	14.30		
Rounded quota	**3**	**6**	**14**	**23**	(too small)

   A divisor between 142 and 149 will work.

	Medina	Alvare	Loranne	Total
Modified divisor				**145**
Modified quota	3.66	6.83	15.45	
Rounded quota	**3**	**6**	**15**	**24**

   (b) With **25 seats**, the apportionment is

	Medina	Alvare	Loranne	Total	
(1000's)	530	990	2240	3760	
Seats				25	
Standard divisor				150.4	(1000)
Standard quota	3.52	6.58	14.89		
Rounded quota	**3**	**6**	**14**	**23**	(too small)

   A divisor less than 142 will work.

	Medina	Alvare	Loranne	Total	
Modified divisor				141	(1000)
Modified quota	3.76	7.02	15.89		
Rounded quota	**3**	**7**	**15**	**25**	

   When the legislature is increased from 24 to 25, no state loses a seat.
   The Alabama Paradox does not occur.

5. We are using Hamilton's method.
   With 24 seats, the original apportionment (from problem 1) is:

	Medina	Alvare	Loranne	Total
(1000's)	530	990	2240	3760
Seats				24
Final Apportion	**4**	**6**	**14**	**24**

   With **24** seats, the apportionment based on the new population is:

	Medina	Alvare	Loranne	Total
(1000's)	680	1250	2570	4500
Seats				24
Standard divisor			187.5	(1000)
Standard quota	3.63	6.67	13.71	
Integer part	3	6	13	22 (2 to be assigned)
Fraction rank		(2)	(1)	
Final Apportion	**3**	**7**	**14**	**24**

	Population change	Original	New Apportionment
Medina	680 ÷ 530: +28%	4	3
Alvare	1250 ÷ 990: +26%	6	7
Loranne	2570 ÷ 2240: +15%	14	14

   Medina (with an increase of 28%) loses a seat, while Alvare (with an increase of 26%) gains a seat. The Population Paradox occurs.

7. We are using Hamilton's method.
   (a) With 50 seats, the apportionment is:

	A	B	C	Total
(1000's)	99	487	214	800
Seats				50
Standard divisor		(800÷50=16)		16 (1000)
Standard quota	6.19	30.44	13.38	
Integer part	6	30	13	49 (1 to be assigned)
Fraction rank		(1)		
Final Apportion	**6**	**31**	**13**	**50**

   (b) With the addition of a new state and corresponding increase in the legislature, the apportionment is

	A	B	C	D	Total
(1000's)	99	487	214	116	916
Seats					57
Standard divisor		(916÷57=16.07)			16.07 (1000)
Standard quota	6.16	30.30	13.32	7.22	
Integer part	6	30	13	7	56 (1 to assign)
Fraction rank			(1)		
Final Apportion	**6**	**30**	**14**	**7**	**57**

   State B loses a seat when the legislature is reapportioned with 57 seats.
   The New States Paradox occurs.

9. We are using Jefferson's method.

   (a) With 50 seats, the apportionment is:

	A	B	C	Total
(1000's)	99	487	214	800
Seats				50
Standard divisor				**16** (1000)
Standard quota	6.19	30.44	13.38	
Rounded quota	**6**	**30**	**13**	**49** (too low)

   A new divisor is needed that will raise the quotas so that one quota will go past the next integer. Divisors between 15.4 and 15.7 will work.

Modified divisor				**15.5** (1000)
Modified quota	6.39	31.42	13.80	
Rounded quota	**6**	**31**	**13**	**50**

   (b) With the addition of a new state and corresponding increase in the legislature, the apportionment is

	A	B	C	D	Total
(1000's)	99	487	214	116	916
Seats					57
Standard divisor					**16.07** (1000)
Standard quota	6.16	30.30	13.32	7.22	
Rounded quota	**6**	**30**	**13**	**7**	**56**

   The two largest fractional parts are about the same, and 30.30 will increase faster than 13.32. Divisors between 15.4 and 15.7 will still work.

Modified divisor					**15.5** (1000)
Modified quota	6.39	31.42	13.80	7.48	
Rounded quota	**6**	**31**	**13**	**7**	**57**

   The number of seats for the original states has not changed.
   The New States Paradox does not occur.

11. The quota rule is that each state's quota is a whole number just below or just above the standard quota. The Jefferson method uses a slightly lower divisor and rounds down. Thus the modified quotients are higher than the standard quotients, so each state's quota is at least as large as the integer part of the standard quota. However, in some cases where several seats are to be assigned, one of the quotas could be increased past a second integer. This happens because one of the populations is relatively small, and one is relatively large. As an example:

	A	B	C	Total
(1000's)	140	950	310	1400
Seats				25
Standard divisor				**56**
Standard quota	2.5	16.69	5.54	
Rounded quota	**2**	**16**	**5**	**23** (too low)

    A new divisor is needed that will raise the quotas so that two changes will occur in the integer parts. The trouble is, they both happen to the quota of the same state. With a divisor of 52, we have:

Modified divisor				**52**
Modified quota	2.69	18.27	5.96	
Rounded quota	**2**	**18**	**5**	**25**

13. Hamilton's method is being used.

	Algebrion	Geometria	Analystia	Stochastica	Total
(1000's)	892	424	664	1162	3142
For **314** seats					
Standard divisor					**10.0064**
Standard quota	89.14	42.37	66.36	116.13	
Integer part	89	42	66	116	313
Fractional rank		(1)			
Final Apportion	**89**	**43**	**66**	**116**	**314**
For **315** seats					
Standard divisor					**9.975**
Standard quota	89.42	42.51	66.57	116.49	
Integer part	89	42	66	116	313
Fractional rank		(2)	(1)		
Final Apportion	**89**	**43**	**67**	**116**	**315**
For **316** seats					
Standard divisor					**9.943**
Standard quota	89.71	42.64	66.78	116.87	
Integer part	89	42	66	116	313
Fractional rank	(3)		(2)	(1)	
Final Apportion	**90**	**42**	**67**	**117**	**316**

SUMMARY	Algebrion	Geometria	Analystia	Stochastica	Total
for 314 seats	89	43	66	116	314
for 315 seats	89	43	67	116	315
for 316 seats	90	42	67	117	316

Yes, the Alabama Paradox occurs. As the legislature is increased from 315 to 316, Geometria loses and both Algebrion and Stochastica gain.

15. Webster's method is being used.

	Algebrion	Geometria	Analystia	Stochastica	Total
(1000's)	892	424	664	1162	3142
For **314** seats					
Standard divisor					**10.0064**
Standard quota	89.14	42.37	66.36	116.13	
Rounded quota	**89**	**42**	**66**	**116**	313
Modified divisor					**9.98**
Modified quota	89.38	42.48	66.53	116.43	
Rounded quota	**89**	**42**	**67**	**116**	314
For **315** seats					
Standard divisor					**9.975**
Standard quota	89.42	42.51	66.57	116.49	
Rounded quota	**89**	**43**	**67**	**116**	315
For **316** seats					
Standard divisor					**9.943**
Standard quota	89.71	42.64	66.78	116.87	
Rounded quota	**90**	**43**	**67**	**117**	317
Modified divisor					**9.97**
Modified quota	89.47	42.53	66.60	116.55	
Rounded quota	**89**	**43**	**67**	**117**	316

SUMMARY	Algebrion	Geometria	Analystia	Stochastica	Total
for 314 seats	89	42	67	116	314
for 315 seats	89	43	67	116	315
for 316 seats	89	43	67	117	316

No, the Alabama Paradox does no occur. As the legislature is increased from 314 to 315 to 316, no state loses a seat.

17. Hamilton's method is being used.

	A	B	C	D	E	Total
(1000's)	1320	1515	4935	1118	1112	10000

For **200** seats

	A	B	C	D	E	Total
Standard divisor						**50**
Standard quota	26.40	30.30	98.70	22.36	22.24	
Integer part	26	30	98	22	22	198
Fractional rank	(2)		(1)			
Final Apportion	**27**	**30**	**99**	**22**	**22**	**200**

For **201** seats

	A	B	C	D	E	Total
Standard divisor						**49.751**
Standard quota	26.53	30.45	99.19	22.47	22.35	
Integer part	26	30	99	22	22	199
Fractional rank	(1)			(2)		
Final Apportion	**27**	**30**	**99**	**23**	**22**	**201**

For **202** seats

	A	B	C	D	E	Total
Standard divisor						**49.505**
Standard quota	26.66	30.60	99.69	22.58	22.46	
Integer part	26	30	99	22	22	199
Fractional rank	(2)	(3)	(1)			
Final Apportion	**27**	**31**	**100**	**22**	**22**	**202**

SUMMARY	A	B	C	D	E	Total
for 200 seats	27	30	99	22	22	200
for 201 seats	27	30	99	23	22	201
for 202 seats	27	31	100	22	22	202

B and C gain, thus the others lose some influence, particularly D, which loses a seat. The Alabama paradox occurs affecting D.

19. Jefferson's method is being used.

	A	B	C	D	E	Total
(1000's)	1320	1515	4935	1118	1112	10000

For **201** seats

	A	B	C	D	E	Total	
Standard divisor						**49.75**	
Standard quota	26.53	30.45	99.20	22.47	22.35		
Rounded quota	**26**	**30**	**99**	**22**	**22**	**199**	(low)
Modified divisor						**48.887**	
Modified quota	27.001	30.99	100.95	22.87	22.75		
Rounded quota	**27**	**30**	**100**	**22**	**22**	**201**	

For **202** seats

Standard divisor						**49.50**

Recall that in looking for a modified divisor for 201 seats that the additional seats weren't generated until the divisor was below 48.9. However, when it was only a bit less than 48.9, too many seats are generated.

Modified divisor (other close values will work)						**48.87**
Modified quota	27.01	31.001	100.98	22.88	22.75	
Rounded quota	**27**	**31**	**100**	**22**	**22**	**202**

SUMMARY	A	B	C	D	E	Total
for 201 seats	27	30	100	22	22	201
for 202 seats	27	31	100	22	22	202

The Alabama Paradox does not occur (it usually doesn't).

21. Hamilton's method is being used.

	A	B	C	D	E	Total
(1000's)	1320	1515	4935	1118	1112	10000
Seats						200
Standard divisor						**50**
Standard quota	26.40	30.30	98.70	22.36	22.24	
Integer part	26	30	98	22	22	198
Fractional rank	(2)		(1)			
Final Apportion	**27**	**30**	**99**	**22**	**22**	**200**

With new population figures, the apportionment is:

	A	B	C	D	E	Total
(1000's)	1370	1565	5035	1218	1212	10400
Standard divisor						**52**
Standard quota	26.35	30.10	96.83	23.42	23.31	
Integer part	26	30	96	23	23	198
Fractional rank			(1)	(2)		
Final Apportion	**26**	**30**	**97**	**24**	**23**	**200**

	Population change	Original	New Apportionment
A	1370 ÷ 1320: +3.8%	27	26
B	1565 ÷ 1515: +3.3%%	30	30
C	5035 ÷ 4935: +2.02%	99	97
D	1218 ÷ 11.18: +8.94%	22	24
E	1212 ÷ 1112: +8.99%	22	23

The Population Paradox occurs. A grew faster than B, but lost a vote.
E grew slightly faster than D, but D gained one on E.

23. Webster's method is being used.

	A	B	C	D	E	Total	
(1000's)	1320	1515	4935	1118	1112	10000	
Seats						200	
Standard divisor						**50**	
Standard quota	26.40	30.30	98.70	22.36	22.24		
Rounded quota	**26**	**30**	**99**	**22**	**22**	**199**	(low)
Modified divisor						**49.8**	
Modified quota	26.51	30.42	99.10	22.45	22.33		
Rounded quota	**27**	**30**	**99**	**22**	**22**	**200**	

With new population figures, the apportionment is:

	A	B	C	D	E	Total	
(1000's)	1370	1565	5035	1218	1212	10400	
Seats						200	
Standard divisor						**52**	
Standard quota	26.35	30.10	96.83	23.42	23.31		
Rounded quota	**26**	**30**	**97**	**23**	**23**	**199**	(low)
Modified divisor						**51.8**	
Modified quota	26.45	30.21	97.20	23.51	23.40		
Rounded quota	**26**	**30**	**97**	**24**	**23**	**200**	

	Population change	Original	New Apportionment
A	1370 ÷ 1320: +3.8%	27	26
B	1565 ÷ 1515: +3.3%%	30	30
C	5035 ÷ 4935: +2.02%	99	97
D	1218 ÷ 11.18: +8.94%	22	24
E	1212 ÷ 1112: +8.99%	22	23

The Population Paradox occurs. A grew faster than B, but lost a vote.
E grew slightly faster than D, but D gained two votes while E gained one.

25. We are using Jefferson's method. For 314 seats, the apportionment is:

	Algebrion	Geometria	Analystia	Stochastica	Total
(1000's)	892	424	664	1162	3142
Standard divisor					**10.0064**
Standard quota	89.14	42.37	66.36	116.13	
Rounded quota	89	42	66	116	313 (low)
Modified divisor	(some smaller ones will also work)				**9.93**
Modified quota	89.83	42.70	66.87	117.02	
Rounded quota	89	42	66	117	314

With the addition of Computvia to the country and 24 seats added to the legislature, we reapportion the legislature using Jefferson's method. The new legislature will have 338 seats.

	Alge.	Geom.	Analys.	Stoch.	Compu.	Total
(1000's)	892	424	664	1162	243	3385
Standard divisor						**10.015**
Standard quota	89.07	42.34	66.30	116.03	24.26	
Rounded quota	89	42	66	116	24	337 (low)
Modified divisor	(some smaller ones will also work)					**9.93**
Modified quota	89.83	42.70	66.87	117.02	24.47	
Rounded quota	89	42	66	117	24	338

SUMMARY

	314 seats	338 seats
Algebrion	89	89
Geometria	42	42
Analystia	66	66
Stochastica	117	117
Computvia	-----	24

The New States Paradox does not occur.

27. The quota is rounded up if it is greater than geometric mean.

State	Quota	Geometric Mean	Rounded Quota
A	10.47	$\sqrt{10 \times 11} = 10.488$	10
B	3.47	$\sqrt{3 \times 4} = 3.464$	4
C	5.47	$\sqrt{5 \times 6} = 5.477$	5
D	7.59	$\sqrt{7 \times 8} = 7.483$	8

29. (a) The following inequalities are equivalent and the first one is obvious.

$$0 \quad < \quad \frac{1}{4} \qquad \text{Next we add } n^2 + n \text{ to both sides.}$$

$$n^2 + n \quad < \quad n^2 + n + \frac{1}{4} \qquad \text{We factor the left side.}$$

$$n(n + 1) \quad < \quad n^2 + n + \frac{1}{4} \qquad \text{The right side can be factored.}$$

$$n(n + 1) \quad < \quad (n + \tfrac{1}{2})^2 \qquad \text{We take the square root.}$$

$$\sqrt{n(n+1)} \quad < \quad n + \frac{1}{2} \qquad \text{The fractional part is less than 0.5.}$$

(b) We begin with the assumption the statement is true.

$$n + 0.41 \quad < \quad \sqrt{n(n + 1)} \qquad \text{Square both sides.}$$

$$(n + 0.41)^2 \quad < \quad n(n + 1) \qquad \text{Then simplify both sides.}$$

$$n^2 + 0.82n + 0.1681 \quad < \quad n^2 + n \qquad \text{Subtract } n^2 + 0.82n \text{ from both sides.}$$

$$0.1681 \quad < \quad 0.18n \qquad \text{Divide by 0.18}$$

$$\frac{0.1681}{0.18} \approx 0.93 \quad < \quad n \qquad \text{This is true for } n \geq 1$$

The last statement is obviously true. Since all steps are reversible, we can begin with it and retrace the steps for the proof.

Chapter 9 Review Problems

Solutions to Odd-numbered Problems

1. Anne wins with plurality: 6 to 4 to 2.

3. It is not possible to tell who would win using the plurality with elimination method. When this method is used, all rankings are adjusted when an alternative is eliminated. The first place votes (if any) are assigned to the person ranked second. We do not know who Claire's voters will support, and if Brad gets both votes, there will be a tie.

5. In the pairwise comparison method, we look at the preferences with respect to the three possible pairings.

Alice is preferred to Bob	7 to 4
Claire is preferred to Alice	6 to 5
Bob is preferred to Claire	9 to 2

    There is no winner using the pairwise comparison method.

7. Switch any two votes of the votes that have Bob - 1, Claire - 2, Alice -3, to Claire - 1, Bob - 2, Alice - 3; or you can switch two other votes from Claire - 1, Alice - 2, Bob - 3 to Bob - 1, Alice - 2, Claire - 3.

9. (a) Largest value for $q = 23$ (unanimous)
   (b) Smallest vale for $q = 12$ (majority)
   (c) $q = 14$; P_1 not a dictator
   (d) $2^5 - 1 = 31$ coalitions

11. Any coalition with two of the first three voters is a winning coalition since the sum of the weights is at least 51. Any two voter coalition with P_4 has fewer votes and is a losing coalition; for example, the coalition with P_1 and P_4 only has 49 votes.

    The Banzhaf Power Index for this system is $\{\frac{1}{3}, \frac{1}{3}, \frac{1}{3}, 0\}$

13. In any 3-voter weighted system using a simple majority as the quota; if there is no dictator, then the power is equally shared. Suppose A and B have the larger two of the three weights. Each one, individually, has less than the quota. The winning coalitions are: {A, B}, {A, C}, and {B, C}. C is a critical voter in each of the last two coalitions. Each of the voters is a critical voter in two coalitions, so power is equally shared.

15. Hamilton's method is being used. The standard divisor and quota for each state were found in problem 14. For 40 seats, the apportionment is:

	A	B	C	D	Total	
(1000's)	275	392	611	724	2002	
Standard divisor					**50.02**	(1000)
Standard quota	5.49	7.83	12.21	14.47		
Integer part	5	7	12	14	38	(2 needed)
Fractional rank	(2)	(1)				
Final Apportion	6	8	12	14	40	

State A	6	State B	8
State C	12	State D	14

17. Jefferson's method is being used.
    For 40 seats, the apportionment is:

	A	B	C	D	Total	
(1000's)	275	392	611	724	2002	
Standard divisor					**50.02**	(1000)
Standard quota	5.49	7.83	12.21	14.47		
Rounded quota	**5**	**7**	**12**	**14**	**38**	(low)
Modified divisor					**48**	(1000)

    Note: Any divisor d with **47.001 ≤ d ≤ 48.266** can be used.

Modified quota	5.73	8.17	12.73	15.08	
Rounded quota	**5**	**8**	**12**	**15**	**40**

State A	5	State B	8
State C	12	State D	15

19. Webster's method is being used.
    For 40 seats, the apportionment is:

	A	B	C	D	Total	
(1000's)	275	392	611	724	2002	
Standard divisor					**50.02**	(1000)
Standard quota	5.49	7.83	12.21	14.47		
Rounded quota	**5**	**8**	**12**	**14**	**39**	(low)

    We want the fractional part of either the first quota or the last quota to increase to 0.5, but not both. A small change is needed.

Modified divisor					**50**	(1000)

    Note: Any divisor d with **49.932 ≤ d ≤ 50.0** can be used.

Modified quota	5.50	7.84	12.22	14.48	
Rounded quota	**6**	**8**	**12**	**14**	**40**

State A	6	State B	8
State C	12	State D	14

21. No, this is not a quota method. To be a quota method the number of seats apportioned to each state must be within 1 of the standard quota. This would be violated if D gets 16 representatives.

10 Critical Thinking, Logical Reasoning, and Problem Solving

Section 10.1 Statements and Logical Connectives

Goals
1. Learn to recognize statements and their negations as defined in logic.
2. Learn about the logical connectives "and," "or," "if...,then," and "...if and only if..." and how they are used in reasoning.
3. Learn about the various forms of "if..., then" statements and how they are used in constructing valid arguments.

Key Ideas and Questions
1. Describe what is meant by a "statement" and how statements may be modified or combined using logical connectives to make new statements.
2. Determine the truth of these combined statements using truth tables.

Vocabulary

Statement	Disjunction (or)	Consequent
Truth Table	Implication	Converse
Truth Values	Conditional	Contrapositive
Compound Statements	Hypothesis	Inverse
Logical Connectives	Antecedent	Logically Equivalent
Conjunction (and)	Conclusion	Biconditional

• •

Overview

We begin the study of mathematical reasoning and problem solving by defining what is meant by a statement and the means by which statements are analyzed and combined.

Statements

In studying logic, the declarative sentence is the most important type of sentence. Declarative sentences that can be classified as true or false are called **statements**. If a sentence cannot be determined to be true or false, then it is not a **statement** as defined in logic. Examples of sentences that are not statements include subjective sentences, exclamations, questions, or paradoxes--sentences that are neither true nor false. Note that it is not necessary for you to know whether the statement is true or false. You only need to be able to determine that it *must* be either true or false.

Examples of **statements**:
1. Based on population, Alaska is the largest state of the U.S. (false)
2. Gasoline is flammable. (true)

Examples of sentences that are **not** statements as defined in logic:
1. Alaska is the most beautiful state in the U. S. (subjective)
2. Are you ready? (a question)
3. The sun in France. (not a sentence)
4. This sentence is false. (neither true nor false; a paradox)

Statements are represented by lower-case letters such as p, q, r and s. For example, the statement "The sun rises in the east" can be represented by p. The statement p can either be True, indicated by T, or False, indicated by F. We call these the possible **truth values** of the statement p. We determine that, since the sun does indeed rise in the east, p is True. The **negation** of p, written as $\sim p$ and read as "not p", is the statement: "The sun does not rise in the east." That is, we usually negate the *verb*. We also know that $\sim p$ is False. When a statement is True, its negation is False; when a statement is False, its negation is True.

Truth Tables

A **truth table** is an organizing device that summarizes and presents the truth values of statements. The **truth table** that summarizes the relationship between a statement and its negation looks like this:

p	$\sim p$
T	F
F	T

Logical Connectives

Two or more statements can be connected to form **compound statements** by using the four common **logical connectives**: "and," "or," "if-then," and "if and only if."

And

If p is the statement "It is snowing" and q is the statement "The moon is shining," then the **conjunction** of p and q is the statement "It is snowing **and** the moon is shining." The conjunction is represented symbolically as "$p \wedge q$". The conjunction of two statements p and q is true only when both p and q are true. The statement "It is snowing and the moon is shining" is only true when "It is snowing" is true and "The moon is shining" is true.

Since the two statements p and q each have two possible truth values--True (T) and False (F)--there are four possible combinations of T and F to consider for both statements.

This truth table displays the truth values of p and q with corresponding values of $p \wedge q$.

p	q	$p \wedge q$
T	T	T
T	F	F
F	T	F
F	F	F

From the truth table for the conjunction it is easy to see that the conjunction of two statements is true only when both statements are true, and false when at least one of the statements is false.

Or

The **disjunction** of statements p and q is the statement "p or q," which is represented symbolically as "$p \vee q$." There are two common uses of "or". The **exclusive "or"** is used when either p is true or q is true, but both cannot be true simultaneously. The statement "Bill will stay or Bill will not stay" is an example of the exclusive "or", since either "Bill will stay" is true, or "Bill will not stay" is true, but both cannot be true at the same time.

The **inclusive "or"** (often referred to as "and/or") includes the possibility that both parts are true. For example, the statement "It will snow or the moon will shine" uses the inclusive "or." The statement is true (1) if it snows, or (2) if the moon shines, or (3) if it snows *and* the moon shines. The inclusive "or" in $p \vee q$ allows for both p and q to be true. It is agreed that in the study of mathematics, the disjunction "p or q", represented as $p \vee q$ uses the **inclusive "or"**. The following truth table summarizes the truth values for the inclusive "or".

p	q	$p \vee q$
T	T	T
T	F	T
F	T	T
F	F	F

From this truth table it is clear that the disjunction of two statements is False only when both statements are False. It also clearly shows that the disjunction is True if at least one of the statements is True.

If-Then

The **conditional statement** is one of the most important compound statements. The statement "If p, then q," represented by "$p \Rightarrow q$", is called a **conditional statement**, or **implication**. The statement p is called the **hypothesis** or **antecedent**. The statement q is called the **conclusion** or **consequent**.

The following truth table gives the truth values for the conditional $p \Rightarrow q$.

p	q	$p \Rightarrow q$
T	T	T
T	F	F
F	T	T
F	F	T

Notice that the only time the conditional is False is when the hypothesis is True and the conclusion is False. Using the alternative terminology, the conditional is False only when the antecedent is True and the consequent is False. Think of the conditional statement as a promise. When you meet the conditions in the hypothesis (p is True) and the consequence doesn't occur (q is False), then the conditional statement is False.

Related Conditionals

Three common forms derived from the conditional and related to it are: the inverse, the converse and the contrapositive.

- The **inverse**: If not p, then not q. The inverse negates the hypothesis and the conclusion. The inverse can also be written $\sim p \Rightarrow \sim q$.

- The **converse**: If q, then p; $q \Rightarrow p$. The converse interchanges the antecedent and consequent of the original conditional. The truth of a converse may or may not agree with the truth of the conditional itself.

- The **contrapositive**: If not q, then not p. The contrapositive interchanges the hypothesis and the conclusion of the original conditional, and it also negates each. The contrapositive can also be written $\sim q \Rightarrow \sim p$.

The table below shows the truth values for the conditional and its related conditionals. Remember that any conditional is false only when the hypothesis is true and the conclusion is false.

Notice that the truth table for the conditional and its contrapositive are the same. Similarly, the converse and inverse have the same truth table. When two statements have the same truth tables, they are considered to be **logically equivalent**. Logically equivalent conditionals are helpful in solving certain problems. Often, mathematical statements in the form of conditionals are proved by using the contrapositive.

				Conditional	Contrapositive	Converse	Inverse
p	q	$\sim p$	$\sim q$	$p \Rightarrow q$	$\sim q \Rightarrow \sim p$	$q \Rightarrow p$	$\sim p \Rightarrow \sim q$
T	T	F	F	T	T	T	T
T	F	F	T	F	F	T	T
F	T	T	F	T	T	F	F
F	F	T	T	T	T	T	T

If and Only If

Another common form derived from the conditional, $p \Rightarrow q$, is the **biconditional** "p if, and only if, q." A biconditional is true when both its hypothesis and conclusion are true, or both are false. The biconditional is "p if and only if q", written $p \Leftrightarrow q$. It is the conjunction of $p \Rightarrow q$ and its converse $q \Rightarrow p$. The truth table for $p \Leftrightarrow q$ is below. Notice that $p \Leftrightarrow q$ is logically equivalent to $(p \Rightarrow q) \wedge (q \Rightarrow p)$.

p	q	$p \Rightarrow q$	$q \Rightarrow p$	$(p \Rightarrow q) \wedge (q \Rightarrow p)$	$p \Leftrightarrow q$
T	T	T	T	T	T
T	F	F	T	F	F
F	T	T	F	F	F
F	F	T	T	T	T

The above truth table shows that the biconditional, $p \Leftrightarrow q$, is true only when p and q have the same truth values. This agrees with the definition as given above: a biconditional is true when both its hypothesis and conclusion are true, or both are false.

· ·

Suggestions and Comments for Odd-numbered problems

33. Since there are two statements, p and q, the truth table will have four lines for truth values. Do the work very systematically.

35. (a) Make separate columns for p $\Rightarrow$ q and q $\Rightarrow$ p.
    (b) Make separate columns for p $\Rightarrow$ q, and then p $\wedge$ (p $\Rightarrow$ q).

41. The language may seem stilted and artificial when it is originally translated.

43. Make a very literal translation of the sentences before changing their form. Remember that if both sentences are not true then at least one of them is false. Conversely if at least one of them is not true, then both of them are false.

45. Apply the laws sequentially; one step at a time.

Solutions to Odd-numbered Problems

1. (a) This is not a statement (it is a question)
   (b) This is a statement (we need more information to know if it is true or false, but it is one or the other)
   (c) This is not a statement (it's an exclamation)
   (d) This is a statement (and it is true)

3. **(a)** This is a statement (although it is false)
   (b) This is not a statement (it's a question)
   (c) This is not a statement (it's a fragment)
   (d) This is a statement (and a true one)

5. **(a)** "A kitten is not teachable." or "No kitten is teachable."
   (b) "A kitten does not have whiskers." or "No kitten has whiskers."

7. **(a)** Phil did not buy a new car.
   (b) The weather is not sunny.

9. **(a)** False
   (b) False
   (c) True (when the antecedent is false, the conditional is true)

11. **(a)** True (a fact of history)
    (b) True (when the antecedent is false, the conditional is true)
    (c) True (if either statement in the disjunction is true, so is the disjunction)

13. **(a)** Roses are red and the sky is blue.
    (b) Roses are red, and either the sky is blue or turtles are green.
    (c) If the sky is blue, then roses are red and turtles are green.
    (d) If turtles are not green and if the sky is not blue, then roses are not red.

15. **(a)** $p \Rightarrow \sim q$
    (b) $\sim q \Rightarrow \sim r$

17. **(a)** $q \Rightarrow \sim p$
    (b) $q \wedge r$
    (c) $r \Leftrightarrow (q \wedge p)$
    (d) $p \vee q$

19. **(a)** $\sim p \Rightarrow q$ or $p \vee q$
    (b) $\sim p \wedge \sim q$

21. **(a)** $p \Rightarrow q$ or $\sim q \Rightarrow \sim p$
    (b) $\sim p \vee \sim q$

23. **(a)** True   (p is true and $\sim q$ is true)
    (b) False   ($p \vee q$ is true, since p is true)
    (c) True   ($\sim p$ is false, so the conditional is true)
    (d) True   ($\sim p$ is false, so $\sim p \wedge r$ is false)

25. **(a)** False   ($\sim q$ is false, so $p \wedge \sim q$ is false)
    (b) False   (p and q are both true, so $p \Leftrightarrow q$ is true)
    (c) True   ($\sim p$ is false, so the conditional is true)
    (d) False   ($\sim p$ is false, so $\sim p \wedge r$ is false; but q is true)

27. (a) True   (if p ⇒ q is false, then p is true and q is false)
   (b) False   (p and q are not both true)
   (c) False   (p and q have different truth values)
   (d) True   (~q is true, and so is p)

29. (a) Antecedent (hypothesis): 'the weather is good'
   Consequent (conclusion): 'we'll go to the game'

   Converse: If we'll go to the game, the weather will be good.
   Inverse: If the weather is not good, we will not go to the game.
   Contrapositive: If we do not go to the game, the weather will not be good.

   (b) Antecedent (hypothesis): 'I don't go to the movie'
   Consequent (conclusion): 'I'll study my math'

   Converse: If I study my math, I won't go to the movie.
   Inverse: If I go to the movie, I will not study my math.
   Contrapositive: If I don't study my math, I'll go to the movie.

   (c) Antecedent (hypothesis): 'I'll get an A on the final'
   Consequent (conclusion): 'I'll get an A for the course'

   Converse: If I get an A for the course, then I'll get an A on the final.
   Inverse: If I don't get an A on the final,
            then I won't get an A for the course.
   Contrapositive: If I don't get an A for the course,
            then I won't get an A on the final.

31. (a) Antecedent (hypothesis): 'You don't have gasoline in the tank'
   Consequent (conclusion): 'Your car won't start'

   Converse: If your car won't start, then you don't have gasoline in the tank.
   Inverse: If you have gasoline in the tank, then your car will start.
   Contrapositive: If your car will start, then you have gasoline in the tank.

   (b) Antecedent (hypothesis): 'I can pass this class'
   Consequent (conclusion): 'I will graduate'

   Converse: If I graduate, I will pass this class.
   Inverse: If I cannot pass this class, then I will not graduate.
   Contrapositive: If I do not graduate, then I cannot pass this class.

33.

p	q	$(\sim p) \vee (\sim q)$	$(\sim p) \vee q$	$\sim(p \wedge q)$
T	T	F	T	F
T	F	T	F	T
F	T	T	T	T
F	F	T	T	T

Because their columns have the same truth values,
$(\sim p) \vee (\sim q)$ and $\sim(p \wedge q)$ are logically equivalent.

35. (a)

p	q	$p \Rightarrow q$	$q \Rightarrow p$	$(p \Rightarrow q) \Rightarrow (q \Rightarrow p)$
T	T	T	T	T
T	F	F	T	T
F	T	T	F	F
F	F	T	T	T

Therefore $(p \Rightarrow q) \Rightarrow (q \Rightarrow p)$ is not always true.

(b)

p	q	$p \Rightarrow q$	$p \wedge (p \Rightarrow q)$	$[p \wedge (p \Rightarrow q)] \Rightarrow q$
T	T	T	T	T
T	F	F	F	T
F	T	T	F	T
F	F	T	F	T

Therefore $[p \wedge (p \Rightarrow q)] \Rightarrow q$ is always true.

37. (a) True   (since p is false, the conditional is true in any case)
   (b) True   (a is true, so $p \vee a$ is true; the conditional is true)
   (c) Unknown (if x is true, the conditional is false; if x is false, the conditional is true)
   (d) True   (a is true, so $a \vee p$ is true)
   (e) False   (the antecedent $a \vee x$ is true, but $p \wedge b$ is false since p is false; therefore, the conditional is false)
   (f) Unknown (the antecedent b is true; if x is true, $b \Rightarrow x$ is true, if x is false, $b \Rightarrow x$ is false)

39. (a) True   (a fact of geometry about triangles)
   (b) False (fish can swim is true; pigs can fly is false)
   (c) False (if a quadrilateral has three right angles, the remaining angle is also a right angle, but the quadrilateral could be a rectangle, and not necessarily a square)

41. (a) "It is not the case that both p and q occur" is equivalent to "either p does not occur or q does not occur."

   "It is not the case that either of p or q occur" is equivalent to "both p does not occur and q does not occur."

   (b) "It is not the case that the moon is dark and the night is cold" has the same meaning as "either the moon is not dark or the night is not cold."

   "It is not the case that either the moon is dark or the night is cold" has the same meaning as "the moon is not dark and the night is not cold."

43. (a) $\sim(p \wedge q)$: It is not the case that unemployment is low and the stock market is going up.
   $\sim(p \vee q)$: It is not the case that unemployment is low or the stock market is going up.
   (b) $\sim(p \wedge q) = \sim p \vee \sim q$: Unemployment is not low, or the stock market is not going up.
   $\sim(p \vee q) = \sim p \wedge \sim q$: Unemployment is not low, and the stock market is not going up.

45. (a) $\sim(p \wedge \sim q) = \sim p \vee \sim(\sim q) = \sim p \vee q$

(b) $\sim(\sim p \vee \sim q) = \sim(\sim p) \wedge \sim(\sim q) = p \wedge q$

47. (a) $p \wedge (q \vee r)$: I plan to work next summer, and I'll enroll for classes in the fall or travel to Asia in the spring.
$(p \wedge q) \vee (p \wedge r)$: I plan to work next summer and enroll for classes in the fall, or I will work next summer and travel to Asia in the spring.

(b) $p \vee (q \wedge r)$: I plan to work next summer or I will enroll for classes next fall and travel to Asia in the spring.
$(p \vee q) \wedge (p \vee r)$: I plan to work next summer or enroll in classes in the fall and I plan to work next summer or travel to Asia in the spring.

Section 10.2 Deduction

Goals

1. Learn to identify valid arguments.
2. Learn to recognize the common forms of fallacies.

Key Ideas and Questions

1. What are the four main argument forms?
2. What are two common forms of argument that are fallacies?

Vocabulary

Argument	Modus Ponens	Disjunctive Syllogism
Rhetoric	Law of Detachment	Argument by Contradiction
Premise	Chain Rule	Indirect Reasoning
Hypothesis	Modus Tollens	Fallacy of Affirming the Consequent
Conclusion	Law of Contraposition	Fallacy of Denying the Antecedent
Valid Deductive Argument	Denying the Consequent	

. . . .

Overview

In this section we consider the basic forms for valid arguments and two types of fallacies (invalid argument forms).

Logical Arguments

An **argument** is a sequence of statements used to persuade someone that some other statement is true. The art of using language to persuade is known as **rhetoric**. In mathematics, persuasion is by means of reason alone. Deductive reasoning proceeds step-by-step from **premises** or **hypotheses** (assumed statements) to a statement called the **conclusion.**

Premises:	If you get 90 or more on the test, you'll get an A.
	You get a 93 on the test
Conclusion:	Therefore, you'll get an A.

The premises usually contain one or more conditionals, and a statement related to one of them. It is often difficult to follow the logic being used in arguments with words, so symbolic notation is used instead. This allows the words to be translated and the logical structure to be revealed.

We look at three valid argument forms involving conditional statements. Each form can be shown to be a valid argument by using truth tables. A **valid deductive argument** is one in which the conclusion *must* be true whenever the premises are true.

Valid Argument Form #1: Modus Ponens.

Modus Ponens		
	Verbally	Symbolically
Premises	if p, then q	$p \Rightarrow q$
	p	p
Conclusion	Therefore, q	$\therefore q$

Modus Ponens is a form of argument that says whenever a conditional statement and its hypotheses are true, the conclusion is also true.

Premises: If you finish the marathon, then I'll buy you dinner.
            You finish the marathon.

Conclusion: I'll buy you dinner.

The following Truth Table shows that Modus Ponens is a valid argument.

		Premises		Conclusion	
p	q	$p \Rightarrow q$	p	q	
T	T	**T**	**T**	**T**	♦ Premises are all true
T	F	F	T	F	
F	T	T	F	T	
F	F	T	F	F	

Valid Argument Form #2: Modus Tollens

Modus Tollens		
	Verbally	Symbolically
Premises	If p, then q.	$p \Rightarrow q$
	Not q.	$\sim q$
Conclusion	Therefore, not p.	$\therefore \sim p$

Modus Tollens is the form of argument that says whenever a conditional statement is true and its conclusion is false, then the hypothesis is also false.

An argument using Modus Tollens:

Premises: If Jamie sells two cars, then she gets a bonus.
Jamie does not get a bonus.
Conclusion: Therefore, Jamie did not sell two cars.

The Truth Table below shows that the conclusion is true when all the premises are true, so Modus Tollens is a valid argument form.

		Premises		Conclusion	
p	q	$p \Rightarrow q$	$\sim q$	$\sim p$	
T	T	T	F	F	
T	F	F	T	F	
F	T	T	F	T	
F	F	**T**	**T**	**T**	♦ Premises are all true

We consider these two forms together since Modus Ponens and Modus Tollens are logically equivalent. Modus Ponens uses the conditional, $p \Rightarrow q$, while Modus Tollens uses the contrapositive, $\sim q \Rightarrow \sim p$.

In Section 10.1 we saw that these two conditionals were equivalent. In terms of mathematical proofs, Modus Tollens is referred to as "indirect" proof. In order to prove that q must be true when p is true, we assume q is *not* true and show that this forces a conclusion that p is not true also - a *contradiction*.

Valid Argument Form #3: The Chain Rule

	Chain Rule	
	Verbally	Symbolically
Premises	If p, then q.	$p \Rightarrow q$
	If q, then r.	$q \Rightarrow r$
Conclusion	Therefore, if p, then r.	$\therefore\ p \Rightarrow r$

Notice that the Chain Rule has a conclusion that is a conditional. The chain rule is stated with only two conditionals, but like a real chain, there can be any number of "links" connecting the first and last statements together.

The validity of the Chain Rule can also be checked by using a Truth Table. Again, if the argument is to be valid the conclusion must be true on every row where all the premises are true.

Truth Table for the Chain Rule

			Premises		Conclusion	
p	q	r	$p \Rightarrow q$	$q \Rightarrow r$	$p \Rightarrow r$	
T	T	T	T	T	T	♦ Premises all true
T	T	F	T	F	F	
T	F	T	F	T	T	
T	F	F	F	T	F	
F	T	T	T	T	T	♦ Premises all true
F	T	F	T	F	T	
F	F	T	T	T	T	♦ Premises all true
F	F	F	T	T	T	♦ Premises all true

Another Valid Argument Form is the **Disjunctive Syllogism**, or Law of the Excluded
Middle. The form arose from the principle of logic requiring each sentence to be either
True or False. So, if the two possibilities are True or False, and one of them is ruled
out, then the other one must hold.

	Disjunctive Syllogism		
	Verbally	Symbolically	
Premises	p or q	$p \vee q$	
	Not q	$\sim q$	Conclusion
Therefore, p		$\therefore p$	

Premises:	Sean earns a commission or Sean earns a salary.
	Sean does not earn a salary.
Conclusion:	Therefore, Sean earns a commission.

The following Truth Table shows the validity of the Disjunctive Syllogism.

		Premises		Conclusion	
p	q	$p \vee q$	$\sim q$	p	
T	T	T	F	T	
T	F	T	T	T	♦ Premises are all true
F	T	T	F	F	
F	F	F	T	F	

Remember: In the inclusive "or", the disjunction is True if at least one of the
statements is True. The disjunction is False only when both statements are
False.

Incorrect Reasoning Patterns

Sometimes arguments are invalid (*not* valid). Such arguments are called logical fallacies. As with other arguments, the lack of validity can be shown by using a Truth Table. When the Truth Table shows all premises to be True, but the conclusion to be False, the argument is not valid.

Fallacy of Affirming the Consequent		
	Verbally	Symbolically
Premises	If p, then q	$p \Rightarrow q$
	q	q
Conclusion	Therefore, p.	$\therefore p$

Premises: If John watches baseball, then he drinks beer.
John is drinking beer.
Conclusion: Therefore, John is watching baseball.

A Truth Table shows that there is a row where the conclusion is false, even though all the premises are true. So, the argument is not valid.

		Premises		Conclusion
p	q	$p \Rightarrow q$	q	p
T	T	T	T	T
T	F	T	F	T
F	T	T	T	F
F	F	F	F	F

◆ Premises are true;
conclusion is false.

The second fallacy involves a negation of one of the premises, although the conclusion is still true.

Fallacy of Denying the Antecedent		
	Verbally	Symbolically
Premises	If p, then q	$p \Rightarrow q$
	not p	$\sim p$
Conclusion	Therefore, not q.	$\therefore \sim q$

Premises: If Sam eats garlic pizza, then he has bad breath.
Sam does not eat garlic pizza.
Conclusion: Therefore, Sam does not have bad breath.

This fallacy negates the hypothesis and conclusion, but it is shown to be invalid by the following Truth Table.

Truth Table for the Fallacy of Denying the Antecedent.

			Premises		Conclusion
p	q		$p \Rightarrow q$	$\sim p$	$\sim q$
T	T		T	F	F
T	F		F	F	T
F	T		T	T	**F** ◆ Premises are all true;
F	F		T	T	T conclusion is false.

Since this Truth Table displays a row where all the premises result in a false conclusion, the argument is invalid.

The Fallacy of Denying the Antecedent is the contrapositive of the Fallacy of Affirming the Consequent. They are logically equivalent.

• •

Suggestions and Comments for Odd-numbered problems

1. (b) Are there any unstated premises?

19. The first two statements are conditionals that can be combined as a single conditional by means of the Chain Rule.

21. There is an unstated premise.

23. "When I ride the bus I am always late" is a statement equivalent to the conditional, "If I ride the bus, then I will be late."

27. Rearrange the conditionals to link up the statements they have in common.

35. through **39.**
Paraphrase the statements, if necessary, to put them in to a more standard form.

41. There is an unstated premise: two lines that are not parallel meet to form an angle.
What would be the measure of the angle in this situation?

Solutions to Odd-numbered Problems

1. **(a)** Premises: If the room is warm, then I'll be uncomfortable.
                          The room is warm.
       Conclusion: I'll be uncomfortable.

(b) Premises: If the weather is bad, I'll go to the movies.
                          It's raining heavily.
             Unstated premise: Heavy rain is bad weather.
       Conclusion: I'll go to the movies.

3. **(a)** Premises: If the weather is good, Barry will paint the house.
                         Barry didn't paint the house.
   Conclusion: The weather was not good.

   (b) Premises: If your average is at least 90% on the tests, you'll get an A.
   You did not get an A.
                         Conclusion: You did not average at least 90% on the tests..

5. $p \Rightarrow q$
   q
   $\therefore p$
   Invalid; this is the fallacy of Affirming the Consequent.

7. $p \Rightarrow q$
   p
   $\therefore q$
   Valid; this is Modus Ponens.

9. $p \Rightarrow q$
   $q \Rightarrow r$
   $\therefore p \Rightarrow r$
   Valid; this is the Chain Rule.

11. $p \Rightarrow q$
    $\sim p$
    $\therefore \sim q$
    Invalid; this is the fallacy of Denying the Antecedent.

13. $p \Rightarrow q$
    $\sim q$
    $\therefore \sim p$
    Valid; this is Modus Tollens.

15. $p \Rightarrow q$
    $\sim p$ (p = I can't go to the movie; $\sim p$ = I can go to the movie)
    $\therefore \sim q$
    Invalid; this is the fallacy of Denying the Antecedent.

17. $p \Rightarrow q$
    $\sim p$
    $\therefore \sim q$
    Invalid; this is the fallacy of Denying the Antecedent.

19. This is an application of the Chain rule followed by Modus Ponens.

21. Modus Ponens.

23. This is an application of the Chain rule followed by Modus Ponens.

25. This is the application of the Disjunctive Syllogism.

27. The original argument is given on the left.

$\sim p \Rightarrow \sim q$ This $u \Rightarrow \sim p$

$\sim r \Rightarrow \sim s$ can be $\sim p \Rightarrow \sim q$

$\sim q \Rightarrow \sim t$ arranged $\sim q \Rightarrow \sim t$

$u \Rightarrow \sim p$ as $\sim t \Rightarrow \sim r$

$\sim t \Rightarrow \sim r$ $\sim r \Rightarrow \sim s$

Conclusion: $u \Rightarrow \sim s$: If a kitten has green eyes (u) , then it (a kitten) will *not* play with a gorilla ($\sim s$).

This a repeated application of the Chain Rule.

29. Conclusion: You did not do your assignments.
Modus Tollens.

31. Conclusion: Michelle finishes her assignment.
Disjunctive Syllogism.

33. Conclusion: Gunder is going to the technical university.
Modus Ponens.

35. The form of the argument is

$p \Rightarrow q$ If it is snowing (p), the bus will be late (q).

q The bus was late (q).

$\therefore p$ It was snowing (p).

Invalid; this is the fallacy of Affirming the Consequent

37. The form of the argument is

$(p \wedge r) \Rightarrow q$ If Tonya works hard (p) and has a good attitude (r) she will be promoted (q).

$p \wedge \sim r$ [$(p \wedge r)$ is F]   Tonya worked hard but had a bad attitude

$\therefore \sim q$ Tonya didn't get promoted.

Invalid; this is the fallacy of Denying the Antecedent

39. The form of the argument is

$p \Rightarrow q$ If we take a class from Smyth, we have to do a term paper.

$\sim p$ We didn't take a class from Smyth.

$\therefore \sim q$ We won't have to do a term paper.

Invalid; this is the fallacy of Denying the Antecedent

41. Fact: If two lines cross another line and if they are not parallel then the three lines form a triangle.

p = "two lines are perpendicular to another line"
q = "two lines are parallel";
unstated premise: "If two lines are not parallel, the two lines will meet (r)".
$(p \wedge {\sim}q) \Rightarrow r$ (This means there will be a triangle formed.)

Fact: The sum of the angles of a triangle is $180°$. Two of the angles are $90°$ so the angle made by the other angle is $0°$. Therefore the two perpendicular lines are the same line and not part of a triangle.
This is a contradiction, so the lines must be parallel.

${\sim}r \Rightarrow {\sim}(p \wedge {\sim}q) \Leftrightarrow {\sim}p \vee {\sim}({\sim}q) \Leftrightarrow {\sim}p \vee q$

Since ${\sim}p$ is false, q must be true.

Section 10.3   Categorical Syllogisms

Goals
1. Learn to use a graphical form to determine the validity of arguments involving categories.

Key Ideas and Questions
1. What are Euler diagrams used for?
2. How are Euler diagrams drawn and used in practice?

Vocabulary

Euler Diagram	Conclusion	Invalid
Syllogism	Categorical Syllogism	Category Fallacy
Premise	Valid	

. . . .

Overview

In this section we use a graphical format and display to test the validity of arguments that deal with categories of "objects" and the relationships between individual objects and the categories.

Euler Diagrams

There is another form of argument called a syllogism that can be analyzed using graphical forms called **Euler Diagrams**. Certain sentences have quantifiers--words such as "all," "some," "every," and equivalents. These types of sentences are true depending upon whether or not some or all things in one category are also in another. An Euler Diagram uses circles to represent interrelationships among statements.

Six types of sentences and the Euler Diagrams that show their truth are presented below:

1. All X are Y. ("All humans (X) are mammals (Y).")

Notice that all of X is included in Y. There is the possibility that some elements in Y are not elements of X.

2. No X are Y. ("No humans (X) have beaks (Y).")

In this case, we separate the two circles as an indication that they share no elements in common.

3. Some X are Y. ("Some pickups (X) are 4-Wheel Drive (Y).")

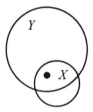

The overlapping of the two circles indicates that the two categories have some elements in common, but we put a dot in the region to make it clear that there is at least one.

4. Some X are not Y. ("Some cars (X) are not All Wheel Drive (Y).")

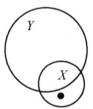

The dot that is in X but not in Y shows that there is at least one element in X that is not in Y.

5. x is a Y--where x is one element. ("My pickup (x) is a 4-Wheel Drive.")

The dot represents the specific person or thing, x.

6. x is not a Y. ("My car (x) is not All-Wheel Drive (Y).")

Valid Category Arguments

A syllogism is an argument that has two statements called **premises** followed by one statement called the **conclusion**. A syllogism with statements about categories is called a **categorical syllogism**.

A syllogism is **valid** if whenever the two premises are true, the conclusion must also be true. To assess the validity of a categorical syllogism check the truth of the premises by making Euler Diagrams of each one. When these two figures are combined and the conclusion is true, then the syllogism is valid.

Sometimes an argument involving categories has more than two premises, and sometimes you may need to draw more than one Euler Diagram. Remember! An argument is valid if the conclusion must be true whenever the premises are true. The fact that the conclusion is true is no assurance that the argument is valid. Truth and validity are different concepts.

Invalid Category Arguments

A categorical syllogism is invalid (*not* valid) if the conclusion is not guaranteed by the truth of the premises. If an argument with categories is not valid, it only takes one Euler diagram to prove it. Any single Euler Diagram in which the premises are true and the conclusion is false will show the syllogism is invalid.

Any invalid argument involving categories is called a **Category Fallacy**. It is often quite easy for such fallacies to slip by you, especially in everyday speech or writing. Whenever you can, take time to sketch an Euler diagram of the situation. If you do this, then you should be able to see the category fallacies. It is very important to consider all possible relationships among the categories that are consistent with the premises before arriving at any conclusions.

· ·

Suggestions and Comments for Odd-numbered problems

5. Draw an Euler diagram for each of the premises and combine the diagrams.

9. Draw Euler diagrams for the premises. Be as general as possible. Are you forced to make the conclusion?

15. Don't make any assumptions regarding names. Just satisfy the premises.

21. and 23.
There are three premises. Draw an Euler diagram for each and combine the drawings. Don't make a conclusion unless you are forced to do so.

35. Draw an Euler diagram for each premise. If Kareem is a sophomore, then he is a "freshman or sophomore".

39. If Tony is short, then he is not tall. If he is not "big or tall", then he is "not big and not tall". The ideas of statements, connectives, and negation carry over from Section 1.1.

Solutions to Odd-numbered Problems

1. **(a)** True   **(b)** True   **(c)** False
   (d) False   **(e)** False

3. **(a)** True   **(b)** False   **(c)** True
   (d) True   **(e)** False   **(f)** True

5. The argument is valid.

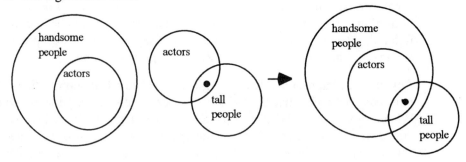

7. The argument is valid.

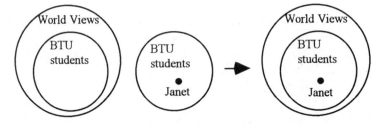

9. The argument is not valid

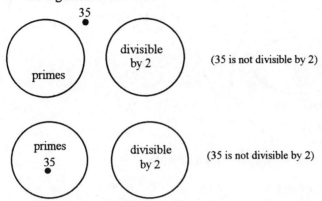

(35 is not divisible by 2)

(35 is not divisible by 2)

Both Euler diagrams above are consistent with the statements. One of them has 35 not a prime. Notice that these statements are not all true.

11. The argument is valid

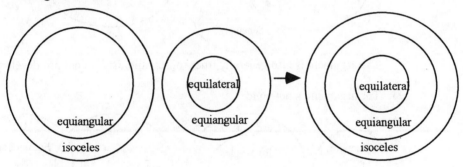

13. The argument is valid.

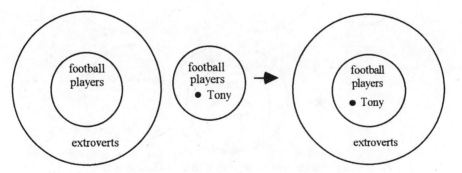

15. The argument is not valid; Sam Jones may be a woman (Samantha?)

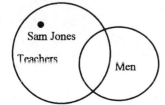

17. The argument is not valid

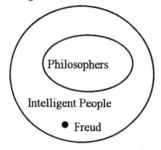

The Venn diagram satisfies the given conditions.

19. The argument is valid.

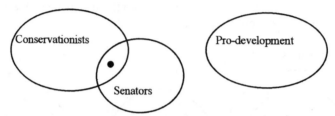

The argument is also generally true; conservationists are not pro-development.

21. The argument is not valid.

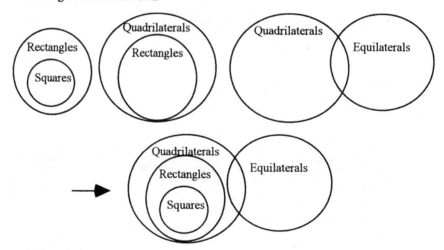

Although the argument is not valid, the conclusion is true.

23. The argument is not valid. The conclusion, however, is probably true.

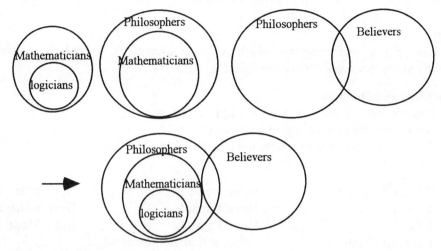

25. Joe Montana is strong.

27. Some extraterrestrials are not carbon based life forms.

29. My roommate is not sane.

31. There are two possible Euler diagrams. There is no valid conclusion.

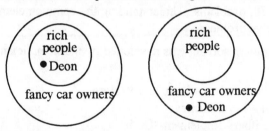

33. Some economists are not trustworthy.

35. Kareem must buy his own lock.

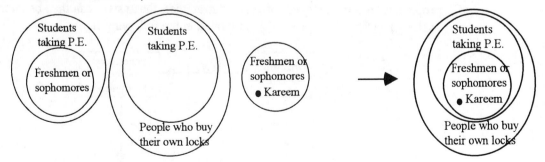

37. Jack is neither a composer nor a musician.

39. Tony is not big.

Section 10.4 Problem Solving

Goals
1. Learn a problem solving framework as well as strategies that can be used throughout the book and in life to solve a variety of problems.

Key Ideas and Questions
1. Describe Pòlya's four step process of problem solving.
2. List at least six problem-solving strategies.
3. Illustrate how various strategies are used to solve a problem.

Vocabulary

Pòlya's Four Steps	Inductive Reasoning	Draw a Graph
Strategy	Use a Variable	Draw a Diagram
Clues	Use a Formula	Use a Model
Guess and Test	Solve an Equation/Inequality	
Look for a Pattern	Draw a Picture	

. . . .

Overview

Problems are encountered by all kinds of people in all kinds of situations. Even though there are diverse problems, there are common elements and an underlying structure for most, if not all, problems. Understanding the common elements and the structure will help you solve problems.

The 4-step process for problem solving was developed by mathematician George Polya. The four steps are:

Step 1: Understand the Problem - State and Restate
Step 2: Devise a Plan - Try a Strategy
Step 3: Carry out the Plan - Implement--Do It!
Step 4: Look Back - Feedback--Check It!

Problems are usually stated in words. It is helpful in solving a problem to translate the words into an equivalent problem using mathematical symbols. After this equivalent problem is solved, using any of various strategies, the answer can then be interpreted back into words. This process is illustrated in the following figure.

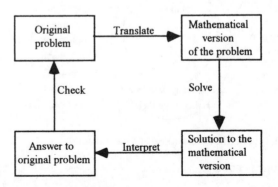

In the problem solving process, Step #2, "Devise a Plan," is very important. Once you thoroughly understand the problem, it is critical to select an appropriate strategy.

The following section will help you learn how and when to use the various strategies. Remember! You must always first completely understand the problem. One good way to test your understanding is to re-state the problem in your own words. The problem statement itself may provide clues to a strategy. You might also find that some problems can be solved in several different ways using different strategies. There is no *one* best way.

Strategy: Guess and Test

You can sometimes begin to solve a problem by "messing around" with it using trial and error. We call this strategy "Guess and Test." Try "Guess and Test" when:

(a) There is a limited number of possible answers to test.
(b) You want to gain better understanding of the problem.
(c) You have a good idea of the answer.
(d) You can systematically try possibilities.
(e) Your choices are narrowed down because other strategies were used earlier.
(f) There's nothing else to try!

Strategy: Look for a Pattern

To use the "Look for a Pattern" strategy, list several specific instances of the problem and look to see if a pattern emerges that suggests a solution. This strategy is often used in a form of reasoning called **inductive reasoning**. With **inductive reasoning** you draw a conclusion based on several observations. Try "Look for a Pattern" when:

(a) A list of data is given.
(b) A sequence of numbers is involved.
(c) Listing special cases helps you deal with complex problems.
(d) You are asked to make a prediction or generalization.
(e) Information can be expressed or viewed in an organized way, such as a table.

Strategy: Use a Variable

A **variable** is a letter or symbol that represents a number. Variables are used in mathematics to solve word problems by restating them as equivalent problems with symbols. Problems stated with variables are easier to solve. Try "Use a Variable" when:

(a) A phrase similar to "for any number" is present or implied.
(b) The problem suggests an equation or inequality.
(c) There is an unknown quantity related to known quantities.
(d) The words "is," "is equal to," or "equals" appear in the problem.
(e) You are trying to develop a general formula, perhaps based on a pattern.

Example 10.21 in the text solves the problem: "What is the greatest number that evenly divides the sum of any three consecutive whole numbers?" As was shown in the solution, the answer is "3". Let's try a related problem.

Example Show that the sum of any five consecutive odd numbers has a factor of 5.

Solution

Step 1: Understand the Problem:
   This is somewhat different from Example 10.21 in that we are talking only about *odd* numbers, such as 3, 5, 11, and 27.
To further understand the problem, try an example:
   $11 + 13 + 15 + 17 + 19 = 75$; this certainly is divisible by 5.
   So also is $13 + 15 + 17 + 19 + 21 = 85$.
However, we need proof that it holds for *any* five consecutive odd numbers.

Step 2: Devise a plan.
   Our strategy will be Use a Variable. We have already used the strategy Look for a Pattern to gain understanding of the problem. The next step will be to represent the five numbers and add them to see if the sum is divisible by 5.

Step 3: Carry out the plan.
   If m is the first of the five odd numbers, the others are $m + 2$, $m + 4$, $m + 6$, and $m + 8$. Their sum is $5m + 20$ (verify).
   Since $5m + 20$ can be written $5 \times (m + 4)$, we see that the sum is divisible by 5. Notice that if m is an even number, the sum is also divisible by 5, as in our first set of five numbers in step 1.

Step 4: Look back.
   Here are some similar statements that can be considered:
   a. The sum of any five consecutive whole numbers has a factor of 5.
   b. The sum of any four consecutive odd numbers has a factor of 4.
   c. The sum of any four consecutive even numbers has a factor of 4.
   d. The sum of any four consecutive whole numbers has a factor of 4.
   The first three of these are true, but the last one is false. After investigating these and several other similar statements, you might be prepared to make a generalization.

Strategy: Draw a Picture

With problems involving a physical situation, drawing a picture can help you understand the problem better so you can devise your plan to solve it. Two strategies are related to "Draw a Picture": **Draw a Graph** and **Draw a Diagram**. Try "Draw a Picture" or related strategies when:

   (a) Geometric figures or measurements are involved.
   (b) A problem can be represented using two variables.
   (c) A visual representation of the problem is possible.
   (d) Finding representations of lines and other geometric figures.

Strategy: Use a Model

The strategy "Use a Model" helps in solving problems involving geometric figures or their applications. We can often gain insight about a problem by seeing a physical model of it. A model is any physical object that resembles the object in the problem. A model can be a simple paper, wood or plastic shape, or it can be a complicated replica. Modeling is common in problem solving. In some applications, modeling is called **simulation**. **Do a Simulation** is a modeling strategy using coins, dice, cards, and other objects in solving probability problems. Try "Use a Model" strategy when:

(a) Physical objects can be used to represent the ideas involved.
(b) A drawing is too complex or inadequate to be of help.
(c) A problem involves 3-dimensional objects.
(d) A problem involves a complicated probability.
(e) A problem has a repeatable process that can be done experimentally.

A Final Note

As an intellectual pursuit, problem solving is somewhere between being an art and a craft. Some people might be more predisposed to being good problem solvers than others, but everyone can improve their skills and benefit from the improvement. What it takes more than anything is a desire to improve, a focusing of energy and attention, and practice, practice, practice. The problem set contains a variety of problems, some of which will challenge anyone. Hints and detailed solutions are provided for nearly all of them. The four-step process is not followed rigidly because it is meant as a guiding paradigm - a way of looking at things. Also, problem solving is not a "straight ahead and down the road you go" kind of thing. For any finished picture you see, there may have been several thrown away. For each model of connected squares that folds into a cube, there was another that folded into something else. Much of what you see are the summaries of the problem solving activity, not the problem solving itself. Problem solving can be satisfying or frustrating. If you have trouble with a problem, be sure you give it adequate time and attention before you resort to the hints and answers. Use unassigned problems for additional examples or practice (practice, practice, practice).

• •

Suggestions and Comments for Odd-numbered problems

1. Review the meaning behind the numbers in the grid in Figure 10.25. How can you extend the pattern to the next row?

3. Try to understand what the problem asks by finding one or more sums of the type discussed. Can you generalize about what happens?

5. Draw pictures and experiment.

7. Make drawings where six squares are connected by edges. The pattern must fold into a cube (a closed "box") with no additional cuts. Try to learn from your mistakes (where else could I have connected that last square?) Patterns drawn on square grid graph paper are the easiest to check. "When in doubt, cut it out."

9. Draw a picture. There are two different sizes of triangles.

11. Try the Guess and Test strategy. Remember the order in which the operations in a calculation are carried out.

13. Use a variable to represent the numbers in the circles. How are the numbers in the squares related to the numbers in the circles? Consider whether the numbers in the squares are even or odd.

15. Draw pictures of the possible paths. Be systematic in the routes you try.

17. Draw a picture. Are different arrangements possible?

19. Draw a series of five blanks (_ _ _ _ _) representing the chairs. Try to fill them using the information in the problem. "Seat" Howard first.

21. Use systematic Guess and Test. Try starting with possibilities for the corners first.

23. Notice that the circle in the upper left has only arrows leading to it. What can you conclude from this? Are there any circles with arrows that only point away?

25. Consider the total amount of wages lost due to the strike and the number of hours she works in a year.

27. Try Systematic Guess and Test. First consider the possible values for the letter P. What must it be, and what are the consequences for the other letters?

29. Run a paper simulation. Make a box representing each pocket and try to fill them.

31. Use Systematic Guess and Test. Since you are asked the man's age in 1949, he must have been born before then. How old could he have been in 1949? This will eliminate some possibilities. For example, he couldn't have been 50 years old since that would mean he was born in 1450 (29×50).

33. Use Guess and Test or Use a Variable.

35. Think about who must ride the elevator first. Drawing a picture might help.

37. Look for a Pattern; try Guess and Test.

39. (a) Look at the difference between successive terms.
    (b) Look at the patterns in the numerator and the denominator separately.
    (c) Look at the ratio between successive terms.
    (d) Each new number is a little more than ten times the previous number, and it has an additional digit.

41. Draw a picture and use Guess and Test. Place numbers in the top circle first.

45. Draw a picture. There will be three different sizes of triangles.

47. Compare successive figures and Look for Patterns in each component in the drawings.

51. Write out the dimensions for each pile of cubes. Find the number of cubes and Look for a Pattern.

53. Make a list. What is the number of squares in each figure, and how many have been added since the last figure? Try writing the number of squares as a sum.

55. Compare the factors in the product with the terms in the sum and with the terms of the Fibonacci sequence.

57. Compare the terms in the difference with the terms in the sum and with the terms of the Fibonacci sequence.

59. Compare the terms in the difference with the terms in the sum and with the terms of the Fibonacci sequence.

63. Make a list with the number of points and line segments at each stage. Count the number of smaller points on each larger point of the star. New points are drawn on each line segment. Is there a pattern to the increase in the number of line segments?

Solutions to Odd-numbered Problems

1. Strategy: Look for a Pattern
   With five squares on each side, the numbers would be as follows:

```
                       *
                  1        1
              1        2        1
          1        3        3        1
      1       4        6        4       1
  1        5       10       10        5        1
      6       15       20       15        6
          21       35       35       21
              56       70       56
                  126      126
                      252
```

   With 4 squares on each side, there are 70 paths.
   With 5 squares on each side, there are 252 paths.

3. Strategy: Use a Variable.
   Let n be the first number. Sums look like this: n + (n+1) + (n+2) + . . .
   (a) n + (n+1) + (n+2) + (n+3) = 4n + 6 = 2(2n + 3)
   Actually, no sum of 4 consecutive whole numbers has a factor of 4.
   The largest factor possible for all sums is 2. Ex: 1 + 2 + 3 + 4 = 10
   Note: An example that "doesn't work" shows that the statement is not true for all
   numbers. This is called a *counter-example*.
   (b) n + (n+1) + (n+2) + (n+3) + (n+4) = 5n + 10. The largest factor is 5.

5. Strategy: Draw a picture (draw lots of pictures)
   Here are some possible solutions exhibiting the maximum number of cuts.

   (a)   **(b)**   **(c)** The maximum is 11 cuts

7. Strategies: Draw a Picture, Guess and Test, Look for a Pattern.
   The following patterns include eleven unique patterns and two duplicates that can be
   folded into the shape of a closed box. Notice that two of the patterns in the second row
   can either be flipped or rotated to correspond to a pattern in the first row. Therefore,
   they do not count as different patterns. Other patterns may be possible.

   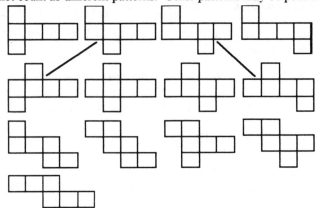

9. Strategy: Draw a Picture. There are a total of 8 triangles

         4 like this
    4 like this

11. Strategy: Guess and Test
    Symbols can be used more than once, and all symbols do not need to be used. Remember
    the order of operation in the calculations. Multiplication and division are done before
    addition or subtraction. The solution can be written as 6 ÷ 6 + 6 + 6 = 13 or 6 + 6 ÷ 6
    + 6 = 13 or 6 + 6 + 6 ÷ 6 = 13

13. Strategy: Use a Variable (three variables, in fact)
Let a, b, and c represent the numbers in the circles as indicated.

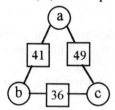

Since a number in a square is the sum of the numbers in the circles on either side of it, we have the following equations:

a + b = 41
a + c = 49
b + c = 36

Since a + b = 41 and a + c = 49, we can write b = 41 - a and c = 49 - a.
When these are substituted into the equation b + c = 36, we have

(41 - a) + (49 - a) = 36
90 - 2a = 36
54 = 2a or a = 27

So b = 41 - 27 = 14 and c = 49 - 27 = 22

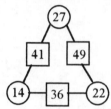

(Be sure to check the totals.)

15. One solution: Systematically draw all the paths she could take from point A to point B. There are 15 paths.

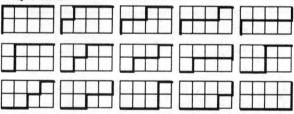

Another solution: Use the strategy from Example 10.19.

1	3	6	10	15
1	2	3	4	5
	1	1	1	1

17. Strategy: Draw a picture and Guess and Test
If you begin with a stool in each corner, there will be six stools left to distribute on the four sides. Since each side must have the same number of stools, the four corners cannot all be occupied. With stools in two of the corners, two possible arrangements are:

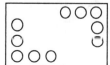 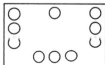

19. Since Gary and Bill sat together and Mike and Tom sat together, Howard must be in the middle chair or at one of the ends.

H x x x x   x x H x x   x x x x H

However, since Bill must sit in the third seat from Howard, we can see that Howard can't sit in the middle seat. Place Howard in an end seat and Bill three seats away.

H x x B x

Gary sat next to Bill, but he has to sit on the right to leave two adjacent seats for Mike and Tom.

H x x B G

Gary is in the third seat from Mike, so Mike is in the seat next to Howard and Tom is in the middle seat with Mike on one side and Bill on the other.

H M T B G

21. There are several possibilities, but first try 7, 8, and 9 in the corners.

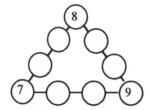

The sums along each side are supposed to be 23.
For the side with 7 and 8, we need 23 - 15 = 8. Either 5 & 3 or 6 & 2.
For the side with 7 and 9, we need 23 - 16 = 7. Either 5 & 2, 4 & 3, or 6 & 1.
For the side with 8 and 9, we need 23 - 17 = 6. Either 5 & 1, or 4 & 2.
Trial and Error produces the following as one solution.

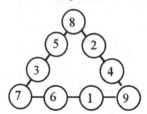

23. Looking at the diagram, we notice the circle in the upper left corner has only arrows leading to it and the one in the lower right has arrows only leading away from it. These circles must contain 9 and 1 respectively. Notice also that the circle below what must be 9 has only one arrow leading away from it and that arrow leads to 9. This circle must contain 8.

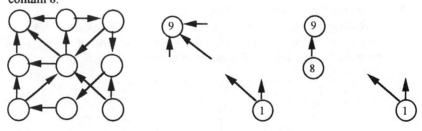

23. Continued   Now try some Guess and Test. With 1 in the lower right hand corner, there are only two possibilities to place the 2. With the 2 in the middle circle, we get the following path.

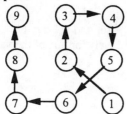

25. Strategy: Use a Variable   Let W = the increase in hourly wage.
With the new contract, she will work 6 hours a day for 240 days. This gives her a total of 1440 hours and an increase of 1440W during one year.
Next, determine the amount of wages lost during the strike. If she worked 6-hour days for 22 days at $9.74 per hour, she lost

(6 hours per day) × (22 days) × ($9.74 per hour) = $1285.68

Since the increase in one year must make up for the wages lost during the strike, we have

1440W = $1285.68   or   W = $1285.68 ÷ 1440 = $0.8928

To make up the lost wages within one year, she must receive at least an additional $ 0.90 per hour.

27. Notice that the P in the sum is alone in the ten thousands place, so we will consider it first. P = 1 since it must be the result of carrying. We conclude also that U = 9 since we could not have carried to get P = 1 if U had been less than 9. Next, E = 0 since E results from U plus a carry. Notice that R plus A = 10 (E = 0, so there must have been a carry) and C must be odd (S + S is even and there is a carry). So far, we have used 0, 1, and 9. R and A could be 2 and 8, 3 and 7, or 6 and 4. If we try R = 8 and A = 2, we see that S must be 3 and C must be 7. We then have the following:

$$\begin{array}{r} 9338 \\ + 932 \\ \hline 10270 \end{array}$$

29. Susan won't be able to arrange the money the way she wants. If 10 pockets have different numbers of bills in each, then the smallest number of bills is 0+1+2+3+4+5+6+7+8+9 = 45. This is impossible since there are only 44 bills.

31. Strategy: Make a Table   From the problem, we know the year of his birth = 29 × (year of his death), and we know he had to be born before 1949. Make a table.

Age at Death	×29	= Year of Birth	Age in1949	
64	29	1856	93	Died at 64 in 1920
65	29	1885	64	Possible
66	29	1914	35	Possible
67	29	1943	6	Will die at 67 in 2010

The man could have been either 64 or 35 years old in 1949.

33. let a, b, c and d represent the missing numbers

$$\begin{array}{ccc} \mathbf{a} & 10 & \mathbf{b} \\ 15 & & 11 \\ \mathbf{c} & 16 & \mathbf{d} \end{array}$$

The following relationships must be true:

$$a + b = 10,\ a + c = 15,\ c + d + 16,\ \text{and}\ b + d = 11$$

Unlike problem 13, we cannot solve uniquely for one of the variables. We try Guess and Test with the following results:

There are several possibilities. Two are;

$$\begin{array}{cccccccc} \mathbf{10} & 10 & \mathbf{0} & & \mathbf{5} & 10 & \mathbf{5} \\ 15 & & 11 & \text{or} & 15 & & 11 \\ \mathbf{5} & 16 & \mathbf{11} & & \mathbf{10} & 16 & \mathbf{6} \end{array}$$

In fact, try any number between 0 and 10 for a.

Then b = 10 - a, c = 11 - b, and d = 16 - c. We see that a + d = 15.

There are a total of 11 solutions using 0 and the whole numbers.

35. Remember that there is a weight restriction of 300 pounds, so the person who weighs 210 pounds must ride the elevator alone, and consequently can't be the first person to ride to the top. Let the 130 and 160 pounders go to the top. The 130 pounder goes down. Then the 210 pounder goes up. The 160 pounder goes down, gets the 130 pounder and they both ride up to the top.

37. Strategy: Look for a Pattern, Guess and Test

$$1234321 * (1 + 2 + 3 + 4 + 3 + 2 + 1) = 4444^2 = 19,749,136.$$

39. Strategy: Look for a Pattern, Guess and Test

There may be more than one acceptable answer for each of these patterns.

(a) The difference between numbers is getting larger.

10 (+7), 17 (+?), ___ (+?), 37 (+13), 50 (+15), 65

The question marks can be replaced with 9 and 11.

The missing number is 26.

(b) Think of 1 as $\frac{1}{1}$. The denominators have the pattern: 1, 2, **?**, 8, 16.

The numerators have the pattern: 1, 3, **?**, 7, 9.

The missing number is $\frac{5}{4}$.

(c) The ratio of the second number to the first is $\frac{324}{243} = \frac{4}{3}$.

The ratio of the third number to the second is $\frac{432}{324} = \frac{4}{3}$.

$\frac{4}{3} \times 432 = 576$, and $\frac{4}{3} \times 576 = 768$.

The missing number is 576.

39. (d) Looking beyond the missing number, we see that each number is increasing by one digit (one position) and the first digits remain unchanged. Consider the third and fourth numbers. How do we get the third number from the fourth?

23481 → 234819

Notice that the new digit is the sum of the preceding two digits.
This is equivalent to multiplying the first number by 10 and then adding the value of the last two digits.
Can the pattern be repeated from the fourth number to the fifth?

234819 → 234819(1+ 9) = 2348190 + 10 = 2348200

Now we go back and verify the pattern with the first three numbers.
2340 + (3 + 4) = 2347
23470 + (4 + 7) = 23481   The pattern works!
The missing number is 2347.

41. Strategy: Guess and Test
We do not want consecutive digits together. The numbers in consecutive circles must differ by at least 2.
Several arrangements are possible. Using the Guess and Test approach, consider the top four numbers. Place the odd digits in these, as they differ by multiples of 2. We could also use the evens.

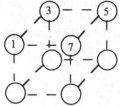

Now we can place the even numbers in order, clockwise, in the bottom circles making sure that the 2 is not under either the 1 or the 3.

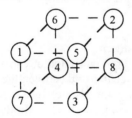

45. Notice that there are three sizes of triangles: 1×1×1, 2×2×2, and 3×3×3.

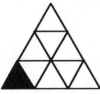 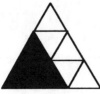 

9 of this size   3 of this size   1 of this size

There are a total of 13 equilateral triangles.

47. Strategy: Look for a Pattern
    (a) Notice that the letters used are in alphabetical order and increase in number by one in each figure. The next figure should contain 5 D's. Notice also that the first figure has one line segment in the center space of the bottom row, the second figure has two line segments, and the third figure has three line segments. The next figure should have four line segments. The line segments in the fourth figure should also form a square. The following figure completes the sequence best.

D	D	D
D	D	
	☐	

    (b) There are three types of objects in the figure. First, there are the circles. The number of circles goes from 1 to 2, then 3, and then 2, as the placement of the circles moves across the diagonals of the grid. The next figure should have one circle in the upper left hand corner of the grid. The triangle in the first three figures is in the upper left corner. It gets smaller and then changes to black. In the fourth figure, there is a larger triangle in the upper right corner. Our best guess is that in the next figure the triangle should get smaller as it did in the second figure. Finally, the arrow moves back and forth in the second row and rotates counter-clockwise (right, up, left, down). The next figure should have an arrow in the right hand column that is pointing to the right. The following figure completes the sequence best.

51. Strategy: Make a list and look for a pattern.
To find the number of cubes in the 100th collection of cubes, Make a List for the 1st, 2nd, 3rd, . . . collections. Look for a relationship between the number of cubes along an edge and the total number in the collection. Notice that in the second collection there are two layers with four cubes each for a total of eight. In the third collection there are three rows of nine cubes each, for a total of 27. The following is the partial list so far.

Dimensions	Number of cubes
$1 \times 1 \times 1$	$1 = 1^3$
$2 \times 2 \times 2$	$8 = 2^3$
$3 \times 3 \times 3$	$27 = 3^3$

For the 100th collection of cubes there should be a total of 100^3 cubes, or 1,000,000 cubes.

53. Strategies: Draw a Picture, Look for a Pattern, Make a List.

    (a)   The next two figures in the sequence are:

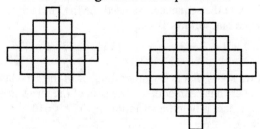

    (b)

Step	Number of Squares	Number of New Squares
1	1	
2	5 = 1 + 4	4 = 4(1)
3	13 = 5 + 8	8 = 4(2)
4	25 = 13 + 12	12 = 4(3)
5	41 = 25 + 16	16 = 4(4)

    (c) The next step should add 20 squares for a total of 61.

    (d) Make a list of the number of squares at each step and look for a pattern.

Step	Number of Squares
1	1 = 1 + 0
2	5 = 4 + 1
3	13 = 9 + 4
4	25 = 16 + 9
5	41 = 15 + 16

At stage n there are $n^2 + (n-1)^2$ squares. For example at stage 6 there
are $6^2 + 5^2 = 61$ squares, which means the pattern seems to hold.

        For the 10th figure, there are $10^2 + 9^2 = 181$ squares.

        For the 20th figure, there are $20^2 + 19^2 = 761$ squares.

        For the 50th figure, there are $50^2 + 49^2 = 4901$ squares.

55. The next 6 terms of the Fibonacci sequence are 34, 55, 89, 144, 233, and 377.
Notice that the sum of the squares of the first n Fibonacci numbers is the product of the
nth and (n-1)st terms of the Fibonacci sequence.

        $1^2 + 1^2 = 2 \times 1$

        $1^2 + 1^2 + 2^2 = 6 = 3 \times 2$

        $1^2 + 1^2 + 2^2 + 3^2 = 15 = 5 \times 3$

        $1^2 + 1^2 + 2^2 + 3^2 + 5^2 = 40 = 8 \times 5$

        $1^2 + 1^2 + 2^2 + 3^2 + 5^2 + 8^2 = 104 = 13 \times 8$

To predict the sum for $1^2 + 1^2 + 2^2 + 3^2 + 5^2 + \ldots 144^2$, we need the number that
follows 144 in the Fibonacci sequence.

        $1^2 + 1^2 + 2^2 + 3^2 + 5^2 + \ldots 144^2 = 233 \times 144 = 33,552$

57. The next six terms of the sequence are 21, 34, 55, 89, 144, and 233.

Notice in the sums that are given that if we want the sum of the first n numbers in the sequence, we need the (n+2)nd term. The second term in the sequence that follows 144 is 377.

$$1 + 1 + 2 + 3 + 5 + \ldots + 144 = 377 - 1 = 376.$$

59. The pattern that we see is that every other number in the Fibonacci sequence (beginning with the first number) has been added. The sum of the terms is 1 less than the next number in the sequence. The number following 377 is 610. $1 + 3 + 8 + 21 + \ldots + 377 = 610 - 1 = 609$.

61.

(a) 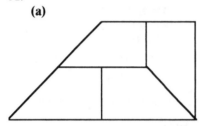 **(b)**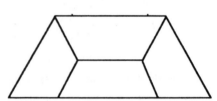

63. Notice that a new point is added to each line segment to form a new star.

(a) For the third star, notice that each of the six points of the second star receives two more points. Where there were two line segments for each point in the second star, there are now 8 line segments in the third star.

Therefore, the third star has $6 \times 3 = 18$.

(b) To form the fourth star, notice that there are 8 line segments for each of the clusters from the third star. Each line segment will receive a new point, making a total of 11 at each of the clusters. The fourth star will have a total of 66 points.

(c) As we go from figure to figure, each line segment receives a point and there are now 4 line segments where there had been 1.

Star	Points	Line Segments
1	3	3
2	6	12
3	18	48
4	66	192

The number of points for the nth star can be expressed as $4^{n-1} + 2$.
The number of points in the fifth star is

$$66 + 192 = 4^4 + 2 = 256 + 2 = 258$$

Chapter 10 Review Problems

Solutions to Odd-numbered Problems

1. Given: p = "snow is cold." p is True
   q = "pigs can fly." q is False
   (a) False   ~p is false when p is true
   (b) False   p ∧ q is false if either p or r is false
   (c) True   p ∨ q is true if either p or q is true
   (d) False   ~q is true; so ~p ∨ ~q is true and ~(~p ∨ ~q) is false
   (e) False   p ⇒ q is false when the antecedent is true and conclusion is not
   (f) True   q ⇒ p is true because the antecedent is false.

3. Conditional: If it rains, the lawn will get wet.   True
   Converse: If the lawn gets wet, it will have rained.   False
   Inverse: If it doesn't rain, the lawn will not get wet. False
   Contrapositive: If the lawn didn't get wet, then it didn't rain. True

5. (a) Invalid;   The fallacy of Denying the Antecedent
   (b) Valid;   Disjunctive Syllogism
   (c) Invalid;   The fallacy of Denying the Antecedent
   (d) Valid;   Modus Ponens
   (e) Valid;   Chain Rule

7. First, with two primary statements, p and q, we will need four rows.
   A separate column should be provided for each modified or combined statement.

p	q	~p	p ⇒ q	~p ∨ q
T	T	F	T	T
T	F	F	F	F
F	T	T	T	T
F	F	T	T	T

Since the truth tables are identical, the statements are logically equivalent.

9. (a)   (b)

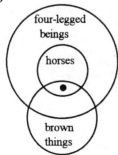

9. (c)   **(d)**

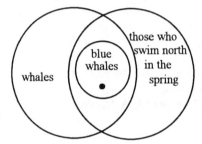

11. (a) True   All of the circle corresponding to R is within the circle for B.
(b) False   The two circles are separated; no elements in common.
(c) False   Part of the circle corresponding to B is outside the circle that
corresponds to A.
(d) False   All of the circle corresponding to Q is contained in the circle
corresponding to A; "All Q are A."

13. (a) The argument is valid.

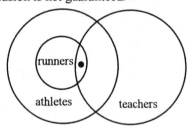

(b) The argument is not valid. We are able to draw an Euler diagram in which the
conclusion is not guaranteed.

15. Strategy: Use a variable.
Let x be your age and y be the age of your brother.
From the problem statement we have:

$$x - y = 8$$
$$\frac{2}{3}\,x - 2 = y \qquad \text{(If a is 2 more than b, then a - 2 = b or a = b + 2)}$$

Substituting for y in the first equation, we have

$$x - (\frac{2}{3}\,x - 2) = 8$$
$$x - \frac{2}{3}\,x + 2 = 8 \quad \text{or} \quad \frac{1}{3}\,x = 6; \quad x = 18 \text{ (your age)}$$

Since x - y = 8, we see that your brother is 10 years old.

17. Strategy: Look for a Pattern, Guess and Test

Since each of the eight possible sums (3 rows, 3 columns, 2 diagonals) must add to 15, we notice that, of the 9 numbers, there are four pairs that add up to 10: (1 and 9, 2 and 8, 3 and 7, and 4 and 6. We put the 5 in the middle square and experiment. There are 8 possible ways. Each may be obtained from the first arrangement by reflections and/or rotations, but the even numbers must all be in the corners.

8	1	6
3	5	7
4	9	2

8	3	4
1	5	9
6	7	2

4	9	2
3	5	7
8	1	6

19. Two cuts: 5 pieces.   Three cuts: 10 pieces.

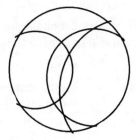

11 Elementary Number Theory

Section 11.1   Numeration Systems

Goals
1. Learn about numbers and the attributes of three historic numeration systems.
2. Understand the Hindu-Arabic system in use today in most of the world.

Key Ideas and Questions
1. Which attributes of the Hindu-Arabic numeration system make it a good system in comparison with other numeration systems discussed in the chapter?

Vocabulary

Tally Numeration System	Multiplicative System	Base
Grouping	Subtractive Principle	Place Value System
Egyptian Numeration System	Positional System	Place Holder
Additive System	Hindu Arabic System	Expanded Form
Roman Numeration System		

. . . .

Overview

A numeration system is a scheme or pattern for representing numbers and communicating information about quantities, patterns and relationships. Throughout history different numeration systems have been developed and played important roles in the advancement of civilization. We will look at various types of these systems.

The Tally Numeration System

The tally numeration system is composed of single strokes, called tallies, where one stroke is used for each object being counted. This system has the advantage of being simple, but it does have two main disadvantages. Very large numbers in this system require lots of tallies, and these large numbers are hard to read. This system was improved by **grouping** the tallies into collections of five tallies.

The Egyptian Numeration System

The **Egyptian Numeration System** was developed about 3400 BC and was based on groupings of ten. It introduced new symbols for powers of ten. This system required fewer symbols than the tally system, once numbers greater than ten were reached. This system is also an example of an **additive** system, since the values for the symbols are added together. The order of symbols is unimportant, but arranging them from left to right makes them easier to read.

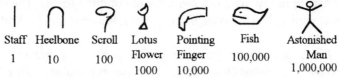

Staff	Heelbone	Scroll	Lotus Flower	Pointing Finger	Fish	Astonished Man
1	10	100	1000	10,000	100,000	1,000,000

The main disadvantage of this system is that even elementary calculations are cumbersome.

The Roman Numeration System

The **Roman numeration system** uses grouping, additivity, and a set of several basic symbols which are used to form the numerals. The basic Roman numerals are listed below.

Roman Numeral	Value
I	1
V	5
X	10
L	50
C	100
D	500
M	1000

Roman numerals are made up of combinations of these basic numerals. To find the values of Roman numerals, you add the values of the basic symbols which comprise the numeral. The Roman system is an additive system.

Two new attributes of numeration systems were introduced with the Roman system: **subtractive principle** and **multiplicative principle.** Both of these principles made it easier to represent numbers, since you could use fewer symbols. Because of the subtractive principle, the Roman system is a **positional** system. The position of the numeral affects the value of the number.

Even though the Roman system required fewer symbols than the Egyptian System, it still requires many more symbols than the Hindu-Arabic system we use today, and is very cumbersome for doing arithmetic.

The Hindu-Arabic Numeration System

The **Hindu-Arabic Numeration System** is the system we use today, and it was first developed about 800 BC. The following points describe the basic features of this system.

1. **Digits**--0, 1, 2, 3, 4, 5, 6, 7, 8, 9,--are the ten symbols that can be used in combination to represent all the non-negative whole numbers.
2. **Grouping by tens** is a basic feature of our system. The number of objects grouped together is called the **base** of the system; thus our Hindu-Arabic system is a base ten system.
3. **Place value (hence positional)**. The position of each digit shows the size of the groups represented by the digit. That size is the **place value**. The digit itself shows how many groups of that size are represented. The digit 0 is important as **a place holder**, so the position of all the non-zero digits can be determined.
4. **Additive and multiplicative.** The value of a Hindu-Arabic numeral is found by *multiplying* each place value by its corresponding digit and then by *adding* all the resulting products. Expressing a numeral as the sum of its digits times their respective place values is called the numeral's **expanded form** or **expanded notation**.

We can identify two advantages of the Hindu-Arabic system. It requires fewer symbols to represent each number, and it is notationally efficient. Also, the Hindu-Arabic system is far superior for performing calculations using paper and pencil methods.

. . . .

Suggestions and Comments for Odd-numbered problems

1. and 3.
The Egyptian system can be treated as a place value system although each symbol has a value and position is actually not important.

$$\text{ⵣⵣ}\cap\cap|| \;=\; \text{ⵣⵣ} \quad \cap\cap \quad || \longleftarrow 2022$$

| 2 x 1000 | 2 x 10 | 2 x 1 |

5. Remember the order of size is $M > D > C > L > X > V > I$
If a lower value symbol precedes a higher value symbol, it is subtracted. If the lower value symbol follows the higher value symbol, it is added. No more than one symbol is ever used in a subtractive mode.

$$DC = D + C \times 600; \quad CD = D - C \times 400$$

9. You can use the Hindu-Arabic system to translate from one numeral system to the other. If you think in terms of *place value*, you can translate directly.

19. and 21.
The Chinese numerals have a type of place value system. If a lower level symbol (for 1 through 9) precedes a higher level symbol (for 10, 100, 1000, etc.) you multiply. If the lower level symbols follow, you add the values.
Look for the symbols for 10, 100, 1000, etc.

25. In the Braille system, one symbol acts as a separator of thousand, millions, etc. The backward "L" precedes every numeral.

Solutions to Odd-numbered Problems

1. (a) 42

∩∩∩∩ ||
4 x 10   2 x 1

(b) 404

𝟵𝟵𝟵𝟵 ||||
4 x 100   4 x 1

3. (a) $9 = 9 \times 1$

|||||||||

(b) $23 = 2 \times 10 + 3 \times 1$

(c) $1231 = 1 \times 1000 + 2 \times 100 + 3 \times 10 + 1 \times 1$

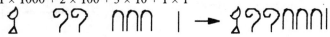

5. (a) MCMXCI = M + CM + XC + I ↔ 1991
         1000 + (1000-100) + (100-10) + 1

(b) CMLXXVI = CM + LXX + VI ↔ 976
        (1000-100) + (50+10+10) + (5+1)

7. (a) 76 = 70 + 6 ↔ LXX + VI = LXXVI
           (50+10+10) + (5+ 1)

(b) 434 = 400 + 30 + 4 ↔ CD + XXX + IV = CDXXXIV
              (500-100) + (10+10+10) + (5-1)

(c) 1999 = 1000 + 900 + 90 + 9 = M + CM + XC + IX = MCMXCIX
             1000 + 900 + 90 + 9

9. (a) MMXCVII = MM + XC + VII ↔ 2000 + 90 + 7
            = 2×1000 + 9×10 + 7×1

9. (b) $\text{MCDLXIV} = \text{M} + \text{CD} + \text{LX} + \text{IV} \quad \leftrightarrow 1000 + 400 + 60 + 4$

$$= 1 \times 1000 + 4 \times 100 + 6 \times 10 + 4 \times 1$$

11. (a) $1993 = 1000 + 900 + 90 + 3 \leftrightarrow \text{MCMXCIII}$

(b) The ad is a cute take-off of the car model's name: the "Mark *Eight*".
When the ad came out (in 1993), the dealer would want readers to know it was the new model (not the 1992 model), and some people can't (or won't) read Roman numerals correctly. He wouldn't be as concerned that they could tell it had 32 valves, since most people don't know how many valves are in their car's engine.

13. (a) $437 = 4 \times 100 + 3 \times 10 + 7$

(b) $5603 = 5 \times 1000 + 6 \times 100 + 0 \times 10 + 3$

15. (a) 105,842

(b) 22,060,300

17. (a) Three hundred forty-five thousand six hundred seventy-eight.

(b) One hundred two million six hundred twenty thousand fifty-seven.

19. (a) 7001   **(b)** 1020

21. (a) $60 + 3$ **(b)** $500 + 80$ **(c)** $2000 + 500 + 40 + 6$

23. The Ionian system is a straight-forward additive system.

(a) $\phi\mu\delta = 500 + 40 + 4 = 544$

(b) $`\delta\sigma\xi = 4000 + 200 + 7 = 4207 \ `\delta = 1000 \times 4$

(c) $\Psi\phi\beta = 900 + 90 + 2 = 992$

25. **(a)** 124,797
    (b) 8,724,640,224

27. **(a)** 156

    ○ ●   ● ○   ● ○   ● ●
    ○ ●   ○ ○   ○ ●   ● ○
    ● ●   ○ ○   ○ ○   ○ ○

(b) 2,450

    ○ ●   ● ○   ○ ○   ● ●   ● ○   ○ ●
    ○ ●   ● ○   ● ○   ○ ●   ○ ●   ● ●
    ● ●   ○ ○   ○ ○   ○ ○   ○ ○   ○ ○

(c) 586,507

    ○ ●   ● ○   ● ○   ● ●   ○ ○   ● ○   ○ ●   ● ●
    ○ ●   ○ ●   ● ●   ● ○   ● ○   ○ ●   ● ●   ● ●
    ● ●   ○ ○   ○ ○   ○ ○   ○ ○   ○ ○   ○ ○   ○ ○

Section 11.2   Divisibility, Factors, and Primes

Goals
1. Learn about prime numbers and composite numbers.
2. Learn how to factor numbers and how to use those factorizations.

Key Ideas and Questions
1. How are divisibility tests used?
2. What is so important about prime numbers?
3. Explain how the GCF and LCM can be used to simplify work with fractions

Vocabulary

Divides	Composite	Greatest Common Factor
Factor	Primality Test	Least Common Multiple
Divisor	Sieve of Eratosthenes	Prime Factorization
Divisible	Factor Trees	Method for Finding
Divisibility Tests	Fundamental Theorem	GCFs and LCMs.
Prime	of Arithmetic	

Overview

The positive integers (the original *counting* numbers) form the basis upon which all other numbers in our number system are defined. In turn, the properties of these numbers have an important effect on the other numbers and operations that combine two or more of them. The most common example is with fractions.

Before the advent of electronic calculators, if you wanted to add two fractions, you needed to find a common denominator (a common multiple of the denominators). One way is simply to multiply all of the denominators together and use that as a common denominator; sometimes it's the best choice.

However, if you had denominators of 12, 14, 18, and 21, you might not want to change all fractions to denominators of 63504. In fact, all denominators can be changed to 252. Finding a least common denominator was a matter of convenience. Also, as a practical matter, since smaller numbers are generally easier to work with, we usually would "reduce" the fraction to lowest terms. To do this, you find the largest common factor in both the numerator and denominator.

The electronic information age has brought new attention to integers and factorization. The need to send sensitive information (political, military, financial) has focused new attention on integers and factors. This section examines the basic properties related to divisibility; in the next section we consider some applications.

Divisibility

Definition

If m and n are positive integers, then m **divides** n if there exists another positive integer q such that

$$n = mq$$

In this case, we also say m is a **factor** of n, that m is a **divisor** of n, or that n is **divisible by** m.   In symbols, we show this as $m \mid n$.

Divisibility Tests

Our ultimate goal will be to find all the possible factors of a given positive integer. We will do this a step at a time, beginning with any recognizable factors and then factoring those factors. Finally, we will express the integer in question as the product of integer factors that are not themselves factorable.

For example, if we want to factor 2760, we would first use $2760 = 276 \times 10$, since any number that ends in zero is divisible by 10. This is one of several divisibility tests that can be used just by looking at the digits that make up the number.

Divisibility Tests

Test for divisibility by 2:   A number is divisibly by 2 if and only if its ones digit is 0, 2, 4, 6, 8.

Test for divisibility by 5:   A number is divisibly by 5 if and only if its ones digit is 0 or 5.

Test for divisibility by 10:   A number is divisibly by 10 if and only if its ones digit is 0.

Test for divisibility by 3:   A number is divisibly by 3 if and only if the sum of its digits is divisible by 3.

Test for divisibility by 9:   A number is divisibly by 9 if and only if the sum of its digits is divisible by 9.

We had $2760 = 276 \times 10$. We recognize that $10 = 2 \times 5$, but what about 276? There are several ways to proceed. You should see recognize that it is divisible by 2, but you might also see that it is divisible by 3. Here are two possible factorization sequences:

$2760 = 276 \times 10$	$2760 = 276 \times 10$
$2760 = 2 \times 138 \times 2 \times 5$	$2760 = 3 \times 92 \times 2 \times 5$
$2760 = 2 \times 2 \times 69 \times 2 \times 5$	$2760 = 3 \times 2 \times 46 \times 2 \times 5$
$2760 = 2 \times 2 \times 3 \times 23 \times 2 \times 5$	$2760 = 3 \times 2 \times 2 \times 23 \times 2 \times 5$

Rearranging factors, we have $2760 = 2^3 \times 3 \times 5 \times 23$

Each of these sequences can be show graphically with a **factor tree**.

When we saw that 2760 was divisible by both 2 and 3, this meant that it was also divisible by 6; $2760 = 460 \times 6$. This is an example of two properties related to divisibility that are quite useful.

Theorem

If $m \mid n$ and $n \mid p$, then $m \mid p$

If $m \mid n$ and $m \mid r$, then $m \mid (n \pm r)$

Prime Numbers

When factoring 2760 in terms of positive integers, we came to a place where no further factorization was possible. Certain integers have only 1 and themselves as positive integer factors.

Prime Numbers

An integer p greater than 1 is called a **prime** if it has exactly two positive divisors, namely 1 and p. A positive integer greater than 1 that is not prime is called a **composite number**. Note: 1 is neither prime nor composite.

The Greek mathematician Euclid (c. 300 B.C.) proved that there were infinitely many primes. Another Greek mathematician, Eratosthenes, devised a simple procedure called a sieve for finding prime numbers.

The primary way to determine whether or not a number is prime is to see if it is divisible by primes that are less than the number. Fortunately, to see if p is prime, you don't have to try any prime that is larger than $\sqrt{p}$. Suppose $p = m \times n$; as one factor gets larger, the other gets smaller. Take 100 for example; $100 = 2 \times 50$, $100 = 4 \times 25$, $100 = 5 \times 20$, $100 = 10 \times 10$, $100 = 20 \times 5$, etc. The factors repeat.

To see if 101 is prime, you need only check 2 (no), 3 (no), 5 (no), and 7 (no). The next prime is 11, but that would make the other factor less than 10, and they've all been checked. 101 is prime.

Greatest Common Factor and Least Common Multiple

The Fundamental Theorem of Arithmetic assures us that there is a level beyond which we cannot (and need not) go when expressing a composite number in terms of factors. Once we have the prime factorization of two (or more) positive integers, we can easily answer two questions that are related to the numbers and divisibility:

What is the largest number that is a factor of both (or all) numbers?

What is the smallest number for which both (or all) of these numbers is a factor?

Greatest Common Factor

The greatest common factor (GCF) of two (or more) positive integers is the largest positive integer that is a factor of both (or all) of the numbers. The greatest common factor of a and b is written **GCF(a, b)**.

Least Common Multiple

The least common multiple (LCM) of two (or more) positive integers is the smallest positive integer that is a multiple of both (or all) of the numbers. The least common multiple of a and b is written **LCM(a, b)**.

For any positive integers a and b, GCF(a, b) $\times$ LCM(a, b) $= a \times b$

To see how this works, suppose we have two number a and b with their respective prime factorizations. To keep the example as simple as, we'll use specific numbers, although the argument could be made abstractly. We'll be looking at the powers of primes as they occur in each factorization. To help make things easier to follow, we'll highlight the power in one number and minimize the power in the other. If the powers of a specific prime are the same, we'll arbitrarily highlight one and minimize the other.

Let $a = 1260$ $1260 = 2^2 \times 3^2 \times 5 \times 7$
    $b = 882$ $882 = 2 \times 3^2 \times 7^2$

The greatest common factor picks out the *smallest* power of each prime that appears in *both* of the numbers: $1260 = 2^2 \times \mathbf{3^2} \times 5 \times \mathbf{7}$; $882 = \mathbf{2} \times 3^2 \times 7^2$

$$\text{GCF}(1260, 882) = \mathbf{2} \times \mathbf{3^2} \times \mathbf{7} = 126$$

The least common multiple picks out the *largest* power of each prime that appears in *either* of the numbers: $1260 = 2^2 \times \mathbf{3^2} \times 5 \times \mathbf{7}$; $882 = \mathbf{2} \times 3^2 \times 7^2$

$$\text{LCM}(1260, 882) = 2^2 \times 3^2 \times 5 \times 7^2 = 8820$$

GCF(1260, 882) $\times$ LCM(1260, 882) $= 126 \times 8820 = 1,111,320 = 1260 \times 882$

. . . .

Suggestions and Comments for Odd-numbered problems

5. If $a \mid b \times c$ is true, then all the factors of a have to be found in b and c.

11. If a statement is true, try to give a general explanation that would apply to similar situations. If a statement is false, a counter-example (an example for which the conclusion of the statement is not true) is sufficient.

15. Begin by crossing out those that are divisible by 2, then those divisible by 3, then those divisible by 5, and so on.

17. through 25.
When looking for factors; unless you "recognize a pair of factors (ex: 240 = 12 × 20) always try 2, 3, 5 since these are easily tested for by the divisibility rules. This will reduce the size of the un-factored numbers. Repeat with testing for 2, 3, 5 if possible.

29. How would you calculate the total for the fruit if you knew the prices (call one price x and one price y)? How can you apply a divisibility test to the total?

31. Assume that the price, P, is a whole number, and x and y are the numbers sold in the two years. The first years sales can be represented by Px and the second by Py. The difference in sales for the two years can be represented by $Py - Px$ or $P(y - x)$.

Solutions to Odd-numbered Problems

1. (a) True $\quad 3 \times 3 = 9$ **(b)** False   6 is a divisor of 12, but 12 is not a divisor of 6

     (c) True $\quad 3 \times 7 = 21$ **(d)** False $3 \times 2 = 6$; 3 is a factor of 6

3. For integers, m and n, we say that m divides n if there is an integer x such that $n = mx$.
     (a) 7 $\quad 49 = 7 \times 7$ **(b)** 10 $\quad 210 = 21 \times 10$ **(c)** $3 \times 18 \quad 9 \times 18 = 3 \times (\mathbf{3 \times 18})$

     (d) $11 \times 5 \times 7 \quad 22 \times 5 \times 7 = 2 \times (\mathbf{11 \times 5 \times 7})$

     (e) $4 \times 32 \times 73 \times 135 \quad 24 \times 32 \times 73 \times 135 = 6 \times (\mathbf{4 \times 32 \times 73 \times 135})$
     There are 3 other possibilities:
     $8 \times 16 \times 73 \times 135$, $24 \times 16 \times 73 \times 45$, or $12 \times 32 \times 73 \times 45$

5. (a) False $\quad 40{,}000$ is even; $27 \times 173 \times 347 \times 501$ is odd

     (b) True $\quad 600 = 2^3 \times 3 \times 5^2 = 24 \times 5^2$;
                 $24 \times 35 \times 55 \times 1043 = \mathbf{24} \times (5 \times 7) \times (5 \times 11) \times 1043$

     (c) True $\quad 147 \times 45 = (\mathbf{21} \times 7) \times 45$; there are other possibilities

     (d) False $\quad r^4$ is a factor of $p^3 q^2 r^4$, but not of $p^6 q^2 r^3$

     (e) True $\quad 15 = 3 \times 5$; $\quad 6 \times 85 \times 45 = (3 \times 2) \times (5 \times 17) \times 45$

7. 1, 3, and 7 $\quad$ If 21 divides m, then any factor of 21 will divide m

9. (a) True $\quad$ The last digit (6) is divisible by 2   **(b)** True $\quad$ The last digit is 0 (or 5)

     (c) True $\quad$ The sum of the digits (12) is divisible by 3

     (d) False $\quad$ The sum of the digits (12) is not divisible by 9

11. (a) False   As a counter-example (an example showing a statement isn't true) use 24. 6 | 24 and 8 | 24, but 48 does not divide 24. The reason is that 6 an 8 share at least one common factor. However, if two positive integers have no factors in common and divide a third positive integer, then their product will also divide the third integer.

(b) False   A counter-example will be sufficient: 4 | 12, but 8 does not divide 12. However, if a positive integer is divisible by 8, then it will be divisible by 4. In general, if m is a divisor of n, and n is a divisor of p, then m is a divisor of p.

13. Several reasons could be given in each case; one of the simplest will be given
   (a) Composite   12 is divisible by 2; the last digit is even

   (b) Composite   123 is divisible by 3; the sum of the digits is 3

   (c) Composite   1234 is divisible by 2; the last digit is even

   (d) Composite   12345 is divisible by 3; the sum of the digits (15) is divisible by 3

15. The primes between 100 and 150 are:
   101, 103, 107, 109, 113, 127, 131, 137, 139, 149

17. (a)

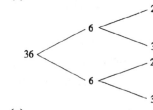

(b)

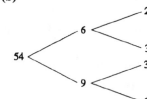

(c)

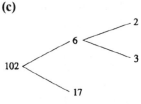

(d)

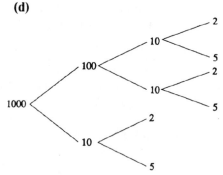

19. Prime Factorizations
   (a) $39 = 3 \times 13$

   (b) $1131 = 3 \times 13 \times 29$

   (c) $55 = 5 \times 11$

21. (a) $8 = 2^3$, $18 = 2 \times 3^2$ GCF(8, 18) = 2

   (b) $36 = 2^2 \times 3^2$, $42 = 2 \times 3 \times 7$ GCF(36, 42) = 6

   (c) $24 = 2^3 \times 3$, $66 = 2 \times 3 \times 11$ GCF(24, 66) = 6

23. **(a)** $15 = 3 \times 5, 21 = 3 \times 7$ LCM$(15, 21) = 3 \times 5 \times 7 = 105$

    (b) $14 = 2 \times 7, 35 = 5 \times 7$ LCM$(36, 42) = 2 \times 5 \times 7 = 70$

    (c) $130 = 2 \times 5 \times 13, 182 = 2 \times 7 \times 13$ LCM$(130, 182) = 2 \times 5 \times 7 \times 13$

25. $72 = 23 \ 32, 108 = 22 \ 33$ GCF$(72, 108) = 22 \ 32 = 36$

    $$\frac{72}{108} = \frac{2 \times 36}{3 \times 36} = \frac{2}{3}$$

27. **(a)** True $6! = 6 \times 5 \times 4 \times 3 \times 2 \times 1$ $n \mid n!$ for all n

    (b) True $6! = 6 \times 5 \times 4 \times 3 \times 2 \times 1$ $(n-k) \mid n!$ for all n and all $k < n$

    (c) False $6! = 6 \times 5 \times 4 \times 3 \times 2 \times 1$ The are 32 numbers $(x \neq 1)$ that divide 6!
    All 32 are composed of factors of 6!

29. Suppose the cantaloupes cost m cents each, and the lemons cost n cents each. Then, the total cost would be $C = 3m + 6n$. But since $3m$ and $6n$ are both divisible by 3, their cost C must also be divisible by 3. However, \$3.52 is *not* divisible by 3.

31. We need to find two positive integers m and n such that $mx = 2567$ and $nx = 4267$, where x is the price of the calculator. To make matters simpler, we assume that x is also a positive integer.

    First, we look at the difference in totals over the last two years:
    $$nx - mx = 1700 \quad \text{or} \quad (n - m)x = 1700 = 2^2 \times 5^2 \times 17$$
    Since x is a divisor of 2567 and 4267, it must be a divisor of the difference.
    By a process of elimination and verification, we conclude that $x = 17$
    151 calculators were sold in the first year, and 251 were sold in the second.

33. **(a)** Between 2 and 4: 3   **(b)** Between 10 and 20: 11, 13, 17, or 19

    (c) Between 100 and 200: 101, 103, 107, 109, 113, 127, 131, 137, 139, 149, and others

35. Answers will vary. To make it a bit of a challenge, AVOID picking an even number. To test a given number from among this group, use your calculator and the list of primes from Figures 11.9 and 11.11 together with those between 100 and 150 listed in problem 33(c).

37. **(a)** 5, 13, 17, 29, 37, 41, 53, 61, 73, 89, and 97

    (b) $5 = 1 + 4$ $13 = 4 + 9$ $17 = 1 + 16$ $29 = 4 + 25$
       $37 = 1 + 36$ $41 = 16 + 25$ $53 = 4 + 49$ $61 = 25 + 36$
       $73 = 9 + 64$ $89 = 25 + 64$ $97 = 16 + 81$

39. Given: $mx = n$ and $ny = p$
    Substituting mx for y, we have $(mx)y = p$, so $m(xy) = p$. Therefore m is a divisor of p.

41. First, since $2 \mid 10$, it also divides any power of 10 and any multiple of powers of 10.
    It follows then, that $2 \mid (a_n \times 10^n + a_{n-1} \times 10^{n-1} + \dots + a_2 \times 10^2 + a_1 \times 10)$ for any n
    Therefore, we know that $2 \mid (a \times 10^3 + b \times 10^2 + c \times 10)$. So, if $2 \mid d$,
    then $2 \mid ((a \times 10^3 + b \times 10^2 + c \times 10) + d)$ or $2 \mid (a \times 10^3 + b \times 10^2 + c \times 10 + d)$

41. Another approach that uses the properties differently and uses a different format.
abcd = $a \times 10^3 + b \times 10^2 + c \times 10 + d$
abcd = abc0 + d   Example: 5236 = 5230 + 6
2 | abc0 since 2 | 10 and 10 | abc0   2 | 5230 since 2 | 10 and 10 | 5230
If 2 | d, then 2 | abc0 + d (= abcd)   Since 2 | 6, 2 | 5230 + 6 (= 5236)

43. The expanded form of n (= abcd) is n = $a \times 10^3 + b \times 10^2 + c \times 10 + d$.
This can be rewritten as a × (999 + 1) + b × (99 + 1) + c × (9 + 1) + d
Expanding: a × 999 + a + b × 99 + b + c × 9 + c + d
Collecting: (a × 999 + b × 99 + c × 9) + (a + b + c + d)
Since   9 | 999, 9 | 99, and 9 | 9; we then know that
         9 | a × 999, 9 | b × 99, and 9 | c × 9   So, 9 | (a × 999 + b × 99 + c × 9)
If 9 | (a + b + c + d), it follows that 9 | ((a × 999 + b × 99 + c × 9) + (a + b + c + d))
This is equivalent to showing that if 9 | (a + b + c + d), then 9 | abcd
The argument can be applied to any number of digits.

Section 11.3   Clock Arithmetic, Cryptography, Congruence

Goals
1. Learn about clock arithmetics and their application to cryptography.
2. Learn about congruence.

Key Ideas and Questions
1. Explain how clock arithmetic can be used to make a code.
2. Learn how basic cipher, substitution, and decimation codes are constructed.
2. How are the 7-clock arithmetic and congruence mod 7 similar but different?

Vocabulary

Clock Addition	Substitution Cipher	Decimation
Clock Subtraction	General System	Congruence mod m
Clock Multiplication	Key	$a \equiv b \bmod m$
Clock Division	Direct Standard	
Caesar Cipher	Alphabet Code	

. . . .

Overview

In this section we deal with the application of mathematics to a problem of communication. Much information has to be transferred from one point to another and contents kept secret. The information could be instructions from the government to one of our foreign embassies, plans for military action sent from the high command to the troops in the field, or instructions regarding a proposed business merger sent to the corporate offices. It could be your financial records. The information could be a video signal for a premium channel that is only supposed to be viewed by subscribers. All this information exchange is performed with various types of codes.

One of the earliest examples of a code used for diplomatic and military purposes was the Caesar cipher, used by Julius Caesar to send messages to the Roman legions. The term *cipher* refers to "a method of transforming a text to conceal its meaning by a systematic substitution or transformation of its letters." It also refers to the *key* to transforming the text. Another term for disguising the information is *encryption*. The sender and receiver know the method used, but anyone else who sees the message can only guess as to its meaning. Code breaking requires a lot of trial and error and special techniques which are made more powerful and adaptable with modern computers. Encryption systems are deemed vital to the national interests, and it is a federal crime to disseminate details regarding some methods of encryption.

The Caesar cipher transformed the text of a message by replacing each letter with the one that comes three places beyond it in the alphabetical order. If the end of the alphabet is reached, the letters "wrap around" and begin again. Each letter has a position, 1 through 26; when 26 is reached, we begin at 1 again. Such a system is referred to as "clock arithmetic". The Caesar cipher is "reversible" in that anyone who knows how to encode a message knows how to decode a message as well.

Sending secure messages is more important than ever, with many people and organizations having the need for privacy. In the mid-1970's, three individuals developed a method known as **public key encryption**. This ingenious method allows anyone who wants to receive secure messages to announce publicly how the messages are to be encoded. However, only the person who receives the message knows how to decode it. The encryption is not "reversible" except for the one who sets it up.

Public key encryption uses a numerical system known as modular arithmetic, in which large prime numbers play a prominent role. Modular arithmetic is closely related to clock arithmetic.

Clock Arithmetic

In clock arithmetic, we have a finite set of integers, 1 through n. When counting in clock arithmetic, we go from 1 to n, and then start over again, in the same way the numbers repeat when you go around the face of a clock in the clock-wise direction.

1 2 3 4 5 6 7 8 9 10 11 **12** 1 2 3 4 ...

In order to create a system in which multiplication (and possibly division) can be performed, it is customary to replace the 12 by 0. Counting proceeds as follows:

1 2 3 4 5 6 7 8 9 10 11 **0** 1 2 3 4 ...

There's a 12 on the 12-hour clock, but there isn't a 12 in 12-clock arithmetic.

Operations in 12-clock Arithmetic

Addition and **subtraction** in 12-clock arithmetic can also be thought of in terms of a physical clock:

$9 \oplus 2 = 11$ "It's 9 o'clock now, what time will it be in 2 more hours?"

$9 \oplus 5 = 2$ "It's 9 o'clock now, what time will it be in 5 more hours?"

$9 \oplus 3 = 0$ ("12")   "It's 9 o'clock now, what time will it be in 3 more hours?"

$9 \oplus 0 = 9$ In general, $k \oplus 0 = k$, 0 is the **additive identity**.

On a physical clock, addition goes "clockwise", but subtraction is "counter-clockwise."

$9 \ominus 3 = 6$ "It's 9 o'clock now, what time was it 3 hours ago?"
$9 \ominus 11 = 10$ "It's 9 o'clock now, what time was it 11 hours ago?"
$9 \ominus 9 = 0$ "It's 9 o'clock now, what time was it 9 hours ago?"

Subtraction is formally defined in terms of addition as follows:

$a \ominus b = c$ if and only if $b \oplus c = a$

If $b \oplus d = 0$, then b and d are **opposites**; *adding d* produces the same result as *subtracting b*. For example, in 12-clock arithmetic:

$3 \ominus 7 = 8$, since $8 + 7 = 3$ in 12-clock, and $3 \oplus 5 = 8$

In terms of 12-clock addition, the opposite of 7 is 5 since $7 \oplus 5 = 12$. We see that adding 5 produces the same result as subtracting 7.

Multiplication in clock arithmetic is defined in terms of repeated additions, adding two at a time.

$5 \otimes 4 = 5 \oplus 5 \oplus 5 \oplus 5 = 5 \oplus 5 \oplus (5 \oplus 5) = 5 \oplus (5 \oplus 10) = 5 \oplus 3 = 8$

A short-cut for clock multiplication uses regular multiplication and then transforms the product to a clock number (by subtracting "excess" multiples of 12).

$5 \otimes 4 = 8$ $5 \times 4 = 20; 20 - 12 = 8$
$8 \otimes 11 = 4$ $8 \times 11 = 88; 88 - (7 \times 12) = 4$
$6 \otimes 4 = 0$ $6 \times 4 = 24; 24 - (2 \times 12) = 12.$
$5 \otimes 1 = 5$ In general, $k \otimes 1 = k$, 1 is the **multiplicative identity**.

Division is defined as it is in regular whole number arithmetic. That is,

$a \oslash b = c$ (*a* divided by *b* equals *c*) if and only if $b \otimes c = a$

However, just like regular whole number arithmetic, division can't always be performed. Unlike whole number division, it can sometimes be performed with numbers other than 1, and in some clock systems (those based on a prime number) it can always be performed.

If $a \otimes b = 1$, then a and b are **multiplicative inverses**. In 12-clock arithmetic, this means that if the regular product is 1 more than a multiple of 12, the numbers are multiplicative inverses. As a consequence, 2, 4, 6, 8, 10 will not have multiplicative inverses, and neither will 3, 6, or 9. That leaves only 5, 7, and 11 as numbers that could possibly have multiplicative inverses in 12-clock. Checking, we see that each number turns out to be its own multiplicative inverse.

$5 \otimes 5 = 1$ "25" $7 \otimes 7 = 1$ "49" $11 \otimes 11 = 1$ "121"

In 12-clock arithmetic, we can generally divide by 5, 7, and 11 only. The exceptions are if all numbers involved are whole numbers less than 12.

$8 \oslash 5 = 8 \otimes 5 = 4$ $10 \oslash 11 = 10 \otimes 11 = 2$ (you should verify these)
$8 \oslash 4 = 2$ $10 \oslash 4$ is not defined

Powers in 12-clock are found by repeated multiplication:

$10^3 = 10 \otimes 10 \otimes 10 = 10 \otimes (\mathbf{10 \otimes 10}) = 10 \otimes \mathbf{4} = 4$

$$100 - 8 \times 12 = 4$$

As a shortcut: $10^3 = 1000 = 83 \times 12 + 4$

With your calculator, you can find the value as follows;
1. Divide 1000 by 12 $1000 \div 12 = 83.333333\ldots$
2. "Discard" the decimal portion (it represents the remainder).
3. Multiply 83 by 12 and subtract from 1000

Multiplication in 12-clock arithmetic can have some interesting results. For example: $10^n = 4$ for all n, while $8^n = 4$ if n is even and $8^n = 8$ if n is odd, and $4^n = 4$ for all n.

Congruence Modulo m

Many of the features of clock arithmetics can be extended to infinite sets of integers and the operations we apply to them. All integers can be associated with the numbers on the 5-clock by thinking about the number line of integers being wrapped around the face clock. We see that certain numbers are associated with each number on the face:

0: ... -10, -5, 0, 5, 10, 15, 20, ...
1: ... -9, -4, 1, 6, 11, 16, 21, ...
2: ... -8, -3, 2, 7, 12, 17, 22, ...
3: ... -7, -2, 3, 8, 13, 18, 23, ...
4: ... -6, -1, 4, 9, 14, 19, 24, ...

There are infinitely many integers associated with each clock number. However, if we consider any two integers associated with the same number on the clack face, we see that they differ by a multiple of 5. The integers that are associated with a given face number are said to be "congruent". This is expressed symbolically as follows:

Congruence Mod *m*

Let a, b, and m be integers, with $m \geq 2$.
Then $a \equiv b \bmod m$ if and only if $m \mid (a - b)$.
This is read "a is congruent to b mod m"; mod is short for *modulus*.

For example: $48 \equiv 3 \bmod 15$, $26 \equiv 2 \bmod 12$, $11 \equiv 23 \bmod 6$, and $53 \equiv 148 \bmod 5$.

In general, for $b < m$, $a \equiv b \bmod m$ if b is the remainder when a is divided by m:
$43 \equiv 3 \bmod 10$, $43 \equiv 1 \bmod 7$, $43 \equiv 7 \bmod 12$, and $43 \equiv 10 \bmod 11$

The congruence relation is much like equality in many ways, and has many similar properties. For simplicity, we will use $a \equiv b$ for $a \equiv b \bmod m$ unless the particular m needs to be specified.
1. $a \equiv a$ for all integers a
2. If $a \equiv b$, then $b \equiv a$
3. If $a \equiv b$ and $b \equiv c$, then $a \equiv c$
4. If $a \equiv b$, then $a + c \equiv b + c$
5. If $a \equiv b$, then $ac \equiv bc$
6. If $a \equiv b$, and $c \equiv d$, then $ac \equiv bd$
7. If $a \equiv b$, then $a^n \equiv b^n$

This last property allows simplification of otherwise complicated powers. Consider 27^{10} mod 12.

First, we note that $27 \equiv 3$ mod 12. Therefore, $27^{10} \equiv 3^{10}$ mod 12

Using a direct approach: $3^{10} = 59,049 \equiv 9$ mod 12 $(59049 = 12 \times 4920 + 9)$

For a number like 43^{19} mod 15, we still have difficulty with this approach.

First, we note that $43 \equiv 13$ mod 15. Therefore, $43^{19} \equiv 13^{19}$ mod 15

Using a direct approach: $13^{19} \approx 1.46192 \times 10^{21}$; the calculator can't display the digits, and only provides an approximate answer.

Calculating large powers for large numbers is a critical step in public key encryption, but poses problems even for a computer. One way we can get around the difficulties is by using the following properties of exponents:

$$A^m \times A^n = A^{m+n} \quad \text{and} \quad (A^m)^n = A^{mn}$$

First, we calculate $13^2 \equiv 4$ mod 15, and then compute successive powers as follows:

$13^4 = (13^2)^2 \equiv 4^2$ mod 15 $\equiv 16$ mod 15 $\equiv 1$ mod 15

$13^8 = (13^4)^2 \equiv 1^2$ mod 15 $\equiv 1$ mod 15

$13^{16} = (13^8)^2 \equiv 1^2$ mod 15 $\equiv 1$ mod 15

Then we note that $13^{19} = 13^{16+2+1} = 13^{16} \times 13^2 \times 13^1$

Note: any integer can be easily expressed as sums of powers of 2.

Thus, we have the following:

$43^{19} \equiv 13^{19}$ mod 15 $\equiv 13^{16} \times 13^2 \times 13^1$ mod 15 $\equiv 1 \times 4 \times 13$ mod 15 $\equiv 7$ mod 15

In mod m arithmetic, the powers of A have to repeat by the time you get to A^m.

Applications to Cryptography

The Caesar Cipher

To see how the Caesar cipher works, we'll consider what happens to messages using a very limited alphabet consisting only of the letters A, B, C, D, and E.

To encode a message, we replace each letter in a message by the letter three places beyond. Think of two sets of letters, extended as follows:

    A B C D E A B C D E

    A B C D E A B C D E

Now, shift the first set three places:

    A B C D E A B C D E   "1" → "1+3"   "3" → "3 + 3 ≡ 1"

A B C D E A B C D E   A → D   C → A

To encode a message, we locate a letter on the top row and replace it by the corresponding letter in the bottom row. To decode a message, we locate the letter on the bottom row and replace it by the letter in the top row.

For example, we encode as follows: BAD → EDB; DEED → BCCB

If we are decoding, we get the following: EDB → BAD and BCCB → DEED

In this example, the underlying numerical system is 5-clock arithmetic, and the type of cipher is called a **substitution cipher**. The number 3 is called the **key**, since you can "unlock" the message if you know the basis for the substitution. If we use all letters of the alphabet and 26-clock arithmetic, it is called a **direct standard alphabet code**.

Decimation

Another type of substitution cipher that is used is called **decimation**. Decimation uses clock multiplication instead of addition. The key can be any number that has a multiplicative inverse in the underlying number system. Using our five letter alphabet, the key can be 2, 3, or 4, since $2 \otimes 3 = 1$ ($2 \times 3 = 6$), and $4 \otimes 4 = 1$ ($4 \times 4 = 16$). To encode a message, we multiply by the key; to decode the message, we multiply by the multiplicative inverse of the key (to code, we multiply; to decode, we "divide").

Suppose our key is 2. We can show the substitutions as follows

	Encoding			Decoding	
$1 \otimes 2 = 2$	A $\rightarrow$ B		$1 \otimes 3 = 3$	A $\rightarrow$ C	
$2 \otimes 2 = 4$	B $\rightarrow$ D		$2 \otimes 3 = 1$	B $\rightarrow$ A	
$3 \otimes 2 = 1$	C $\rightarrow$ A		$3 \otimes 3 = 4$	C $\rightarrow$ D	
$4 \otimes 2 = 3$	D $\rightarrow$ C		$4 \otimes 3 = 2$	D $\rightarrow$ B	
$5 \otimes 2 = 5$	E $\rightarrow$ E		$5 \otimes 3 = 5$	E $\rightarrow$ E	

Encoding: CAB $\rightarrow$ ABD   Decoding: ABD $\rightarrow$ CAB

. . . .

Suggestions and Comments for Odd-numbered problems

1. through 7.
If necessary, review the definitions for each of the operations and the definitions for opposites and multiplicative inverses.

11. Follow the order of operations indicated with the parentheses, and compare the results. What do the results indicate in terms of general operations and properties?

15. and **17.**
Use Table 11.2 for problem 15, and make a similar table for problem 17.

19. Using decimation with key 3 and 26-clock arithmetic, "I" codes as "A" as follows: "I" is the 9th letter; $9 \otimes 3 = 1$; the 9th letter is coded as the 1st letter. Multiply with whole numbers, then remove multiples of 26 when the product is greater than 26..

21. Two numbers are congruent (mod m) if their difference is divisible by m.

27. Since we only want the last two digits, we will use mod 100 arithmetic:
Ex. $3^9 = 19683 \equiv 83$ mod 100.   Use the properties of exponents as in the example that precedes "Applications to Cryptography" in this section.

Solutions to Odd-numbered Problems

1. (a) 7 $8 + 11 = 7$ (19 - 12)   **(b)** 1 $4 + 9 = 1$ (13 - 12)

(c) 1 $7 \otimes 6 = 6$ (42 - 36)   **(d)** 7 $5 \otimes 11 = 7$ (55 - 48)

3. (a) 1 $9 \oplus 10 = 1$ (19 - 18)   **(b)** 3 $3 \otimes 5 = 2$ (15 - 13)

(c) 33 $33 \oplus 5 = 1$ (38 - 37)   **(d)** 1 $3 \otimes 9 = 1$ (27 - 26)

5. (a) opposite = 4, $4 + 3 = 0$ (7);   multiplicative inverse = 5, $5 \otimes 3 = 1$ (15 - 14)

(b) opposite = 7, $7 + 5 = 0$ (12);   multiplicative inverse = 5, $5 \otimes 5 = 1$ (25 - 24)

(c) opposite = 1, $1 + 7 = 0$ (8);   multiplicative inverse = 7, $7 \otimes 7 = 1$ (49 - 48)

(d) opposite = 4, $4 + 4 = 0$ (8);   there is no multiplicative inverse

7. (a) 7   Step-by-step: $7^3 = 7 \otimes 7 \otimes 7 = 7 \otimes (7 \otimes 7) = 7 \otimes 1$ (49 - 48) $= 7$
"Short-cut": $7^3 = 7$ ($7^3 = 343 = 8 \times 42 + 7$ in ordinary arithmetic)

(b) 4 $4^5 = 4 \otimes 4 \otimes 4 \otimes 4 \otimes 4 = 4 \otimes 4 \otimes 4 \otimes (4 \otimes 4) = 4 \otimes 4 \otimes 4 \otimes 1$ (16 - 15)
    $= 4 \otimes 4 \otimes (4 \otimes 1) = 4 \otimes 4 \otimes 4 = 4 \otimes (4 \otimes 4) = 4 \otimes 1 = 4$
"Short-cut": $4^5 = 4$ ($4^5 = 1024 = 5 \times 204 + 4$ in ordinary arithmetic)

Note: In part (a), we found that $7^2 = 1$ and $7^3 = 7$ in 8-clock arithmetic. You can easily verify that $7^4 = 1$ in 8-clock arithmetic and that, in general, in 8-clock arithmetic:
$$7^n = 1 \text{ if n is even, and } 7^n = 7 \text{ if n is odd}$$
In part (b), we found that $4^2 = 1$ in 5-clock arithmetic and $4^5 = 4$ in 5-clock arithmetic. What do you think is the value of 9^{45} in 10-clock arithmetic? Could you generalize?

9. The addition table for 7-clock arithmetic

$\oplus$	0	1	2	3	4	5	6
0	0	1	2	3	4	5	6
1	1	2	3	4	5	6	0
2	2	3	4	5	6	0	1
3	3	4	5	6	0	1	2
4	4	5	6	0	1	2	3
5	5	6	0	1	2	3	4
6	6	0	1	2	3	4	5

The table is symmetric with respect to the diagonal line that runs diagonally from the upper left to the lower right hand corner; this shows that the addition operation is commutative. That is, $m \oplus n = n \oplus m$ for all m and n.

To find the opposite of any number, look for a 0 in the number's row, Then the number in the column of that 0 is the opposite of the original number. For example, to find the opposite of 4, we find a 0 in the 3 column. That means 3 is the opposite of 4; $3 \oplus 4 = 0$ Notice that no number is its own opposite. That's because the base of the clock arithmetic is an odd number.

11. **(a)** $3 \otimes (4 \oplus 5) = 3 \otimes 2 = 6$,
    $(3 \otimes 4) \oplus (3 \otimes 5) = 5 \oplus 1 = 6$

    (b) $2 \otimes (3 \oplus 6) = 2 \otimes 9 = 6$,
    $(2 \otimes 3) \oplus (2 \otimes 6) = 6 \oplus 0 = 6$

    (c) $5 \otimes (7 \ominus 3) = 5 \otimes 4 = 2$,
    $(5 \otimes 7) \ominus (5 \otimes 3) = 8 \ominus 6 = 2$

    (d) $4 \otimes (3 \ominus 5) = 4 \otimes 4 = 4$,   Note: $3 \ominus 5 = 4$ since $5 \oplus 4 = 3$
    $(4 \otimes 3) \ominus (4 \otimes 5) = 2 \ominus 8 = 4$

    (e) Distributivity of multiplication over addition and subtraction

13. In 12-clock arithmetic, the product of two numbers is 0 if their ordinary product is a multiple of 12. The pairs for which this is true are 1 and 11, 2 and 10, 3 and 9, 4 and 8, 5 and 7; also, 6 added to itself equals 0. Further, in any clock arithmetic for which the base is a composite number, there will be non-zero numbers that have a product of 0.

15. The Caesar Cipher displaces each letter three places (and "wraps around", if needed. $M \rightarrow P$, $A \rightarrow D$, $K \rightarrow N$, $E \rightarrow H$, and so on.

    MAKE MY DAY $\rightarrow$ PDNH PB GDB

17. With a direct standard alphabet code and a key of 13, we have the following:
    $B \rightarrow 0$ B is the 2nd letter, it codes as the 15th letter
    $I \rightarrow V$ (9th to 22nd); $G \rightarrow T$ (7th to 20th); $E \rightarrow R$ (5th to 18th); $A \rightarrow N$ (1st to 14th);
    $S \rightarrow F$ (19th to 6th: $19 + 13 = 32 \equiv 6 \pmod{26}$); $Y \rightarrow L$ (25th to 12th)

    BIG EASY $\rightarrow$ OVT RNFL

19. In decimation with key 3, we multiply the usual position of a letter in the alphabet by 3 and express the result from using 26-clock arithmetic.

    (a) $H \rightarrow X$ ($H \leftrightarrow 3$, $3 \otimes 8 = 24 \leftrightarrow X$)
    $O \rightarrow S$ ($O \leftrightarrow 15$, $3 \otimes 15 = 19 \leftrightarrow S$)
    $T \rightarrow H$ ($T \leftrightarrow 20$, $3 \otimes 20 = 8 \leftrightarrow H$)
        etc.
    HOT DOG $\rightarrow$ XSH LSU

    (b) Decoding using decimation requires that we work backward - instead of multiplying, we will "divide" by multiplying using the multiplicative inverse. In order to have an inverse in 26-clock multiplication, the key must be a prime number. We need to find m such that $m \otimes 3 = 1$. Since $27 - 26 = 1$, we see that the multiplicative inverse of 3 is 9. We will decode by multiplying by 9

    $U \leftrightarrow G$ ($U \leftrightarrow 21$, $9 \otimes 21 = 7$ ($189 = 7 \times 26 + 7$), $7 \leftrightarrow G$
    $S \leftrightarrow O$ ($S \leftrightarrow 19$, $9 \otimes 19 = 15$ ($171 = 6 \times 26 + 15$), $15 \leftrightarrow O$
        etc.
    USSL FWO $\rightarrow$ GOOD BYE

21. **(a)** False $14 \equiv 2 \bmod 3$ **(b)** False $-3 \equiv 1 \bmod 4$, $7 \equiv 3 \bmod 4$

    (c) True   **(d)** True   **(e)** True

    (f) False   If $a \equiv b \bmod 8$, the $a - b$ is a multiple of 8; but $-7 + 11 = 4 \equiv 4 \bmod 8$
               Also, $-7 \equiv 1 \bmod 8$, and $-11 \equiv -3 \bmod 8 \equiv 5 \bmod 8$

23. If $a + c \equiv b + c$, then $(a + c) + (-c) \equiv (b + c) + (-c) \Leftrightarrow a + (c + (-c)) \equiv b + (c + (-c))$
then $a \equiv b$

25. If $a \equiv b$, then $a + (-b) \equiv b + (-b)$ or $a - b \equiv 0$
That means there $(a - b)c \equiv 0$; $m \mid (a - b)c$
Since $(a - b)c = ac - bc$, we have $ac - bc \equiv 0$
Finally, $(ac - bc) + bc \equiv 0 + bc$, which leads to our result: If $a \equiv b$, then $ac \equiv bc$

27. The last two digits of 3^{48} can be found as follows:
Since we only want the last two digits, we will use mod 100 arithmetic:
Ex. $3^9 = 19683 \equiv 83 \bmod 100$

$3^{48} = 3^{45} \times 3^3 = (3^9)^5 \times 27 \equiv (83)^5 \times 27 \equiv 43 \times 27 \equiv 61$
$(3^9)^5 \otimes 27 \equiv (83)^5 \otimes 27 \equiv 43 \otimes 27 \equiv 61$
$3^{49} = 3^{48} \times 3 = \ldots \ldots 61 \times 3$; the last two digits will be 83

Chapter 11 Review Problems

Solutions to Problems

1. Properties of Numeral Systems

   (a) Roman: Grouping, Additive, Subtractive, Multiplicative, Positional

   (b) Egyptian: Grouping, Additive

   (c) Hindu-Arabic: Grouping, Additive, Multiplicative, Positional, Place Value

2. **(a)** $1123 = 1 \times 1000 + 1 \times 100 + 2 \times 10 + 3 \times 1$

   (b) MCCIVII = M   CC   XL II $\rightarrow$ 1242
   1000 200   40   2

   (c) $239 = 200 + 30 + 9 \rightarrow CC + XXX + IX = CCXXXIX$

3. The expanded form of 34,719
$3 \times 10^4 + 4 \times 10^3 + 7 \times 10^2 + 1 \times 10 + 9$.
Thirty four thousand seven hundred nineteen.

4. Prime numbers between 150 and 160: 151, 157

5. The divisors of 48: 1, 2, 3, 4, 6, 8, 12, 16, 24, 48

6. The factorization tree for 72 (one possibility based on $72 = 8 \times 9$)

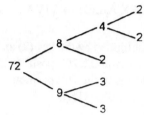

$72 = 2^3 \times 3^2$

7. $48 = 2^4 \times 3 \qquad 72 = 2^3 \times 3^2$

     $\text{LCM} = 2^3 \times 3 = 24 \qquad 48 = 2^4 \times \mathbf{3}, \quad 72 = \mathbf{2^3} \times 3^2$

     $\text{GCD} = 144 = 2^4 \times 3^2 \qquad 48 = \mathbf{2^4} \times 3, \quad 72 = 2^3 \times \mathbf{3^2}$

8. Given $\; 2^3 \times 3^5 \;$ and $\; 3^2 \times 7^4$

     $\text{LCM} = 3^2 \qquad\qquad\qquad 2^3 \times 3^5, \quad \mathbf{3^2} \times 7^4$

     $\text{GCD} = 2^3 \times 3^5 \times 7^4 \qquad \mathbf{2^3} \times \mathbf{3^5}, \quad 3^2 \times \mathbf{7^4}$

9. (a) In 11-clock, $5 \oplus 9 = 3 \quad (14 - 11)$

     (b) In 9-clock, $8 \otimes 8 = 1 \quad (64 - 63)$

     (c) In 10-clock, $3 \ominus 7 = 6 \quad 6 \oplus 7 = 3 \; (13 - 10)$

     (d) In 13-clock, $4 \oplus 9 = 12 \quad 9 \otimes 12 = 4 \; (108 - 104; \, 8 \times 13 = 104)$

10. (a) $3 \oplus (9 \oplus 7) = 3 \oplus (7 \oplus 9) = (3 \oplus 7) \oplus 9 = 9$ in 10-clock

     (b) $(8 \otimes 3) \otimes 4 = 8 \otimes (3 \otimes 4) = 8 \otimes 1 = 8$ in 11-clock

     (c) $(5 \otimes 4) \oplus (5 \otimes 11) = 5 \otimes (4 \oplus 11) = 5 \otimes 0 = 0$ in 15-clock

     (d) $(6 \otimes 3) \oplus (3 \otimes 4) \oplus (3 \otimes 3) = 3 \otimes (6 \oplus 4 \oplus 3) = 3 \otimes 0 = 0$ in 13-clock

11. Opposites and multiplicative inverses

     (a) 4 in 7-clock   Opposite = 3 $\;(3 + 4 = 7)$ Inverse = 2 $\;(4 \times 2 = 8 \rightarrow 1)$

     (b) 4 in 8-clock   Opposite = 4 $\;(4 + 4 = 8)$ Inverse: does not exist

     (c) 0 in 5-clock   Opposite = 0 $\;(0 + 0 = 0)$ This is true in any clock system
                              Inverse: does not exist   0 *never* has a mult. inverse

     (d) 5 in 12-clock   Opposite = 7 $\;(7 + 5 = 12)$ Inverse = 5 $\;(5 \times 5 = 25 \rightarrow 1)$

12. **(a)** ELVIS LIVES (in the Caesar Cipher; letters translate 3 places)

    E→H, L→O, V→Y, etc   ELVIS LIVES → HOYLV OLYHV

    (b) ANAF QFX AJLFX (direct standard alphabet with key 5; translate 5 places *back*)

    A→V, N→I, F→A, etc   ANAF QFX AJLFX → VIVA LAS VEGAS

    (c) TITANIC (decimation with key 7)

    In decimation, each letters position is multiplied by the key (7) and then converted via 26-clock to the new letter's position: T is in the 20th position, $20 \times 7 = 140$. In 26-clock, $140 \equiv 10$ $(140 - 5 \times 26)$ and J is in the 10th position.

    T →J, I→K, A→G, etc   TITANIC → JKJGTKU

13. **(a)** Find n such that $n \equiv 4$ mod 9, $-15 \le n \le 15$

    -14, -5, 4, 13 $n = 9 \times k + 4$ is required; $k = -2, -1, 0, 1$

    (b) Find n such that $15 \equiv 3$ mod n, $1 < n < 20$

    $n = 12$ $15 = k \times n + 3$ is required, and so is $n < 15$.

    (c) Find n such that $8 = n$ mod 7

    { ..., -13, -6, 1, 8, 15, 22, ...} This is the set of numbers with the form $7k + 1$, where k is an integer.

14. $a \equiv a$ mod m since $m \mid (a - a)$.

15. Assume there are 365 days in a year. Then, $365 \equiv 1$ mod 7. This implies that the next year will start 1 day later, namely, on Tuesday. If January 1 is on Monday in a leap year, the next year will start on Wednesday ($366 \equiv 2$ mod 7).

12 Geometry

Section 12.1 Tilings

Goals
1. Determine whether a regular or semi-regular tiling of the plane is possible.

Key Ideas and Questions
1. Why are there no regular tilings for n-gons where n is greater than 6?
2. How do you test to see if a tiling of the plane is semiregular?
3. Suppose you have measured the sides of a triangle. What must be true for the triangle to be a right triangle?

Vocabulary

Polygon	Tessellation	Edge-to-Edge Tiling
Side	Vertex Angle	Vertex Figure
Vertex/Vertices	Diagonals	Semiregular Tiling
n-gon	Regular Polygon	Convex Polygonal Region
Polygonal Region	Irregular Polygon	Concave Polygonal Region
Tiling	Regular Tiling	Pythagorean Theorem

. . . .

Overview

Patterns made by using tilings originated in early civilization. Tilings became works of art as well as being used for walls and floors. In this section, tiling patterns are classified in some of the more elementary cases.

Polygons

Mathematically, a tiling is a collection of polygonal regions. A **polygon** is a figure consisting of line segments lying in a plane that can be traced so that the starting and ending points are the same, and the path never crosses itself or is retraced. The line segments of a polygon are called its **sides**, and the endpoints of the sides are called **vertices** (singular is vertex). A polygon with n sides, hence also n angles, is called an **n-gon**. There are familiar names for some of the smaller n-gons such as triangle for a 3-gon and quadrilateral for a 4-gon. A **polygonal region** is a polygon together with the portion of the plane that is enclosed by the polygon.

Polygonal regions form a **tiling** or **tessellation**, if

    (1) The entire plane is covered without gaps.
    (2) No two polygonal regions overlap.
        (The only points common to the polygonal
        region are points on their common sides.)

A tiling composed of triangles with angle measures a, b, and c shows that $a + b + c = 180°$. This relationship is true for any triangle.

Angle Measures in a Triangle

The sum of the measures of the angles in a triangle is $180°$.

The angles in a polygon are called its **vertex angles**. The symbol "$\angle$" indicates an angle. Line segments are indicated by listing the endpoints with a line drawn above. Line segments joining nonadjacent vertices in a polygon are called **diagonals**.

The sum of the measures of the vertex angles of a polygon can be found by subdividing the polygon into triangles using diagonals. The sum of all the vertex angles is then found by adding the angles in the triangles. This technique is generalized to find the sum of the measures of the vertex angles in <u>any</u> polygon.

Sum of the Angle Measures in a Polygon

The sum of the measures of the vertex angles in a polygon with

n sides is $(n - 2)180°$.

Regular Polygons

Regular polygons are polygons in which all sides have the same length and all vertex angles have the same measure. Polygons that are not regular are called **irregular polygons**. Squares and equilateral triangles are examples of regular polygons.

In a regular n-gon there are n angles. The sum of all the vertex angles of an n-gon is $(n - 2)180°$, and the measure of any one of the vertex angles in a *regular* n-gon must be:

Vertex Angle Measure in a Regular Polygon

The measure of a vertex angle in a *regular* n-gon is $\dfrac{(n-2)180°}{n}$.

We can calculate the measure of a vertex angle in any regular polygon. The following table contains a list of several regular n-gons with the measure of their vertex angles.

n-gon	n	Measure of a vertex angle in regular n-gon
Triangle	3	$(3-2)180°/3 = 60°$
Quadrilateral	4	$(4-2)180°/4 = 90°$
Pentagon	5	$(5-2)180°/5 = 108°$
Hexagon	6	$(6-2)180°/6 = 120°$
Heptagon	7	$(7-2)180°/7 = 128\frac{4}{7}°$
Octagon	8	$(8-2)180°/8 = 135°$
Nonagon	9	$(9-2)180°/9 = 140°$
Decagon	10	$(10-2)180°/10 = 144°$

Regular and Semiregular Tilings

A **regular tiling** is a tiling composed of regular polygons in which all the polygons are the same size and shape. If the polygonal regions in the tilings have entire sides in common, they are called **edge-to-edge** tilings. Tilings do not need to be edge-to-edge. For <u>any</u> regular tiling, the vertex angle is a <u>factor</u> of 360°. For regular n-gons:

(1) There must be at least 3 vertex angles at each point.
(2) The vertex angle measures for n-gons where $n > 6$ must exceed 120°.

The sum of the vertex angle measures in any n-gon tiling for $n > 6$ must exceed $360°$, causing an overlap. By our definition of a tiling, this is not possible. Thus, the only regular edge-to-edge tilings are triangles, squares or hexagons. We will restrict our attention to tilings that are edge-to-edge.

Regular Tilings

The only regular tilings of the plane are those consisting of triangles, squares, or hexagons.

A **vertex figure** of a tiling is the polygon formed when line segments join consecutive midpoints of the sides of the polygons sharing that vertex. The following figure illustrates the vertex figure when two squares and three equilateral triangles meet at a common vertex.

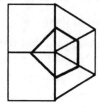

A **semiregular tiling** is an edge-to-edge tiling of two or more regular polygons such that their vertex figures are the same size and shape.

+---+
| **Semiregular Tilings** |
| There are only eight semiregular tilings. |
+---+

Miscellaneous Tilings

We will now discuss tilings made up of irregular figures of the same size and shape.

3-gons (triangles): Since the sum of the angles in a triangle is 180°, any triangle can form a tiling by forming infinite strips. The fact that the angle sum is 180° is what allows us to bring all three angles together at a vertex with two edges falling on a line. The strips can then be used to tile the plane.

4-gons (quadrilaterals): The sum of the angles in a quadrilateral is $360°$; thus the four angles of a quadrilateral will fit around a point. For a triangle, it takes six copies to surround a point.

Triangles, as well as most quadrilaterals, are examples of convex polygons. A polygonal region is called **convex** if, for any two points in the region, the line segment having the two points as endpoints is also in the region. Otherwise, the region is called **concave**.

The text showed how all convex quadrilaterals tile the plane, and showed the start of a tiling by a concave quadrilateral. All concave quadrilaterals will also tile the plane.

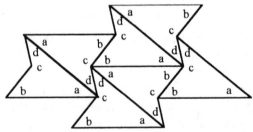

For both kinds of quadrilaterals, the tiling is accomplished by taking four copies of the quadrilateral and rotating two of them $180°$.

5-gons (pentagons): Earlier we saw that regular pentagons do not form a tiling. However, polygonal regions shaped like a baseball plate do tile the plane. The tiling shown was just one of at least 14 general types with irregular pentagons that tile the plane. It is unknown if there are any more.

This type of pentagon can tile the plane because there is a pair of sides that are parallel and of equal length.

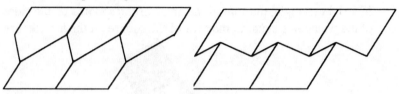

6-gons (hexagons): Regular hexagons produce a regular tiling of the plane, and if a pair of opposite sides of a regular hexagon are extended and kept at the same length, such an irregular hexagon can also tile the plane.

There are exactly three types of convex hexagons that tile the plane. These three types are illustrated in the problem set. One of the types also has the property that a pair of sides are parallel and of equal length.

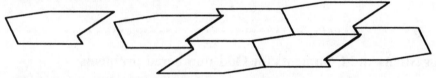

***n*-gons for $n \geq 7$:** It was proved in 1927 that no convex polygon with more than 6 sides can tile the plane. This also holds for concave polygons.

The following table summarizes the tilings discussed above:

Tilings by Irregular Polygonal Regions of the same Size and Shape	
Number of Sides	**Number of Possible Tilings**
3	All are possible
4	All are possible
5	Unknown, but ≥ 14
6	3
7 and more	None are possible

The Pythagorean Theorem

The Pythagorean Theorem is one of the oldest and most famous theorems in mathematics. Most remember it from algebra as "$a^2 + b^2 = c^2$". Actually, the theorem is really one about geometry, since it deals with sides of right triangles. The general case of the Pythagorean Theorem is usually proved algebraically in algebra (of course) but it can be verified geometrically using tiles. You might want to review the presentation in the text and one or more of the methods of proof. The graphical representation of the theorem is also useful to remember as part of the statement.

The Pythagorean Theorem

In a right triangle, the sum of the areas of the squares on the sides
of the triangle is equal to the area of the square on the hypotenuse.

The converse of the Pythagorean theorem is also true. The converse states that if a
triangle has sides of lengths a, b, and c, and $a^2 + b^2 = c^2$; then the triangle is a right
triangle.

· ·

Suggestions and Comments for Odd-numbered problems

9. Vertex angles and exterior angles have the following relationship.

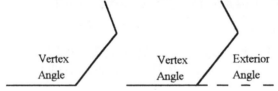

11. n × (measure of exterior angle) = 360°.

19. Consider the combinations of 108 and 60 (at least one of each) that could be used to get
a total of 360.

27. What is required in the definition for a tiling to be called semiregular?

35. If you know two sides of a right triangle, you can use the Pythagorean Theorem to find
the remaining side.

41. The dimensions of four squares are known, and three of these can be used to find the
height of the rectangle. Two of them can be also be used to find the unknown
dimension of one of the larger squares. Since the left and right edges of the rectangle
have the same length, and the top and bottom edges have the same length, you can then
find the missing dimensions by subtracting the known lengths from the total length.

43. The center of the hexagon will be the common vertex for the figures.

45. Carefully count the number of sides for different size polygons. Are there any that can't
be used in a tiling?

47. (a) Given $\frac{1}{3} + \frac{1}{b} + \frac{1}{c} = \frac{1}{2}$, clear the equation of fractions by multiplying by the least common denominator, which is 6bc.

$$6bc \times (\frac{1}{3} + \frac{1}{b} + \frac{1}{c}) = 6bc \times \frac{1}{2}$$
$$2bc + 6c + 6b = 3bc$$

This can be simplified to $b \times (c - 6) = 6c$. Substitute different values for c, and see if the result leads to a whole number for b. Why must c be greater than 6?

49. When four polygons surround a point, the sum of the vertex angles must be 360°. From our previous work, we can conclude that only angles whose measure is a multiple of 15° are possible choices, and the smallest possible angle is 60°. The only possible angles are 60°, 90°, 105°, 120°, 135°, and 150°.

Solutions to Odd-numbered Problems

1. For each of these n-gons, we use the formula $S = (n - 2)180°$

(a) 1800° $(12 - 2) \times 180°$

(b) 1440° $(10 - 2) \times 180°$

(c) 2340° $(15 - 2) \times 180°$

3. (a) 2520° $(16 - 2) \times 180°$

(b) 3960° $(24 - 2) \times 180°$

5. Each vertex angle in a regular nonagon has a measure of 140°.
The sum of the vertex angles is $(9 - 2) \times 180 = 1260$; $1260 \div 9 = 140$.

7. Each vertex angle in a regular icosagon has a measure of 162°.
The sum of the vertex angles is $(20 - 2) \times 180 = 3240$; $3240 \div 20 = 162$.

9. First, the sum of a vertex angle and related exterior angle is 180°. Since each vertex angle is 144°, each exterior angle is 36°.
The sum of all exterior angles is 360°, which means there are 10 exterior angles, and therefore 10 sides.

11. The sum of all exterior angles is 360°. Since each exterior angle is 24°, this means there are 15 exterior angles ($360 \div 24$), and therefore 15 sides.

13. (a) Yes   The sliding of the rows will not cause any "breaks" or "gaps".

(b) Yes   All that is required in a regular tiling is that the figures are all regular polygons of the same shape and size.

(c) A regular tiling does not have to be edge-to-edge.

15. Two possibilities are the following:

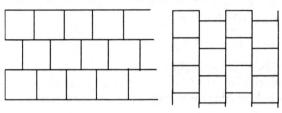

17. The vertex angle measure of a regular 7-gon does not divide into 360° evenly.

19. There are 108° in a vertex angle of a regular pentagon. There are 60° in a vertex angle of a equilateral triangle. There is no way to add multiples of these two angles to sum to 360°.

$$3 \times 108 + 1 \times 60 = 384$$
$$2 \times 108 + 2 \times 60 = 336$$
$$2 \times 108 + 3 \times 60 = 396$$
$$1 \times 108 + 4 \times 60 = 348$$
$$1 \times 108 + 5 \times 60 = 408$$

21. The following shows two types of tiling that could be made. Even smaller triangles can be used in the first tiling, but the small triangles must be able to tile one of the larger ones.

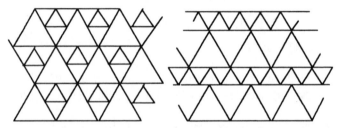

23. A tiling cannot be made of equilateral triangles and regular octagons. There are no combinations of multiples of the vertex angles (60° and 135°) that add up to 360°.

$$2 \times 135 + 1 \times 60 = 330$$
$$2 \times 135 + 2 \times 60 = 390$$
$$1 \times 135 + 3 \times 60 = 315$$
$$1 \times 135 + 4 \times 60 = 375$$

25.

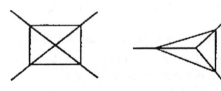

27. The tiling is not semiregular because the vertex figures are different.

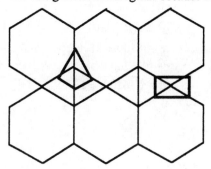

29. (a) Yes, this pentagon will tile the plane.

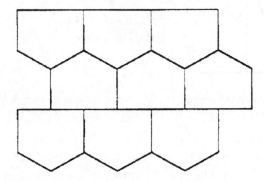

(b) No, this one will not tile the plane.

31. (a) 13 $5^2 + 12^2 = 25 + 144 = 169$; $\sqrt{169} = 13$

   (b) 10 $6^2 + 8^2 = 36 + 64 = 100$; $\sqrt{100} = 10$

   (c) $\sqrt{586} \approx 24.2$ $15^2 + 19^2 = 225 + 361 = 586$

33. The longest side has to be the hypotenuse if it is a right triangle.

   (a) Yes   Does $10^2 + 24^2 = 26^2$?; $100 + 576 = 676$

   (b) Yes   Does $(\sqrt{2})^2 + (\sqrt{3})^2 = (\sqrt{5})^2$?; $2 + 3 = 5$

   (c) No   Does $6^2 + 8^2 = 12^2$?; $36 + 64 \neq 144$

35. Since x is the hypotenuse of a right triangle for which we know two sides, we can find its value:
$$x^2 = (8.1)^2 + (7.4)^2 = 65.61 + 54.76 = 120.37$$
We also know that x is the hypotenuse of the other triangle, so
$$x^2 = 4^2 + y^2 \text{ or } 120.37 = 16 + y^2. \text{ This means that } y^2 = 104.37$$
$x \approx 11.0, y \approx 10.2$

37. Approximately 127.3 ft

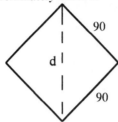

$$d^2 = 90^2 + 90^2$$
$$d^2 = 8100 + 8100$$
$$d^2 = 16200$$
$$d = \sqrt{16200} \approx 127.3$$

39. 15.5 feet

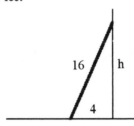

$$h^2 + 4^2 = 16^2$$
$$h^2 = 16^2 - 4^2$$
$$h^2 = 256 - 16 = 240$$
$$h = \sqrt{240} \approx 15.5$$

41. First, we notice that we can get a height of 33 for the rectangle: $14 + 4 + 15 = 33$. This allows us to find the dimensions for the squares that help form the borders on the left and right sides

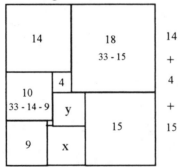

Since the width is now known to be 32, the square marked x must satisfy the equation $9 + x + 15 = 32$. Also, the square marked y must satisfy the equation $x + y = 15$. From these facts, we conclude that $x = 8$ and $y = 7$. The little square must have a side length of 1.

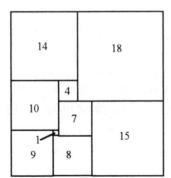

43. (a) 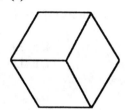 **(b)**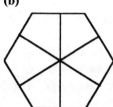

45. The figure shows a vertex surrounded by a regular 5-gon, a regular 6-gon, and a regular 7-gon.

The sum of a vertex angle measure of these is $108° + 120° + 128\frac{4}{7}°$, which is *not* 360°.

At another vertex, there are a regular 5-gon, 6-gon, and 8-gon. The sum of a vertex angle measure of these is $108° + 120° + 135°$; again, this is *not* 360°.

47. (a) (3, 7, 42), (3, 8, 24), (3, 9, 18), (3, 10, 15), (3, 12, 12)

Given $\frac{1}{3} + \frac{1}{b} + \frac{1}{c} = \frac{1}{2}$, we clear of fractions by multiplying by the least common denominator, which is 6bc.

$$6bc \times (\frac{1}{3} + \frac{1}{b} + \frac{1}{c}) = 6bc \times \frac{1}{2}$$
$$2bc + 6c + 6b = 3bc$$
$$6c + 6b = bc \text{ or } bc = 6c + 6b$$
$$bc - 6c = 6b$$
$$c \times (b - 6) = 6b \quad [\text{We could also have } b \times (c - 6) = 6c]$$

Since b and c are both whole numbers, with b > 6, we try different values for b

If b = 7, we have c × (7 - 6) = 42, or c = 42   (3, 7, 42)
If b = 8, we have c × (8 - 6) = 48, or c = 24   (3, 8, 24)
If b = 9, we have c × (9 - 6) = 54, or c = 18   (3, 9, 18)
If b = 10, we have c × (10 - 6) = 60, or c = 15   (3,10,15)
If b = 11, we have c × (11 - 6) = 66, or c = 13.2   No solution
If b = 12, we have c × (12 - 6) = 72, or c = 12   (3,12,12)

For values of b greater than 12, we will get a value for c that is less than 12; these values have already been considered since the roles of b and c are interchangeable.

(b) (4, 5, 20), (4, 6, 12), (4, 8, 8)

Given $\frac{1}{4} + \frac{1}{b} + \frac{1}{c} = \frac{1}{2}$, we clear of fractions by multiplying by the least common denominator, which is 8bc.

$$8bc \times (\frac{1}{4} + \frac{1}{b} + \frac{1}{c}) = 8bc \times \frac{1}{2}$$
$$2bc + 8c + 8b = 4bc$$
$$8c + 8b = 2bc \qquad \text{switch sides and divide through by 2}$$
$$bc = 4c + 4b \text{ or } bc - 4c = 4b$$
$$c \times (b - 4) = 4b \quad [\text{We could also have } b \times (c - 4) = 4c]$$

Continued on next page

(b) Continued

Since b and c are both whole numbers, with b > 4, we can try different
values for b.

If b = 5, we have c × (5 - 4) = 20, or c = 20	(4, 5, 20)
If b = 6, we have c × (6 - 4) = 24, or c = 12	(4, 6, 12)
If b = 7, we have c × (7 - 4) = 28, or c = 9.3333	
If b = 8, we have c × (8 - 4) = 32, or c = 8	(4, 8, 8)

Once the values for b and c are the same, we can quit.

(c) (5, 4, 10), (5, 5, 20)

Given $\frac{1}{5} + \frac{1}{b} + \frac{1}{c} = \frac{1}{2}$, we clear of fractions by multiplying by the least common
denominator, which is 10bc, and simplify the equation as before.

$$10bc \times (\frac{1}{5} + \frac{1}{b} + \frac{1}{c}) = 10bc \times \frac{1}{2}$$

$$2bc + 10c + 10b = 5bc \qquad \text{subtract 2bc and switch the sides}$$
$$3bc = 10c + 10b \quad \text{or} \quad 3bc - 10c = 10b$$
$$c \times (3b - 10) = 10b$$

Since b and c are both whole numbers, with b > 3, we try different values for c.

If b = 4, we have c × (12 - 10) = 40, or c = 20	(5, 4, 20)
If b = 5, we have c × (15 - 10) = 50, or c = 10	(5, 5, 10)
If b = 6, we have c × (18 - 10) = 60, or c = 7.5	
If b = 7, we have c × (21 - 10) = 70, or c = 6.36	

Once again, we've come to a place where we can quit.

Note: (5, 4, 20) is equivalent to (4, 5, 20) since the order of the polygons is not
important.

(d) (6, 6, 6), (6, 4, 12)

The equation used to generate the solutions is c × (b - 3) = 3b
Note: (6, 4, 12) is equivalent to (4, 6, 12).

(e) After accounting for duplications, there are 10 different sets of values that
satisfy the equation.

49. (a) (4, 4, 4, 4), (3, 3, 4, 12), (3, 3, 6, 6), (3, 4, 4, 6)

When four polygons surround a point, the sum of the vertex angles must be 360°. From
our previous work, we can conclude that only angles whose measure is a multiple of 15°
are possible choices, and the smallest possible angle is 60°. The only possible angles are
60°, 90°, 105°, 120°, 135°, and 150°.

The possible combinations of angles that sum to 360° and the corresponding
combination of whole numbers (a, b, c, d) are:

90° + 90° + 90° + 90°	(4, 4, 4, 4)
60° + 60° + 90° + 150°	(3, 3, 4, 12)
60° + 60° + 120° + 120°	(3, 3, 6, 6)
60° + 90° + 90° + 120°	(3, 4, 4, 6)

49. (b) (4, 4, 4, 4)

This is a regular tiling with squares.

(c) (3, 3, 4, 12), (3, 4, 3, 12), (3, 3, 6, 6), (3, 4, 4, 6)

These combinations cannot be extended to semiregular tilings.

(d) (3, 6, 3, 6) and (3, 4, 6, 4) can be extended to semiregular tilings.

These correspond to the two left-most tilings in the bottom row of Figure 12.14.

51. (a) (3, 3, 3, 3, 3, 3)

The smallest number that can be used is 3, which corresponds to an equilateral triangle. To have one of these numbers be larger, we would have to also have one of them smaller.

(b) No, the angle measures would be less than 60° and no regular polygon has this vertex angle measure.

Section 12.2 Motion, Symmetry, and Escher Patterns

Goals
1. Describe rigid motions and symmetries.

Key Ideas and Questions
1. Describe, in terms of rigid motions, the possible types of symmetry that a pattern may have.
2. Give examples of each of type of rigid motion.

Vocabulary

Strip Pattern	Rigid Motion	Directed Angle
One-dimensional Pattern	Isometry	Center
Symmetry	Motion Geometry	Rotation
Reflection Symmetry	Reflection in a Line	Fixed Point
Rotation Symmetry	Vector	Glide Reflection
Translation Symmetry	Translation	

Overview

When we looked at tilings, we classified them according to how the polygonal regions could be arranged around a point. Another way to look at tilings and other patterns in art is through symmetries.

Strip Patterns and Symmetry

The figure below shows a **strip** or **one-dimensional pattern.** A figure has **symmetry** if it can be moved so that the resulting figure looks identical to the original figure. When the pattern in the figure is flipped or **reflected** across the vertical line or the horizontal line, the pattern looks the same after the flipping or reflection as it did originally.

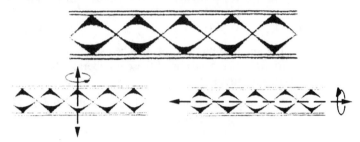

When this occurs, the pattern is said to have **reflection symmetry**.

Now consider the same pattern, but turn or **rotate** it a half-turn (180°) around a given point. The same pattern is produced, so it is said to have **rotation symmetry**. (See figure below) If the only rotation symmetry of a pattern requires a full 360° turn, then it has <u>no</u> rotation symmetry.

When the pattern extends indefinitely in two directions or wraps around an object such as a cylinder or pot, three dots are put at either end of the pattern to show that it continues. If you slide or **translate** the pattern to the right as shown by the dashed arrow in the figure below, the same pattern results. This pattern is thus said to have a **translation symmetry.**

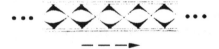

More complex patterns can be made by combining reflections, rotations, and translations that are performed one after the other on a basic figure with copies left in place. In this way, the transformations are built into the pattern.

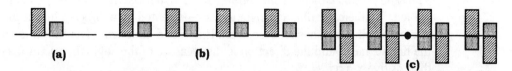

(a)   **(b)**   **(c)**

In part (a), we have a basic figure. In part (b) the basic figure is **translated**, with copies of the figure left in place. In part (c), the image following the **translation** is **rotated**, with a copy left in place. Notice that part (c) can be made with any two adjacent rectangles in the same line using the **translation** and **rotation** as described.

It is common for one-dimensional strip patterns to have translation symmetry, but not all of them do. There are horizontal patterns that do not have translation symmetry, even though they might have both reflection and rotation symmetry.

Rigid Motions

Any combination of reflections in lines, rotations around a point, and translations is called a **rigid motion** or **isometry** (which means "same measure"). When rigid motions are involved, the shapes, lengths, and areas associated with figures do not change. The study of reflections, rotations, translations and combinations of them is called **motion geometry**.

We will now define the terms **reflection**, **translation**, and **rotation** in precise mathematical terms.

A **reflection with respect to line** l, is defined by describing the location of the image of each point of the plane as follows:
(A' represents the **image of** A with respect to the motion)

(1) If A is a point on l, then $A = A'$ (This says that a point on the line of reflection is its own image)

(2) If A is not on l, then l is the perpendicular bisector of $\overline{AA'}$. See figure (b). This says that A and A' are the same distance away from l.

Reflections are often thought of as mirror images with respect to the line of reflection.

The rigid motion called a *translation* can be visualized as a hockey puck sliding along the ice. Mathematically, a translation is represented by a <u>directed</u> line segment, or **vector**. Each vector has an associated length and direction. The length is the length of the line segment, and the direction is the measure of the angle the vector makes with the positive x-axis.

A vector is denoted as υ or as $\overline{AA'}$, where A is the initial point of the arrow and A' is the tip of the arrowhead. A **translation** is defined by describing the location of the image of each point of the plane as follows:

A vector, υ, assigns to every point A in a plane, a point A' which is determined by the length and direction of υ.

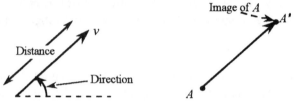

The rigid motion called a *rotation* involves turning a figure clockwise or counterclockwise. A rotation is determined by a point O , and a directed angle. A **directed angle** is an angle where one side is the initial side and its second side is the terminal side. An angle can be directed either clockwise or counterclockwise. Angles directed counterclockwise are assigned a positive number for their measure. The point O is called the **center** of the rotation.

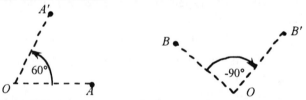

A is rotated 60° counterclockwise about O B is rotated 90° clockwise about O

A **rotation** is defined by describing the location of the image of each point of the plane as follows: The image of a point X under the rotation determined by the directed angle $\angle AOB$ in figure (a) below is the point X' where

(1) $OX = OX'$ and   (2) $\angle XOX' = \angle AOB$ as directed angles (figure b)

Once a rotation is defined by its center and its directed angle (whose vertex is the center of rotation), the image of every point in the plane is determined. This means we can find the image A' of any point A in the plane.

The center, O, of a rotation always corresponds to itself, so it is called a **fixed point**. A point A is called a **fixed point** under a transformation if A and its image A' are the same point. So, the center of any rotation is a fixed point. In a reflection in a line, the fixed points are the points of the line of reflection. Translations have no fixed points.

Rigid motions are often composed of a sequence of basic transformations.

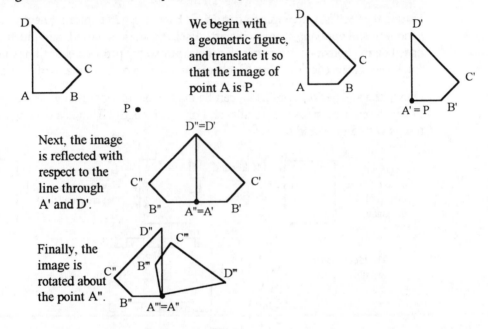

We begin with a geometric figure, and translate it so that the image of point A is P.

Next, the image is reflected with respect to the line through A' and D'.

Finally, the image is rotated about the point A".

A **glide reflection** is a rigid motion that is a reflection combined with a translation. We will usually assume that the translation is in a direction parallel to line of reflection, although this assumption is not necessary.

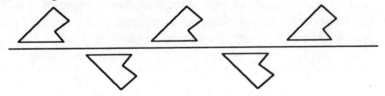

Escher Patterns

So far we have looked at tiling the plane and using Euclidean geometry. If the surface to be covered is not a plane, but is a sphere or other form, new types of tilings can be created. More interesting types of tilings can be created if the shapes used are not polygons, but are some other figures that cover the surface with no gaps or overlaps.

During the 1930's, the graphic artist M.C. Escher began exploring new concepts in geometry and applying them to his art. Escher's work is varied, and much of it is based on mathematics. In particular, he explored the properties of tilings in two non-Euclidean geometries. Escher used rigid motions to prepare many of his patterns.

One of the simpler types of tiling can be made from any parallelogram, a quadrilateral in which opposite sides are parallel and of equal length. This is a property that led to tilings with 5-gons and 6-gons.

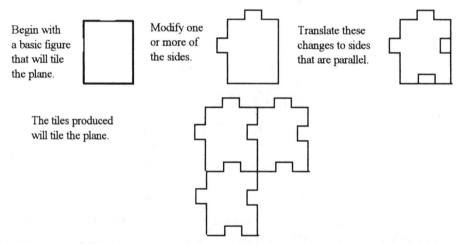

Begin with a basic figure that will tile the plane.

Modify one or more of the sides.

Translate these changes to sides that are parallel.

The tiles produced will tile the plane.

Another type of tiling is based on the rotational symmetries of an equilateral triangle or a regular hexagon. These tilings are investigated in the extended problems. They will help give insight into how and why certain tilings work.

- -

Suggestions and Comments for Odd-numbered problems

17. Consider the background of thick diagonal lines. What symmetry does it have?

19. through **23.**
   If a figure has a line of symmetry, the line has to go through the center of the figure.

27. In a glide reflection, the line of reflection cannot also be a line of reflection symmetry for the original figure.

Coordinate Geometry

Recall that we can use a grid of horizontal and vertical lines to cover the plane in such a way that every point can be assigned a unique pair of numbers called its coordinates. These numbers are assigned with respect to horizontal and vertical reference lines called the axes. The place where the axes meet is called the origin, or center. The first number in the pair tells the horizontal displacement from the center (positive to the right and negative to the left). The second number of the pair tells the vertical displacement from the center (positive upward and negative downward). The origin has coordinates of (0,0).

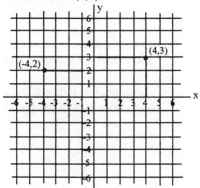

Solutions to Odd-numbered Problems

1. Symmetries

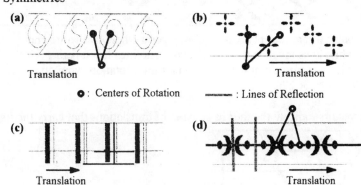

3. Lines of reflection symmetry

Any line that goes through the midline of an "arrow" is a line of reflection symmetry.

There are three such types of lines.

5. Lines of reflection symmetry

Any line through the long diagonal of a rhombus or a kite is a line of reflection symmetry.

There are six such types of lines.

7. Centers for rotation symmetry

Any point at the center of a white or black "propeller" is a center of rotation symmetry.

The figure has rotation symmetries of 120° or 240° about each of the points

9. Centers for rotation symmetry

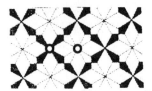 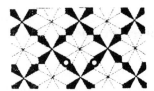

Any point where four black or four white kites touch is the center for a rotation symmetry of 90°, 180°, or 270°.

Any point where two white and two black kites touch is the center for a rotation symmetry of 180°.

11. The vectors indicate the four basic types of translations that carry the figure onto itself.

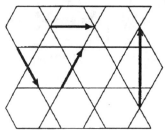

13. The vectors indicate the four basic types of translations that carry the figure onto itself.

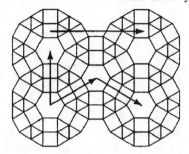

15. Reflections across horizontal, vertical and diagonal lines; rotations by 90°, 180°, or 270° about the center.

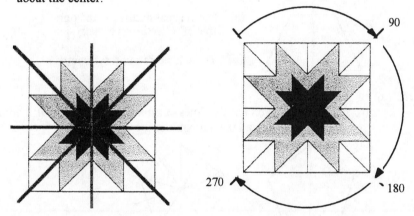

17. Reflections across diagonal lines only; rotations by 180° about the center.

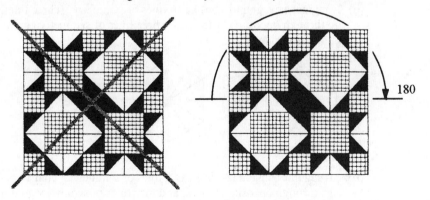

19. Symmetries in a rectangle

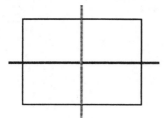
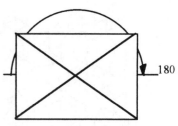

(a) The rectangle has two lines of reflection symmetry.

(b) There is one rotation symmetry of 180° about the intersection of the diagonals.

21. Symmetry in an isosceles trapezoid.

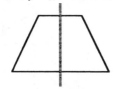

(a) The vertical midline is the only line of reflection symmetry.

(b) There is no rotation symmetry

23. (a)

(i) A regular hexagon has five lines of symmetry. Each line of symmetry goes through a vertex and the midpoint of the opposite side.

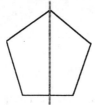
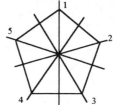

(ii) A regular hexagon (6-gon) has 6 lines of symmetry. Three lines of symmetry go through opposite vertices, and three lines of symmetry go through midpoints of opposite sides.

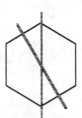
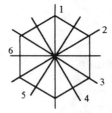

(iii) A regular octagon (8-gon) has 8 lines of symmetry. Four lines of symmetry go through opposite vertices, and four lines of symmetry go through midpoints of opposite sides.

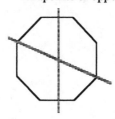
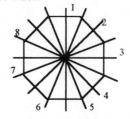

23. (b) A regular n-gon will have n lines of symmetry. Moreover, if n is odd, each line of symmetry will go through a vertex and the midpoint of the opposite side.

If n is even, half the lines will go through opposite vertices and half will go through midpoints of opposite sides.

25.

 (a)

Reflection Symmetry

 (b)

Reflection Symmetry through both diagonals.
180 degree rotation symmetry about the center.

 (c)

Reflection symmetry

27. A glide reflection that produces the original pattern can be any combination of a reflection in one of the vertical lines followed by a translation that moves the reflected image either up or down an odd number of rows indicated by the arrow on the right.

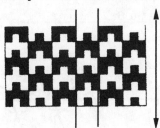

Translation in either direction

Reflection in a vertical line separating two "columns".

29. Five types of translations that map the tiling onto itself are from D to C, from C to B, from A to B, from A to E, and from A to F.

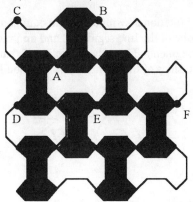

31. **(a)**   **(b)**   **(c)**

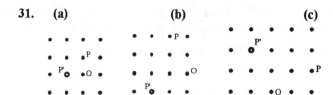

33. **(a)**

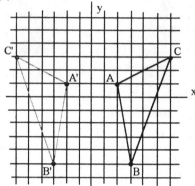

(b) A' = (-2, 1), B' = (-3, -5), C' = (-6, 3)

(c) If P = (a, b), then P' = (- a, b)

35. **(a)**   **(b)**

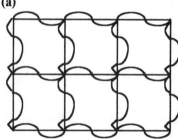
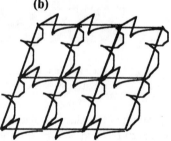

37. These tilings utilize the 60° rotational symmetry of an equilateral triangle. A modification is made to one of the line segments, and an image is rotated 60° to coincide with another line segment. The only restriction is that the modification doesn't overlap one of the other two sides of the triangle, and the modification and its image don't overlap.

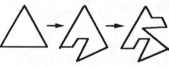

Many tessellations are possible.

Section 12.3   Conic Sections - Parabolas

Goals
1. Define conic sections in terms of a cone and a plane.
2. Write an equation and draw the graph for a parabola.

Key Ideas and Questions
1. How are the conic sections described in terms of cross-sections of cones?
2. How are parabolas described in terms of a point, a line, and distances?
3. Suppose that a light source is put at the focus of a parabola. What happens to the rays of light?

Vocabulary

Cone	Conic Sections	Directrix of a Parabola
Right Circular Cone	Parabola	Focus of a Parabola
Vertex of a Cone	Ellipse	Axis of a Parabola
Nappes of a Cone	Hyperbola	Vertex of a Parabola
Oblique Circular Cone	Circle	Paraboloid

· ·

Overview

Like tilings, conic sections have a long history dating back to the Greeks. Unlike tilings, however, little has been added to our knowledge of conic sections in the past 2000 years except for their representation in algebraic form and their application to advances in science and technology. Our understanding of the paths of projectiles, the orbits of planets, and the receivers and projectors of light and sound have been greatly enhanced by knowledge of the conic sections.

The Conic Sections

The common mathematical definition of a **cone** is the surface formed as follows: Consider a circle with center C on a plane and a point, V, above the plane so that $\overline{VC}$ is perpendicular to the plane. The figure formed by all lines passing through V and the circle is called a **right circular cone**. Point V is called the **vertex** of the cone, and the two parts above V and below V are called the **nappes** of the cone; the two nappes are infinite in extent because they are composed of lines. If $\overline{VC}$ is *not* perpendicular to the plane, then the figure generated is called an **oblique circular cone.**

Slicing a cone with a plane produces several types of curves in the plane. These curves are called **conic sections.** The three types of conic sections are the parabola, the ellipse, and the hyperbola. Circles are a special case of an ellipse.

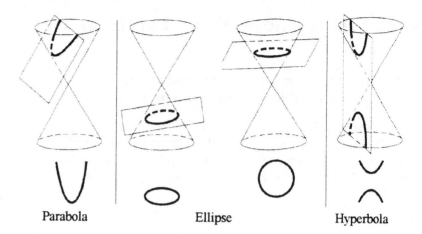

| Parabola | Ellipse | Hyperbola |

A **parabola** is formed when the intersecting plane is parallel to the side of the cone. An **ellipse** is formed when the intersecting plane intersects only one nappe of the cone but is not parallel to the side of the cone. A **hyperbola** is formed when the intersecting plane intersects both nappes of the cone. A **circle**, which is a special ellipse, is formed in the case when the intersecting plane is parallel to the original plane.

The Parabola

A parabola is defined as follows:

> **Parabola**
>
> A **parabola** is a figure in a plane determined by a fixed line and fixed point, not on the line, as follows: A point is on the parabola if it is the same distance from the fixed line and the fixed point.

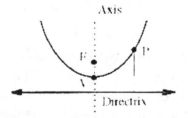

The figure above illustrates the definition of a parabola. The line that determines the parabola is called its **directrix**. The point that determines the parabola is called its **focus**. In the figure above, notice that each of the points P and V is equidistant from the focus F and the directrix, so by definition, each one is on the parabola.

The line through the focus, F, and perpendicular to the directrix is called the **axis**. The axis is the line of symmetry of the parabola.

The **vertex**, V, is the intersection of the parabola and the axis. The vertex is the point of the parabola closest to the focus. By definition, the vertex is midway between the focus and the directrix.

Applications of the Parabola

One property of the parabola is that all rays parallel to the axis that hit the parabola from within and are reflected off the parabola will pass through the focus.

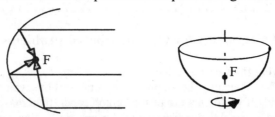

If a parabola is rotated around its axis, the three-dimensional shell that is formed is called a **paraboloid**. See the figure above. By definition, a paraboloid has the property that every plane cross-section that contains the axis of the paraboloid is a parabola. The **focus** and **vertex** of the paraboloid is the same as the focus and vertex of each of these parabolas.

A television dish antenna that is a portion of a paraboloid is used to collect signals from satellites. Satellites send out a signal in all directions. By the time they reach the earth, they are quite weak. The signals hit the dish in parallel rays and are then reflected to a receiver placed at the focus of the paraboloid. This produces a stronger, reinforced signal. Large parabolic reflectors are able to obtain "readable" signals from deep space.

Conversely, flashlights and headlights are constructed using portions of paraboloids as reflectors. The light source is placed at the focus and the light is bounced off the reflector producing parallel light rays which form a cylinder of light. The reflection property of the parabola and paraboloid are used in many other applications, such as a whispering gallery.

Equation of a Parabola

The equation of a two-dimensional parabola whose axis is the y-axis, with vertex on the x-axis at the origin and focus on the y-axis, is derived algebraically.

Equation of a Parabola

The equation of a parabola whose axis is the y-axis, whose focus is $F(0, b)$, and whose vertex is $V(0, 0)$ is

$$y = \frac{1}{4b}x^2, \text{ and conversely.}$$

In general, the equation of a parabola with a vertical axis is

$y = Ax^2 + Bx + C$, where A, B, and C are constants, and $A > 0$.

If the parabola has a horizontal axis, the general equation is

$\underline{x} = Ay^2 + By + C$.

The additional terms change the location of the focus and vertex, but the shape of the parabola is completely determined by the squared term.

· ·

Suggestions and Comments for Odd-numbered problems

Graphing the Conic Sections

When sketching the graphs of the conic sections, you should always try to use a good grid. If you do not use printed graph paper, then make a grid carefully using a pencil and a straight-edge (ruler, folded paper, etc.). You can also use copies of the grid below. The shape of the graph can be determined by the form of the equation or the description of the relationship between the variables. You should try to learn to visualize the shape from the form of the equation. Only as a last resort should you make a long table of values and plot them point-by-point.

After determining the general shape of the graph, you need to sketch it with the correct placement and shape. In order to get the correct placement of the graph, use information about the foci and vertices. To get the correct shape (not all parabolas are the same), substitute one or more values for x and find the corresponding values for y. The curves should be sketched so that they:

1. go through the vertices,
2. go through all the plotted points,
3. are "smooth", and
4. are symmetrical with respect to their axes.

The last condition is a visual double-check.

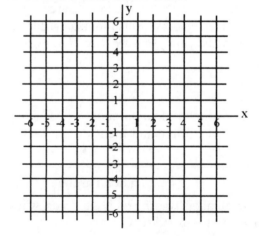

21. How does the distance between the vertex and focus affect the equation of a parabola?

23. If the vertex was at the origin and the focus was still the same distance and direction from the vertex, what would the equation be?

33. Draw a sketch of the cross-section of the dish with the vertex at the origin. What form of equation is associated with the graph? How is the distance between vertex and focus used in the equation?

35. The distance between the focus and vertex can be found from the equation when it is in the standard form $y = \frac{1}{4b} x^2$. First, use the form $y = Ax^2$ and the fact that the coordinates of any point on the graph have to satisfy the equation.
Draw a sketch of the cross-section. What would be the coordinates of a point on the rim of the dish?

39. Draw a sketch. What path will the bullet follow?
In order to find the maximum height, you can use the symmetry of the parabola. What is the height of the bullet when the gun is fired? When will it be that height again? What is the height of the bullet when it hits the ground? When will that happen? In order to find the time associated with a given height, substitute the height for y in the equation and solve for t.

41. Follow the hints given in the text. Use the form $y = Ax^2$ and the fact that the coordinates of any point on the graph have to satisfy the equation.

Solutions to Odd-numbered Problems

1. $y = x^2$

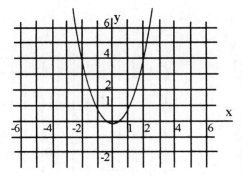

3. $y = 3x^2$

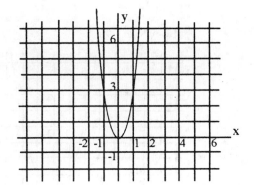

5. $y = -2x^2$

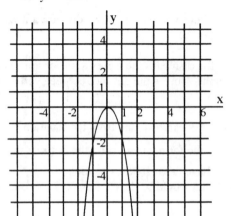

7. $y = -\frac{1}{4}x^2$

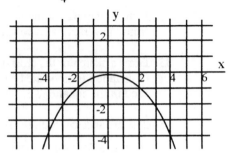

9. $y = x^2 + 2x$

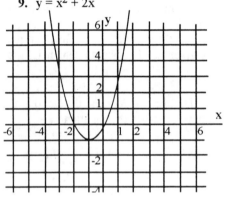

11. $y = x^2 + 5x$

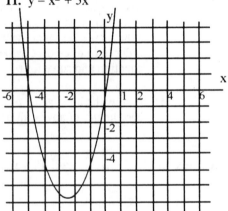

13. $y = x^2 - 4x$

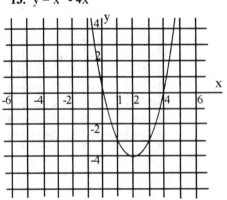

15. $y = x^2 - 2$

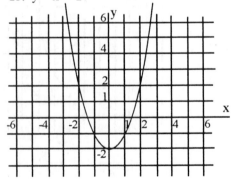

17. $y = x^2 + 3$

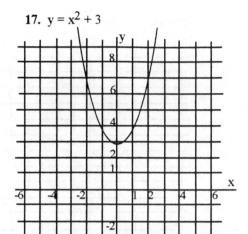

19. $y = x^2 - 5$

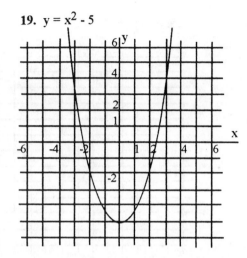

21. Since the vertex is at the origin and the focus is on the y-axis, the parabola is in a standard position, and the equation has the form $y = \frac{1}{4b} x^2$, where b is the distance from the vertex to the focus. We can see that b = 2, and $y = \frac{1}{8} x^2$.

23. Refer to problems 15 - 19 and problems 20 and 21.

The coefficient of x^2 is determined by the distance between the vertex and the focus. Since vertex and focus are on the y-axis and the distance from the focus to the vertex is $b = 2$ (so that 4b = 8), the equation has the form $y = \frac{1}{8} x^2 + C$.

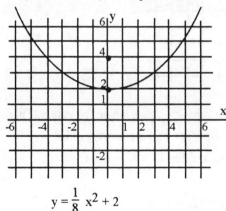

$$y = \frac{1}{8} x^2 + 2$$

25. $x = \frac{1}{4} y^2$

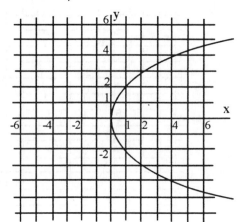

27. $x = y^2 + 2y$

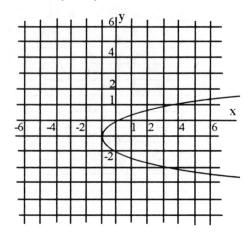

29. $x = y^2 + 3$

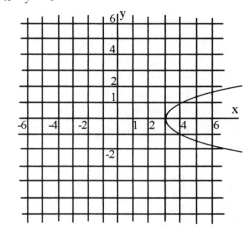

31. All of the equations have the form $y = \frac{1}{4b} x^2$, where the vertex is at the origin, b is the distance from the vertex to the focus, and the focus is (0,b).

   (a) $\frac{1}{4b} = 0.25 = \frac{1}{4}$. This means that $4b = 4$, or $b = 1$

   The focus is $(0, 1)$

   (b) $\frac{1}{4b} = 3$. This means that $12b = 1$, or $b = \frac{1}{12}$.

   The focus is $(0, \frac{1}{12})$

   (c) $\frac{1}{4b} = -4$. This means that $-16b = 1$, or $b = -\frac{1}{16}$

   The focus is $(0, -\frac{1}{16})$

33. If we assume the vertex is at the origin and the focus is directly above it, then the equation has the form $y = \frac{1}{4b} x^2$, where b is the distance from the vertex to the focus. Since the focus is 4 feet from the vertex, the equation is $y = \frac{1}{16} x^2$.

35. First, we draw a sketch matching the description, with the cross-section of the paraboloid centered at the origin of the coordinate system. All dimensions have been changed to feet. The satellite dish has a diameter of 3 feet, so a point on the rim of a cross-section of the paraboloid would be $\frac{3}{2}$ feet from the y-axis (the axis of the paraboloid) and $\frac{1}{2}$ foot above the x-axis.

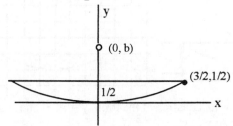

The equation has the standard form $y = \frac{1}{4b} x^2$. The coordinates of the point have to satisfy the equation, so $\frac{1}{2} = \frac{1}{4b} \left(\frac{3}{2}\right)^2$. This simplifies to $4b = \frac{9}{2}$, or $b = \frac{9}{8}$.

The receiver (focus) should be in the center $1\frac{1}{8}$ ft. (or $13\frac{1}{2}$ in) above the vertex.

37. Because the parabolas are congruent (the same size and shape), the distance from the vertex to the focus is the same for both parabolas; that is, two meters. The woman should stand at the focus of the parabola on the right, which is two meters to the left of the vertex of the parabola.

39. A rough sketch of the situation is as follows:

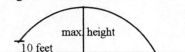

The equation of the path is $y = -16t^2 + 96t + 10$

(a) When the bullet is fired, t = 0. The initial height is 10 ft.

(b) The maximum point on the curve will occur halfway between the time it is 10 feet high on the way up and when it is 10 feet high on the way down. To find the times when this happens, we set y = 10 in the equation and solve for t.

$$10 = -16t^2 + 96t + 10$$

This reduces to $16t^2 = 96t$; either t = 0, or t = 6.
The maximum height will be reached when t = 3.
Max. height $= -16(3)^2 + 96(3) + 10 = 154$ ft

39. **(c)** We want to know "When does y = 0?" We substitute in the equation and solve for t: 0 = -16t^2 + 96t +10. The quadratic equation is used, taking the positive value.

$$t = \frac{-b \pm \sqrt{b^2 - 4ac}}{2a} = \frac{-96 \pm \sqrt{96^2 - 4(-16)(10)}}{2(-16)}$$

Approximately 6.1 sec

41. First, we draw a rough sketch of the bridge. Beside it we place the graph of a parabola centered at the origin and having the same measurements as the horizontal and vertical components of the path of the cable.

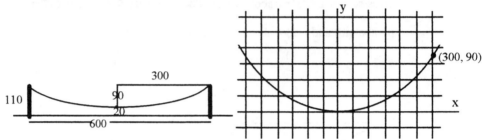

The parabola at the right has an equation of the form y = Ax^2, and is satisfied by y = 90 when x = 300. Solving for A, we get

90 = A × (300)^2, or A = 0.001

The equation of the parabola on the right is y = 0.001x^2.

The equation of the cable (with the vertex raised 20 feet) is y = 0.001x^2 + 20. The height of the cable 100 feet from the center of the span would correspond to the value of y when x = 100 or -100. The height of the cable 100 feet from the center of the span is about 30 feet. This is an approximation, since the parabola is an approximation.

45. **(a)** (x + 2)^2 = 12(y - 3)

$$12y = x^2 + 4x + 40 \text{ or } y = \frac{1}{12} x^2 + \frac{1}{3} x + \frac{10}{3}$$

(b) (x - 6)^2 = - 12(y +2)

$$12y = - x^2 + 12x - 60 \text{ or } y = -\frac{1}{12} x^2 + x - 5$$

47. **(a)** (y - 3)^2 = 16(x - 2)

$$\text{or } 16x = y^2 - 6y + 41 \text{ or } x = \frac{1}{16} y^2 - \frac{3}{8} y + \frac{41}{16}$$

(b) (y + 2)^2 = - 28(x - 6)

$$\text{or } - 28x = y^2 + 4y - 164 \text{ or } x = -\frac{1}{28} y^2 - \frac{1}{7} y + \frac{41}{7}$$

Section 12.4   Conic Sections - Ellipses and Hyperbolas

Goals
1. Write an equation and draw the graph for an ellipse.
2. Write an equation and draw the graph for a hyperbola.

Key Ideas and Questions
1. How are ellipses and hyperbolas described in terms of points and distances?
2. Suppose that a light source is put at a focus of an ellipse. What happens to the rays of light?

Vocabulary

Ellipse	Minor Axis of an Ellipse	Foci of a Hyperbola
Foci of an Ellipse	Mean Distance	Axis of a Hyperbola
Center of an Ellipse	Eccentricity	Center of a Hyperbola
Major Axis of an Ellipse	Hyperbola	

. . . .

Overview

To philosophers in many cultures, both ancient and modern, the circle epitomizes perfection. The circle, however, is a special case of the conic section known as an ellipse. We now study the ellipse as well as the hyperbola. These curves have important modern applications.

The Ellipse

An ellipse is formed when a plane intersects one nappe of a cone. The working definition follows.

The Ellipse

An **ellipse** is a figure in a plane determined by two fixed points as follows: A point is on the ellipse if the sum of its distances from the two fixed points in the plane is a constant.

The fixed points in the definition are called **foci** (singular is **focus**) of the ellipse. The **center** of an ellipse is the midpoint of the segment whose endpoints are the foci. The line segment that contains the foci and has its endpoints on the ellipse is called the **major axis** of the ellipse. The segment that is the perpendicular bisector of the major axis and has its endpoints on the ellipse is called the **minor axis** of the ellipse. Each axis is a line of symmetry for the ellipse.

Terminology related to an ellipse and examples of the distance property:

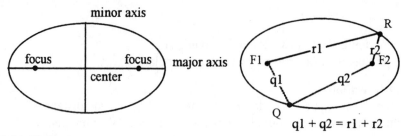

Applications of the Ellipse

Elliptical shapes appear in many situations in the physical world. The planets move in elliptical orbits around the sun with the sun at one focus. Comets have elliptical orbits as well. Whispering galleries have usually been constructed in the elliptical shape.

The Equation of an Ellipse

Suppose P is any point of the ellipse as shown in the following figure.

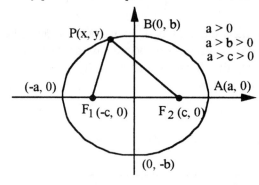

The relation among a, b, and c plays an important role in the following.

Equation of an Ellipse

An **ellipse** whose center is the origin and whose foci are at $(c, 0)$ and $(-c, 0)$ on the x-axis is the set of all points satisfying the equation

$$\frac{x^2}{a^2} + \frac{y^2}{b^2} = 1,$$

where $(a, 0)$, $(-a, 0)$, $(0, b)$, and $(0, -b)$ are points of the ellipse, $a > b$, and

$$c^2 = a^2 - b^2.$$

If the foci of an ellipse are located on the vertical axis, then b will be greater than a, although the form of the equation is essentially the same.

An ellipse with foci on the vertical axis:

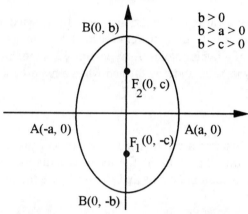

$b > 0$
$b > a > 0$
$b > c > 0$

Equation of an Ellipse

An **ellipse** whose center is the origin and whose foci are at $(0, c)$ and $(0, -c)$ on the y-axis is the set of all points satisfying the equation

$$\frac{x^2}{a^2} + \frac{y^2}{b^2} = 1,$$

where $(a, 0)$, $(-a, 0)$, $(0, b)$, and $(0, -b)$ are points of the ellipse, $b > a$, and

$$c^2 = b^2 - a^2.$$

The Hyperbola

The hyperbola, although composed of two disjoint parts of infinite extent, has a definition similar to the ellipse.

Hyperbola

A **hyperbola** is a figure in a plane determined by two fixed points as follows: A point is on the hyperbola if the (positive) difference of its distances from the two fixed points is a constant.

The figure below illustrates this definition.

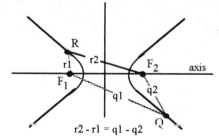

The fixed points F_1 and F_2 are called the foci of the hyperbola.

The line determined by the foci is called the axis of the hyperbola.

Applications of the Hyperbola

Hyperbolas occur in a variety of situations. A radio assisted navigational system is based on the differences. Cooling towers for nuclear reactors are approximately in the shape of a portion of a hyperboloid. A **hyperboloid** is the three dimensional counterpart of a hyperbola, and it is formed when a hyperbola is revolved around an axis.

The Equation of a Hyperbola

The definition of a hyperbola can be used to derive an equation for a hyperbola which is similar to that of the ellipse. In the figures below, the foci of the hyperbolas are the points F_1 and F_2 on a coordinate axis and P_1 and P_2 are the two points of the hyperbola that are on the axis.

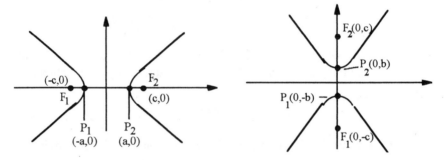

The form of the equation depends on whether the foci are on the x-axis or the y-axis.

If the foci are on the x-axis, we have the following:

Equation of a Hyperbola

A **hyperbola** whose center is the origin and whose foci are at $(c, 0)$ and $(-c, 0)$ on the x-axis is the set of all points satisfying the equation

$$\frac{x^2}{a^2} - \frac{y^2}{b^2} = 1,$$

where $(a, 0)$ and $(-a, 0)$ are points of the hyperbola and

$$c^2 = a^2 + b^2.$$

If the foci are on the y-axis, then the equation is:

> ### Equation of a Hyperbola
>
> A **hyperbola** whose center is the origin and whose foci are at $(0, c)$ and $(0, -c)$ on the y-axis is the set of all points satisfying the equation
>
> $$\frac{y^2}{b^2} - \frac{x^2}{a^2} = 1,$$
>
> where $(0, b)$ and $(0, -b)$ are points of the hyperbola and
> $$c^2 = a^2 + b^2.$$

The **center** of a hyperbola is the midpoint of the line segment whose endpoints are the foci of the hyperbola.

. . . .

Suggestions and Comments for Odd-numbered problems

1. through 9.
See comments in Graphing the Conic Section from the Hints in Section 12.3.

11. through 17.
Draw a sketch of the conic section described. Label any points or distances given with the problem. How do these help you find values for a, b, or c?

19. Follow the hints given in the text. If a point on the ellipse was on the x-axis with coordinates (a, 0), how would you express the sum of the distances to the two foci? What would this sum have to equal?

21. through 25.
Write each equation in a standard form by dividing the equation by the number on the right hand side of the equation. From these forms you can get the values of a and b needed to find x- and y-intercepts (if any) and foci.

25. A fraction such as $\frac{m}{n}$ can be re-written as $\frac{1}{\frac{n}{m}}$. Example: $\frac{2}{3} = \frac{1}{\frac{3}{2}}$.

29. Complete the sketch with a coordinate system. How can you get the values for a and b needed to write the equation?

31. Follow the hints in the text. Mark a point where the lightning could strike. Call the distances from Hal and Bob d_1 and d_2, respectively. What is the difference in the distances (what is the difference in times)? In terms of the geometry of the hyperbola the difference between distances is 2a, which is the distance between the vertices.

33. When working the problem, you may find it easier to use 93, rather than 93,000,000 in the calculation. However, in the final equation the denominators should be in terms of (miles)^2, although we won't assign units of measurement to the values. (Note: The concept of area is not involved; that is, the units are not "square miles".)

37. In an ellipse, the relationship between the distances between center, foci, and vertices is $0 < c < a$. What is the definition of e in terms of the distances?

49. You need b^2 in order to write the equation in the general form, and $c^2 = a^2 + b^2$ in the equation of a hyperbola. Since $b^2 = c^2 - a^2$, try to look for ways to factor $c^2 - a^2$ from common terms when you've removed all parentheses as suggested by the instructions. For example: $Ax^2 - Bx^2 = (A - B)x^2$.

Solutions to Odd-numbered Problems

1. **(i)** $\frac{x^2}{4} + \frac{y^2}{25} = 1$; this is the equation of an ellipse.

   $a^2 = 4$ and $b^2 = 25$; the major axis is vertical.
   The foci are $(0, \pm c)$, where $c^2 = b^2 - a^2 = 25 - 4 = 21$; $c = \sqrt{21}$.

   (ii) The foci are $(0, -\sqrt{21})$ and $(0, \sqrt{21})$. $\sqrt{21} \approx 4.58$

   (iii) x-intercepts: (2, 0), (-2, 0);
   y-intercepts: (0, 5), (0, -5)

   (iv)

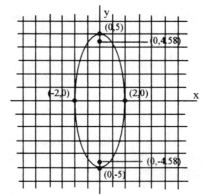

3. **(i)** $\frac{x^2}{16} + \frac{y^2}{25} = 1$; this is the equation of an ellipse.

   $a^2 = 16$ and $b^2 = 25$; the major axis is vertical.
   The foci are $(0, \pm c)$, where $c^2 = b^2 - a^2 = 25 - 16 = 9$; $c = 3$.

   (ii) The foci are $(0, -3)$ and $(0, 3)$.

   (iii) x-intercepts: (4, 0), (-4, 0);
   y-intercepts: (0, 5), (0, -5)

   Continued on next page

3. (iv)

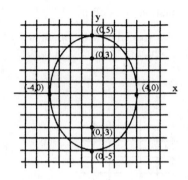

5. (i) $\frac{x^2}{16} - \frac{y^2}{9} = 1$; this is the equation of a hyperbola.

$a^2 = 16$ and $b^2 = 9$; the axis is horizontal..

The foci are $(\pm c, 0)$, where $c^2 = a^2 + b^2 = 16 + 9 = 25$; $c = 5$

(ii) The foci are $(5, 0)$ and $(-5, 0)$.

(iii) x-intercepts: $(4, 0)$, $(-4, 0)$;

y-intercepts: None

(iv)

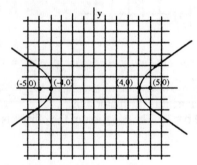

7. (i) $\frac{y^2}{4} - \frac{x^2}{10} = 1$; this is the equation of a hyperbola.

$a^2 = 10$ and $b^2 = 4$; the axis is vertical.

The foci are $(0, \pm c)$, where $c^2 = a^2 + b^2 = 10 + 4 = 14$; $c = \sqrt{14} \approx 3.74$

(ii) The foci are $(0, \sqrt{14})$ and $(0, -\sqrt{14})$.

(iii) x-intercepts: None;

y-intercepts: $(0, 2)$ and $(0, -2)$

(iv)

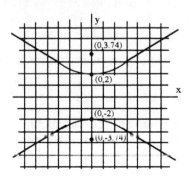

9. (i) $\dfrac{x^2}{16} + \dfrac{y^2}{16} = 1$; this is the equation of an ellipse.

$a^2 = 16$ and $b^2 = 16$; This is a special case of an ellipse, a circle.

(ii) There is a single focus, the center of the circle, at $(0, 0)$.

(iii) x-intercepts: $(4, 0)$, $(-4, 0)$;

y-intercepts: $(0, 4)$, $(0, -4)$

(iv)

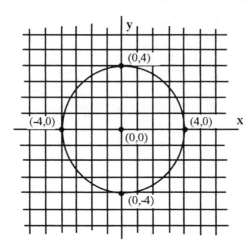

11. The equation for an ellipse centered at the origin is $\dfrac{x^2}{a^2} + \dfrac{y^2}{b^2} = 1$.

The major axis has a length of 20; the minor axis has a length of 12.
Since the foci are on the x-axis, it is the major axis.
This means that $a = 10$ and $b = 6$. The equation of the ellipse is:

$$\dfrac{x^2}{100} + \dfrac{y^2}{36} = 1$$

13. The equation for an ellipse centered at the origin is $\dfrac{x^2}{a^2} + \dfrac{y^2}{b^2} = 1$.

The foci are on the y-axis, so it is the major axis.
The major axis has a length of 8, so $b = 4$.

The minor axis has a length of 5, so $a = \dfrac{5}{2}$.

The equation is $\dfrac{x^2}{\frac{25}{4}} + \dfrac{y^2}{16} = 1$. This can be written as $\dfrac{4x^2}{25} + \dfrac{y^2}{16} = 1$

15. A hyperbola centered at the origin with foci on the y-axis has an equation of the form $\dfrac{y^2}{b^2}$

$- \dfrac{x^2}{a^2} = 1$. We know that $(0,-3)$ is on the y-axis, so $b = 3$.

The distance between the two foci is equal to $2c$. Since this is 12, $c = 6$.
In a hyperbola, $c^2 = a^2 + b^2$, or $a^2 = c^2 - b^2$. Therefore, $a^2 = 36 - 9 = 27$.

The equation of the hyperbola is $\dfrac{y^2}{9} - \dfrac{x^2}{27} = 1$

17. Two points and a focus lie on the x-axis, so it is the axis of the hyperbola. The center is the midpoint between the two intercepts on the axis, so the center has to be the origin, (0,0). The equation of the form $\frac{x^2}{a^2} - \frac{y^2}{b^2} = 1$.

    x-intercepts are (5,0) and (-5,0), so a = 5. The given focus is (7,0), so c = 7.
    In a hyperbola, $c^2 = a^2 + b^2$, or $b^2 = c^2 - a^2$. Therefore, $b^2 = 49 - 25 = 24$.
    The equation is $\frac{x^2}{25} - \frac{y^2}{24} = 1$

19. In an ellipse, the sum of the distance from a point on the ellipse to the foci is equal to the length of the major axis.

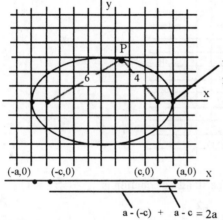

    What if P is the x-intercept (a,0)?

    Then, the distances to the foci are a - c and a + c.

    The sum equals 2a.

    $a - (-c) + a - c = 2a$

    This means that 2a = 6 + 4 = 10. This is length of the major axis; a = 5.
    For the ellipse, $a^2 = b^2 + c^2$, or $b^2 = a^2 - c^2$. Therefore, $b^2 = 25 - 16 = 9$.
    The equation is $\frac{x^2}{25} + \frac{y^2}{9} = 1$

21. Given $16x^2 - 20y^2 = 320$, the standard form of the equation is $\frac{x^2}{20} - \frac{y^2}{16} = 1$.

    (i) This a hyperbola centered at the origin with foci on the x-axis.
    (ii) The foci are (±c, 0) where $c^2 = a^2 + b^2 = 20 + 16 = 36$.
    The foci are (6, 0) and (-6, 0).
    (iii) x-intercepts: $(\sqrt{20}, 0)$, $(-\sqrt{20}, 0)$ $\sqrt{20} \approx 4.47$
    (iv)

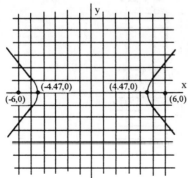

23. Given $25x^2 + 16y^2 = 400$, the standard equation is $\dfrac{x^2}{16} + \dfrac{y^2}{25} = 1$

    (i) This is the equation of an ellipse centered at the origin with the y-axis as the major axis.

    (ii) The foci are $(0, \pm c)$ where $c^2 = b^2 - a^2 = 25 - 16 = 9$; $c = 3$.
        The foci are $(0, -3)$ and $(0, 3)$.

    (iii) x-intercepts: $(4, 0)$, $(-4, 0)$;
        y-intercepts: $(0, 5)$, $(0, -5)$

    (iv)

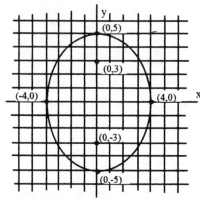

25. Given $4x^2 + 16y^2 = 100$, the standard equation is $\dfrac{x^2}{25} + \dfrac{4y^2}{25} = 1$.

    This can be written as $\dfrac{x^2}{25} + \dfrac{y^2}{\frac{25}{4}} = 1$.

    (i) This is the equation of an ellipse centered at the origin with the x-axis as the major axis.

    (ii) The foci are $(\pm c, 0)$, where $c^2 = a^2 - b^2 = 25 - \dfrac{25}{4} = \dfrac{75}{4}$; $c = \dfrac{5}{2}\sqrt{3} \approx 4.33$
        The foci are $(\dfrac{5}{2}\sqrt{3}, 0)$ and $(-\dfrac{5}{2}\sqrt{3}, 0)$.

    (iii) x-intercepts: $(5, 0)$, $(-5, 0)$;
        y-intercepts: $(0, \dfrac{5}{2})$, $(0, -\dfrac{5}{2})$ or $(0, 2.5)$, $(0, -2.5)$

    (iv)

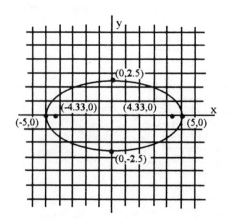

27. Since the dimensions of the gallery are 30 m by 20 m, these are the lengths of the major axis and minor axis, respectively. That is, a = 15 and b = 10.
The people should stand at the foci on a line through the vertices. The foci are located at (±c, 0), where $c^2 = a^2 - b^2 = 225 - 100 = 125$; $c \approx 11.18$.
They should stand about 11.18 m from the center, on a line through the vertices; or 3.82 m to the right of the left vertex and about 3.82 m. to the left of the right vertex.

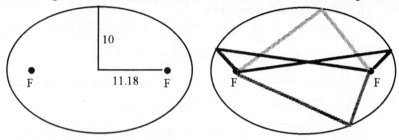

29. First we draw a sketch of the bridge.

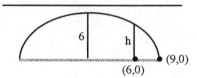

The major axis is 18 m, so a = 9. The semi-minor axis is 6 m, so b = 6.

If the center is the origin, the equation of the ellipse is $\frac{x^2}{81} + \frac{y^2}{36} = 1$ and a point 3 m from the end corresponds to x = 6.

$\frac{x^2}{81} + \frac{y^2}{36} = 1$ is equivalent to $36x^2 + 81y^2 = 2916$

When x = 6, we have $36(6)^2 + 81y^2 = 2916$, or $81y^2 = 1620$; $y^2 = 20$.

The height above the water 3 feet from the end is $h = \sqrt{20} \approx 4.47$ m.

31. First, we will draw a sketch. Since we are measuring the difference between the times Hal and Bob hear the thunder, this is equivalent to measuring the distances the sound travels until they hear it. This suggests a hyperbola with Hal and Bob at the foci. Since they are 8800 ft apart, the foci are (4400, 0) and (-4400, 0).

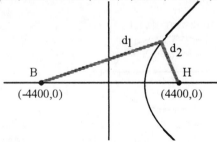

Continued on next page

31. Continued

Since Hal hears the thunder 4 seconds before Bob, the difference in distances the sound travels is $4 \times 1100 = 4400$ feet. In terms of the geometry of the hyperbola the differences between distances is 2a, which is the distance between the vertices. That is, $2a = 4400$, or $a = 2200$.

For the hyperbola, $b^2 = c^2 - a^2 = (4400)^2 - (2200)^2$.

This can be simplified as follows:

$$b^2 = (4 \times 1100)^2 - (2 \times 1100)^2 = 16 \times (1100)^2 - 4 \times (1100)^2 = 12 \times (1100)^2$$

The equation of the hyperbola is $\dfrac{x^2}{4(1100)^2} - \dfrac{y^2}{12(1100)^2} = 1$.

The lightning could have struck anywhere along the part of the hyperbola closest to Hal.

33. First, we draw a sketch to show the relationships.

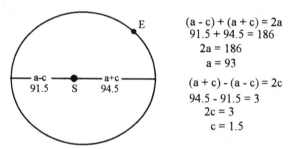

$$(a - c) + (a + c) = 2a$$
$$91.5 + 94.5 = 186$$
$$2a = 186$$
$$a = 93$$

$$(a + c) - (a - c) = 2c$$
$$94.5 - 91.5 = 3$$
$$2c = 3$$
$$c = 1.5$$

The major axis has a length of 186 million miles. The semi-major axis has a length of 93 million miles. The difference between the longest and shortest distances is 2c, so c = 1.5 million miles.

For the ellipse, $a^2 = b^2 + c^2$, or $b^2 = a^2 - c^2 = (93)^2 - (1.5)^2 = 8646.75$
$a = 93,000,000$, and $b = 92,988,000$; the orbit is nearly circular.
The equation of the orbit is

$$\frac{x^2}{8.649 \times 10^{15}} + \frac{y^2}{8.64675 \times 10^{15}} = 1$$

35. The eccentricity of an ellipse is defined as $e = \dfrac{c}{a}$.

For the Earth's orbit around the sun, we have c = 1.5 and a = 93.

Therefore, $e = \dfrac{1.5}{93} \approx 0.016$.

37. Consider the following sketch:

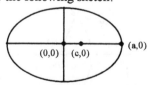

The focus is closer to the origin than the vertex, so $0 < c < a$.
We divide all terms by a; since a is positive, the inequalities still hold.

That is, $\dfrac{0}{a} < \dfrac{c}{a} < \dfrac{a}{a}$ or $0 < e = \dfrac{c}{a} < 1$.

39. Refer to the sketch in problem 37.

In the ellipse, $a^2 = b^2 + c^2$; as c gets closer to 0, b gets closer to a.
The ellipse approaches a circle. When c = 0, the ellipse becomes a circle.
The eccentricity is 0 if, and only if, a = b.

Note: As c gets closer to a, the ellipse "collapses" onto the line segment between (-a, c) and (a, c).

43. With the center at (3,4), we have h = 3 and k = 4.
The major axis is horizontal with length 12. This means a = 6.
The minor axis has length 8; b = 4.

The equation of the ellipse is $\dfrac{(x - 3)^2}{36} + \dfrac{(y - 4)^2}{16} = 1$

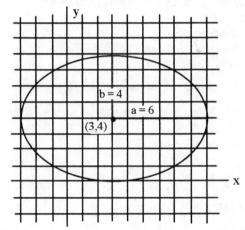

45. With the center at (2,-3), we have h = 2 and k = -3.
The major axis is vertical with length 12. This means b = 6.
The minor axis has length 8; a = 4.

The equation of the ellipse is $\dfrac{(x - 2)^2}{16} + \dfrac{(y + 3)^2}{36} = 1$. Note: y - (-3) = y + 3

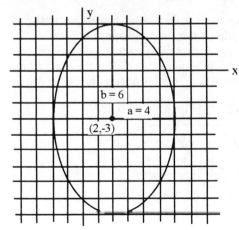

49. We begin with the given equation: $\sqrt{(x + c)^2 + y^2} - \sqrt{(x - c)^2 + y^2} = 2a$.
We add 2a to both sides: $\sqrt{(x + c)^2 + y^2} = 2a + \sqrt{(x - c)^2 + y^2}$

Squaring both sides yields
$$(x + c)^2 + y^2 = 4a^2 + 4a\sqrt{(x - c)^2 + y^2} + (x - c)^2 + y^2.$$
Next, we isolate the radical
$$(x + c)^2 - (x - c)^2 - 4a^2 = 4a\sqrt{(x - c)^2 + y^2}$$
which simplifies to $4cx - 4a^2 = 4a\sqrt{(x - c)^2 + y^2}$.
Then, we divide both side s by 4: $cx - a^2 = a\sqrt{(x - c)^2 + y^2}$.

Squaring both sides yields $c^2x^2 - 2a^2cx + a^4 = a^2[(x - c)^2 + y^2]$.

Simplifying the right hand side yields
$$c^2x^2 - 2a^2cx + a^4 = a^2x^2 - 2a^2xc + a^2c^2 + a^2y^2,$$
which can be rearranged as
$$c^2x^2 - a^2x^2 - a^2y^2 = a^2c^2 - a^4 \text{ or } (c^2 - a^2)x^2 - a^2y^2 = a^2(c^2 - a^2).$$

In a hyperbola, $c^2 - a^2 = b^2$. Substitution gives us $b^2x^2 - a^2y^2 = a^2b^2$.

Dividing both sides by a^2b^2 we obtain $\dfrac{x^2}{a^2} - \dfrac{y^2}{b^2} = 1$.

Chapter 12 Review Problems

Solutions to Odd-numbered Problems

1. The formula for the sum of the vertex angles in an n-gon is $(n - 2) \times 180°$.
For a 9-gon, the sum of the vertex angles is $(9 - 2) \times 180° = 1260°$.
Each of the vertex angles in a regular 9-gon is $1260° \div 9 = 140°$.

3. There are two possibilities using squares and equilateral triangles, both of which are included in Figure 12.14 (top row, second and fourth figures).
These are semi-regular tilings because all vertex figures are the same. The vertex figure for the fourth tiling is shown in Example 12.4. For the other tiling, the vertex figure is shown with the tiling below, which is equivalent to the one in Figure 12.14.

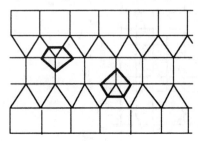

5. First, we draw a sketch.

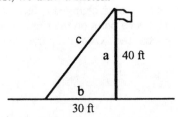

The distance from the top of the flagpole to the tip of the shadow is the hypotenuse of a right triangle:

$$c^2 = a^2 + b^2 = (40)^2 + (30)^2 = 1600 + 900 = 2500. \quad c^2 = 2500; \; c = 50.$$

The distance from the top of the flagpole to the tip of the shadow is 50 feet.

7. There are 5 lines of symmetry in a regular pentagon.

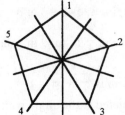

9. Reflect about a vertical axis and translate one unit upward.

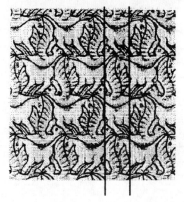

11. The pentagon is reflected across the line $\overline{AB}$.

Lines of reflection are at 90° to $\overline{AB}$.

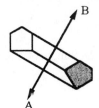

13. First, we draw a sketch, with the cross-section of the paraboloid centered at the origin of the coordinate system. All dimensions have been changed to feet. The satellite dish has a diameter of 6 feet and a depth of 3 feet, so a point on the rim of a cross-section of the paraboloid would be 3 feet from the y-axis (the axis of the paraboloid) and 3 foot above the x-axis.

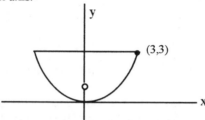

The equation has the standard form $y = \frac{1}{4b} x^2$. The coordinates of the point have to satisfy the equation, so $3 = \frac{1}{4b}(3)^2$. This simplifies to $4b = 3$, or $b = \frac{3}{4}$.

The receiver (focus) should be in the center $\frac{3}{4}$ ft. (or 9 in.) above the vertex.

15. Given $8y^2 - 12x^2 = 72$, the standard form is $\frac{y^2}{9} - \frac{x^2}{6} = 1$.

This is the equation of a hyperbola centered at the origin with a vertical axis.
This means that $b^2 = 9$ and $a^2 = 6$.
In a hyperbola, $c^2 = a^2 + b^2 = 6 + 9 = 15$. $c^2 = 15$; $c = \sqrt{15} \approx 3.87$
The foci are $(0, \sqrt{15})$ and $(0, -\sqrt{15})$.

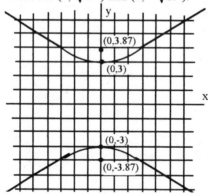

17. Since the vertices and foci are on the x-axis and equidistant from the origin, the form of

the equation is $\frac{x^2}{a^2} - \frac{y^2}{b^2} = 1$.

Since the vertices are $(\pm 2, 0)$, we see that $a = 2$. Similarly, $c = 3$.
For a hyperbola, $b^2 = c^2 - a^2 = 9 - 4 = 5$. $b^2 = 5$

The equation of the hyperbola is $\frac{x^2}{4} - \frac{y^2}{5} = 1$.

13 Growth and Scaling

Section 13.1 Scaling of Length and Area

Goals
1. Solve problems involving similar triangles.

Key Ideas and Questions
1. Describe similitude and scaling factors.
2. How are length and area of similar figures related to the scaling factor in a similitude that transforms one figure into the other?

Vocabulary

Similar Triangle	Translation	Contraction
Scaling Factor	Rotation	Similitude
Scaling of Perimeter	Size Transformation	
Scaling of Area	Expansion	

. . . .

Overview

This chapter examines the concept of similar figures and the way in which length, area, and volume are affected by changes in size.

Similar Triangles

We begin our study of **geometric similarity** with the example of **similar triangles**. Two triangles $\triangle ABC$ and $\triangle DEF$ are **similar** if there is a correspondence of points $A\leftrightarrow D$, $B\leftrightarrow E$, and $C\leftrightarrow F$ such that corresponding angles are equal (1) $\angle A = \angle D$ (2) $\angle B = \angle E$ (3) $\angle C = \angle F$, and the ratios of lengths of corresponding sides are equal (4) $\dfrac{AB}{DE} = \dfrac{BC}{EF} = \dfrac{CA}{FD}$. By convention, the notation $\overline{AB}$ indicates the line segment connecting A and B, and AB indicates the *length* of the line segment. Also, a pair of similar triangles is indicated with the corresponding points in the same order:

$$\triangle ABC \sim \triangle DEF$$

This notation means that the triangles are similar.

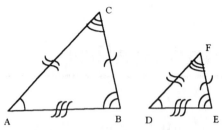

In the above figure, $\triangle ABC \sim \triangle DEF$ because $\angle A = \angle D$, $\angle B = \angle E$, $\angle C = \angle F$, and $\dfrac{AB}{DE} = \dfrac{BC}{EF} = \dfrac{CA}{FD}$.

Another way to state this similarity is that $\triangle ABC \sim \triangle DEF$ since the corresponding angles have the same measure and the corresponding sides are proportional (in the same ratio). A fundamental property of similar triangles relies on the fact that the sum of the angles in any triangle is $180°$.

Angle-Angle Property of Similar Triangles

If two angles of one triangle are equal to two angles of another, then the triangles are similar.

Consider similar triangles, $\triangle ABC \sim \triangle DEF$. Since by definition, the ratios of the lengths of corresponding sides are equal, $\dfrac{AB}{DE} = \dfrac{BC}{EF} = \dfrac{CA}{FD} = r$. The value of r represents the common ratio in the triangles and it is called the **scaling factor** of $\triangle ABC$ with respect to $\triangle DEF$. Also, the scaling factor of $\triangle DEF$ with respect to $\triangle ABC$ is $\dfrac{1}{r}$.

When $\triangle ABC \sim \triangle DEF$, the perimeter of $\triangle ABC$ can be obtained by multiplying the perimeter of $\triangle DEF$ by the scaling factor of $\triangle ABC$ with respect to $\triangle DEF$.

Scaling of Perimeter

If $\triangle ABC \sim \triangle DEF$ and r is the scaling factor of $\triangle ABC$ with respect to $\triangle DEF$, then perimeter of $\triangle ABC$ = r × perimeter of $\triangle DEF$

The scaling factor between corresponding sides of similar triangles is also used in figuring the area of a triangle.

Scaling of Area

If $\triangle ABC \sim \triangle DEF$ and r is the scaling factor of $\triangle ABC$ with respect to $\triangle DEF$, then area of $\triangle ABC$ = r^2 × area of $\triangle DEF$

Similitudes

A triangle can be transformed into a similar triangle by applying a series of transformations. To do this, you need a translation, rotation about a point, reflection in a line, and a size transformation. A **size transformation** is defined as follows:

For a fixed point, C, called the **center**, and any nonnegative number, k, called the **scaling factor**, any point P (other than C) corresponds to the point P' where P' is on the ray from C through P with $\frac{CP'}{CP} = k$.

Equivalently, $CP' = k \times CP$. See figure below.

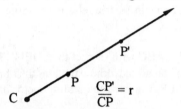

A size transformation is also known as a dilation, a magnification, or a dilatation. When $r > 1$, the result of a size transformation is a larger similar figure. This is called an **expansion**. When $r < 1$, the result of a size transformation is a smaller similar figure. This size transformation is called a **contraction**. The term size transformation refers to either possibility.

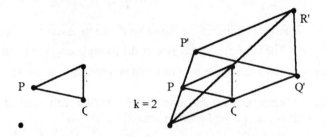

Any combination of the four basic transformations; that is, translations, rotations, reflections or size transformations, is called a **similitude**.

When similitudes are applied to objects in the plane, shapes are preserved, but sizes may not be. If one similitude is followed by another similitude, then the combination is also a similitude and the scaling factor of the combination is the product of the scaling factors.

Suggestions and Comments for Odd-numbered problems

1. through **7.**

When asked to find the scaling factor of $\triangle ABC$ with respect to $\triangle DEF$, you need to find the ratio of the sides in $\triangle ABC$ compared to the corresponding sides in $\triangle DEF$. In part (b), when you want to find the missing sides or perimeter in $\triangle DEF$, you will need to know the scaling factor of $\triangle DEF$ with respect to $\triangle ABC$. This is the reciprocal of the scaling factor in part (a).

To find the perimeter of $\triangle DEF$, you can either find each of the sides in $\triangle DEF$ using the scaling factor of $\triangle DEF$ with respect to $\triangle ABC$ and then find their sum, or you can use the fact that the perimeter has the same scaling factor as the lengths. Conversely, the ratio of the perimeters in two similar triangles is the same as the scaling factor for the sides.

9. What is the scaling factor of $\triangle ABC$ with respect to $\triangle DEF$? This can be used to find the perimeter of $\triangle ABC$, and you already know the lengths of two sides in $\triangle ABC$. To find the missing sides in $\triangle DEF$, you need the scaling factor of $\triangle DEF$ with respect to $\triangle ABC$.

11. Two triangle are similar if corresponding angles are equal. However, since the sum of the angles in a triangle is 180°, if you know the measures of two of the angles in a triangle, you can find the third by subtraction. As a result, if you can show that two pairs of corresponding angle are equal, this will mean the triangles are similar. How does the midpoint help you find the scaling factor?

13. and **15.**

The area of a triangle is found by taking one-half the product of the base and the height, or Area $= \frac{1}{2}$ bh. The base can be any side of the triangle, and the height is the length of the line segment from the opposite vertex that is perpendicular to the side.

17. Use the scaling factor between the perimeters to find the scaling factor between the triangles. What is the scaling factor for area?

19. Assume that the cost of materials is proportional to the area of the decking.

29. through **33.**

When there is a sequence of transformations, make a sketch and show the intermediate transformations.

37. What is the length of the base in the larger of the two triangles?

39. and **41.**

Think of physically moving or altering one triangle to change it into the size and location of the other, and then describe the moves in terms of basic transformations. To determine the factor for a size transformation, take a pair of corresponding sides and find the ratio of their lengths. Reflections can be made with respect to any line. Each problem can be solved in many ways.

Solutions to Odd-numbered Problems

1. (a) Since A↔D and B↔E, $\overline{AB}$ and $\overline{DE}$ are corresponding sides.

The scaling factor of △ABC with respect to △DEF is $r = \dfrac{AB}{DE} = \dfrac{12}{7}$.

(b) To find EF and DF we need the scaling factor of △DEF with respect to △ABC. This scaling factor is $\dfrac{1}{r} = \dfrac{7}{12}$. Each side of △ABC is multiplied by $\dfrac{7}{12}$ to find the length of the corresponding side in △DEF.

$EF = \dfrac{7}{12} \times BC = \dfrac{7}{12} \times 9 = \dfrac{63}{12} = \dfrac{21}{4} = 5.25$

$DF = \dfrac{7}{12} \times AC = \dfrac{7}{12} \times 4 = \dfrac{28}{12} = \dfrac{7}{3} = 2.333$

3. (a) Since A×D and B×E, $\overline{AB}$ and $\overline{DE}$ are corresponding sides.

The scaling factor of △ABC with respect to △DEF is $r = \dfrac{AB}{DE} = \dfrac{7}{10}$.

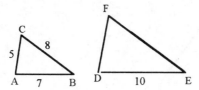

(b) The scaling factor of △DEF with respect to △ABC is $\dfrac{1}{r} = \dfrac{10}{7}$.

The perimeter has the same scaling factor as the sides:
perimeter(△ABC) = 5 + 8 + 7 = 20

perimeter(△DEF) = $\dfrac{10}{7} \times 20 = \dfrac{200}{7} = 28.571$

Note: $EF = \dfrac{10}{7} \times 8 = 11.429$; $DF = \dfrac{10}{7} \times 5 = \dfrac{50}{7} = 7.143$

5. (a) Since A↔D and C↔F, $\overline{AC}$ and $\overline{DF}$ are corresponding sides.

The scaling factor of △ABC with respect to △DEF is $r = \dfrac{AC}{DF} = \dfrac{5}{10} = \dfrac{1}{2}$.

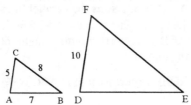

(b) The scaling factor of △DEF with respect to △ABC is $\dfrac{1}{r} = 2$.

The perimeter has the same scaling factor as the sides:
perimeter(△ABC) = 5 + 8 + 7 = 20
perimeter(△DEF) = 2 × 20 = 40
Note: DE = 2 × 7 = 14; EF = 2 × 8 = 16

7. **(a)** The scaling factor for perimeter is the same as the scaling factor for the sides. Since $\triangle ABC$ has a perimeter of $8 + 7 + 5 = 20$ and the perimeter of $\triangle DEF$ is 30, the scaling factor of $\triangle ABC$ with respect to $\triangle DEF$ is $\frac{2}{3}$.

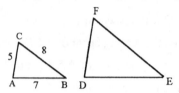

 (b) The scaling factor of $\triangle DEF$ with respect to $\triangle ABC$ is $\frac{1}{r} = \frac{3}{2} = 1.5$.

  $DF = 1.5 \times 5 = 7.5$, $EF = 1.5 \times 8 = 12$, and $DE = 1.5 \times 7 = 10.5$
  Note: As a double-check, $7.5 + 12 + 10.5 = 30$

9. The scaling factor for perimeter is the same as the scaling factor for the sides.

 Since $A \leftrightarrow D$ and $C \leftrightarrow F$, $\overline{AC}$ and $\overline{DF}$ are corresponding sides.

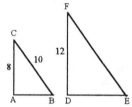

 The scaling factor of $\triangle ABC$ with respect to $\triangle DEF$ is $r = \frac{AC}{DF} = \frac{8}{12} = \frac{2}{3}$.

 Since the perimeter of $\triangle DEF$ is 36, $\triangle ABC$ has a perimeter of $\frac{2}{3} \times 36 = 24$.

 Since we know the lengths of two of the sides of $\triangle ABC$, we see that $AB = 6$.

 The scaling factor of $\triangle DEF$ with respect to $\triangle ABC$ is $\frac{1}{r} = \frac{3}{2} = 1.5$.

  Since $AB = 6$, $DE = 1.5 \times 6 = 9$, and $EF = 1.5 \times BC = 1.5 \times 10 = 15$
 Check: perimeter$(\triangle DEF) = 12 = 9 + 15 = 36$

11. **(a)** Since D and E are midpoints, from the first fact preceding the problem we can see that $\overline{DE}$ is parallel to $\overline{BC}$.

  Since the line through A and B intersects the parallel lines through D and E and B and C, the second fact tells us $\triangle ADE = \triangle ABC$. Similarly, $\triangle AED = \triangle ACB$. The corresponding angles of the two triangles are equal, so the triangles are similar.

 (b) The scaling factor for $\triangle DEF$ with respect to $\triangle ABC$ is $\frac{AD}{AB} = \frac{1}{2}$.

  $DE = \frac{1}{2} \times 16 = 8$

 (c) perimeter$(\triangle DEF) = AD + DE + AE = 4 + 8 + 6 = 18$

 (d) The scaling factor for the perimeter is the same as the scaling factor for the sides. Since the scaling factor is $\frac{1}{2}$ and the perimeter of $\triangle ADE$ is $8 + 16 + 12 = 36$,

  perimeter$(\triangle DEF) = \frac{1}{2} \times 36 = 18$.

13. **(a)** Based on the correspondence of points, $\overline{AB} \leftrightarrow \overline{DE}$.

The scaling factor for $\triangle ABC$ with respect to $\triangle DEF = \dfrac{AB}{DE} = \dfrac{7}{12} = 0.583$

(b) Area($\triangle ABC$) = $\frac{1}{2} \times b \times h = \frac{1}{2} \times 8 \times 4.43 = 17.72$

The scaling factor for $\triangle DEF$ with respect to $\triangle ABC = \dfrac{12}{7} = 1.714$

The area scales by the square of the scaling factor.
Area($\triangle DEF$) = $r^2 \times$ Area($\triangle ABC$) = $(1.714)^2 \times 17.72 = 52.08$

15. **(a)** Based on the correspondence of points, $\overline{AC} \leftrightarrow \overline{DF}$.

The scaling factor for $\triangle ABC$ with respect to $\triangle DEF = \dfrac{AC}{DF} = \dfrac{5}{15} = \dfrac{1}{3} = 0.333$.

(b) Area($\triangle ABC$) = $\frac{1}{2} \times b \times h = \frac{1}{2} \times 8 \times 4.43 = 17.72$

The area scales by the square of the scaling factor.

The scaling factor for $\triangle DEF$ with respect to $\triangle ABC = \dfrac{15}{5} = 3$

Area($\triangle DEF$) = $r^2 \times$ Area($\triangle ABC$) = $9 \times 17.72 = 159.48$

17. First, we need to find the scaling factor for the sides. This is the same as the scaling factor for the perimeter. We are given perimeter($\triangle DEF$) = 54, and we find perimeter($\triangle ABC$) = 5 + 10 + 12 = 27. The scaling factor of $\triangle DEF$ with respect to $\triangle ABC$ is $\dfrac{54}{27} = 2$. The scaling factor for area is $r^2 = 4$.

Area($\triangle DEF$) = $r^2 \times$ Area($\triangle ABC$) = $4 \times 26 = 104$

19. The two decks are similar rectangles: $r = \dfrac{12}{8} = \dfrac{15}{10} = 1.5$.

The scaling factor for area is $r^2 = 2.25$.
We must make an assumption that the cost of the decking materials will be proportional to the area.
Cost(larger deck) = $r^2 \times$ Cost(smaller deck) = $2.25 \times \$150 = \$337.50$

21.

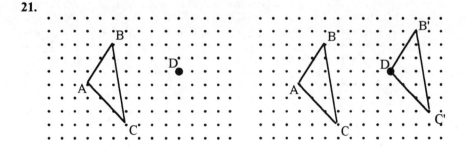

23.

25.

27.

29.

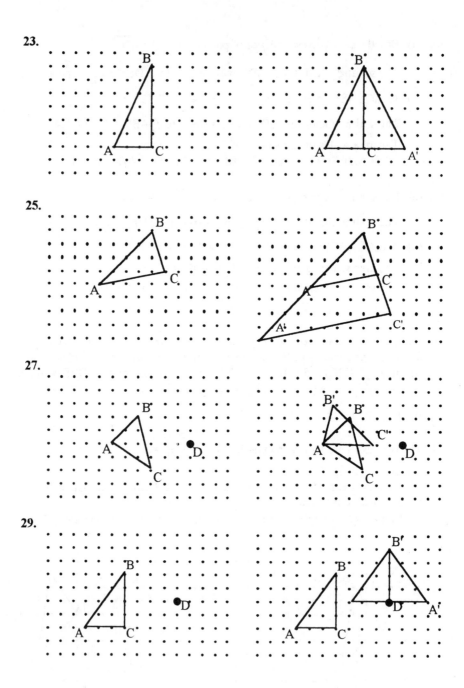

31.

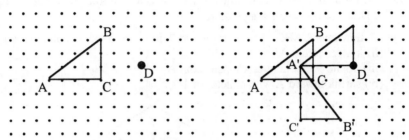

33.

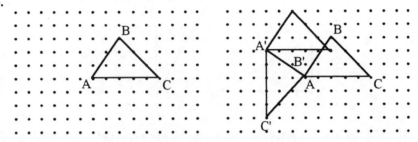

35. First, we need to look at the essentials from the sketch.

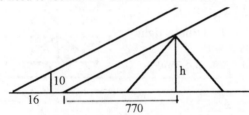

Since the sun's rays are essentially parallel, the angles they make with the ground are equal, and the two right triangles are similar (they have 2 corresponding angles equal). The scaling factor for the large triangle with respect to the smaller triangle is $r = \frac{770}{16}$.

Since h ↔ 10, we have h = r × 10 = $\frac{7700}{16}$ ≈ 481 feet

The height of the great pyramid would be calculated as approximately 481 feet.

37. First we draw the sketch slightly differently to emphasize the lengths.

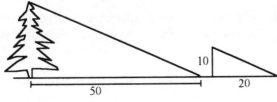

The scaling factor is $r = \frac{50}{20}$ = 2.5.

The height of the tree is r × 10 = 25 feet.

39. The following sequence of transformations is a similitude that maps △ABC into △DEF. Many others are possible.

(i) translate △ABC 11 units right and 7 units up so that A coincides with D.
(ii) reflect across AB
(iii) rotate 90° counterclockwise about A
(iv) shrink toward A with a factor of 1/2.

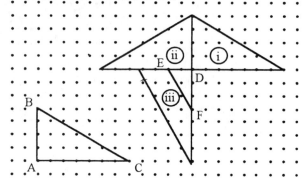

41. The following sequence of transformations is a similitude that maps △ABC into △DEF. Many others are possible.

(i) reflect across AB
(ii) translate A to D; 11 units right and 3 units down
(iii) dilate by a factor of 3 away from A

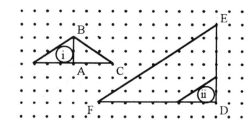

Section 13.2   Similarity and Scaling

Goals
1. Transform geometric objects into larger or smaller objects of similar shape through scaling.
2. Compute quantities such as perimeter, area, volume and weight of scaled objects.

Key Ideas and Questions
1. How do lengths, areas and volumes transform when the scaling factor is r?
2. What are the fundamental similitudes?

Vocabulary
Similar Plane Figures   Scaling Factor of Length   Scaling Factor of Volume
Similar 3-Dimensional Figures   Scaling Factor of Area

Overview

We will now look at changes in the sizes of all geometric objects and the way in which area and volume are affected.

Plane Figures

As it is in the case of triangles, two plane figures are similar if their corresponding angles are equal and their corresponding sides are proportional. We will use the transformation definition of similarity. Two plane figures P and Q are **similar** if there is a similitude that transforms one figure into the other. Recall that a similitude is a combination of four basic types of transformations: translation, rotation, reflection and size transformation. This definition also holds for triangles.

Scaling Length and Area

The following general rule applies to all one-dimensional figures:

Scaling of the Length of Plane Figures

If P and Q are one-dimensional plane figures, $P \sim Q$, and the similitude that transforms P into Q has scaling factor r, then

$$\text{length of } Q = r \times \text{length of } P$$

The following general rule applies to all two-dimensional figures:

Scaling of the Area of Plane Regions

If S and T are two-dimensional plane regions, $S \sim T$, and the similitude that transforms S into T has scaling factor r, then

$$\text{area of } T = r^2 \times \text{area of } S$$

Scaling Volume

If two objects in three-dimensional space are similar, then there is a **similitude** that transforms one into the other. If P and Q are similar three dimensional objects, we will write it as P $\sim$ Q. What happens to the volume of an object when it is transformed by a similitude? Once again, the critical piece of information is the scaling factor of the expansion or contraction involved. The effect, as generalized below, is that the volume is multiplied by the cube of the scaling factor.

Scaling of the Volume of Three-Dimensional Objects

If P and Q are three-dimensional objects, $P \sim Q$, and the similitude that transforms P into Q has scaling factor r, then

$$\text{volume of } Q = r^3 \times \text{volume of } P.$$

Suggestions and Comments for Odd-numbered problems

1. through 9.
The scaling factors for area and volume are the square and cube, respectively, of the scaling factor for length. Conversely, the scaling factor for length is the square root and cube root, respectively, of the scaling factors for area and volume.

11. There are two ways you can do this problem: Use the scaling factors related to area and volume of similar figures, or modify the formula through substitution. The first way is preferred, but you should do both and compare the results.

13. What are the scaling factors for area and volume?

17. and 19.
What is the scaling factor for the volume? If the scaling factor for the volume is the cube of the scaling factor for length, how do you find the scaling factor for length from the scaling factor for volume?

23. As the spiral is continued toward the inside, the semicircles added will have diameters that are $\frac{3}{4}$ of the previous semicircles. As you continue the spiral outward, the semicircles added will have diameters that are $\frac{4}{3}$ of the previous semicircles.

Solutions to Odd-numbered Problems

1. The scaling factor for the sides is $\frac{4}{2} = 2$, and the scaling factor for area is $2^2 = 4$.

The area of the larger regular hexagon is $4 \times 6\sqrt{3} = 24\sqrt{3}$.

3. The scaling factor for the sides is $\frac{4}{3} \div 2 = \frac{2}{3}$. Therefore, the scaling factor for area is $(\frac{2}{3})^2 = \frac{4}{9}$.

The area of the larger hexagon is $\frac{4}{9} \times 6\sqrt{3} = \frac{8}{3}\sqrt{3}$

5. If r is the scaling factor for length, then the scaling factor for area is r^2.
From the areas of the two hexagons, we see that $r^2 = \frac{20}{10.4} = 1.923$.

Therefore, $r = \sqrt{1.923} = 1.387$
The side length of the larger hexagon is $1.387 \times 2 = 2.774$.

7. The scaling factor for perimeter is the same as the scaling factor the size transformation, while the scaling factor for the area is the square of the scaling factor for the size transformation.
The perimeter is 3 times larger and the area is 9 times larger.

9. The scaling factor for volume is the cube of the scaling factor of the size transformation. $r^3 = (2.5)^3 = 15.265$.
The volume of the larger region is $15.265 \times 45 = 703.125$.

11. There are two ways to approach this problem:

(i) Since the scaling factor for the size transformation is $\frac{2s}{s} = 2$, the scaling factor for surface area is $2^2 = 4$ and the scaling factor for volume is $2^3 = 8$.

For the larger cube: surface area = $4 \times 6s^2 = 24s^2$, volume = $8 \times s^3 = 8s^3$.

(ii) We can also substitute 2s in place of s in each formula:

$$\text{Area} = 6(2s)^2 = 6 \times 4s^2 = 24s^2$$
$$\text{Volume} = (2s)^3 = 8s^3$$

13. Since the scaling factor for the size transformation is 1.5, the scaling factor for surface area is $(1.5)^2 = 2.25$ and the scaling factor for volume is $(1.5)^3 = 3.375$. For the larger region: surface area = $2.25 \times 76 = 171$ in^2, and volume = $3.375 \times 40 = 135$ in^3.

15. The scaling factor for volume is the cube of the scaling factor for the size transformation, so $r^3 = (\frac{1}{2})^3 = \frac{1}{8}$.

The volume of the sports arena would be $\frac{1}{8} \times 91,400,000$ ft^3.

Volume = 11,425,000 ft^3

17. If r is the scaling factor for the size transformation and the volume of the scaled replica is 1,000,000 ft^3, then the scaling factor for volume is

$r^3 = \frac{1000000}{91400000} = 0.010941.$ Therefore, $r = \sqrt[3]{0.010941} = 0.222.$

19. (a) If all the dimensions were doubled, this is a size transformation with a scaling factor of 2. The scaling factor for volume is $r^3 = 2^3 = 8$.
The volume of the new altar would be 8 times as large as the original.

(b) If the volume is twice the original, then $r^3 = 2$.

The scaling factor should be $r = \sqrt[3]{2}$, or about 1.26.

21. A self-similar figure composed of squares.   23. A self-similar figure composed of half-circles.

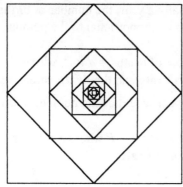
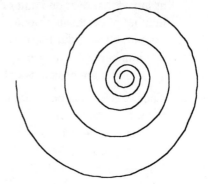

The figures keep expanding outward and shrinking inward

25. Many natural objects have a structure that exhibits self-similarity. These include many ferns, some trees, and many seashells.

Section 13.3   Applications to Population Growth and Radioactive Decay

Goals

1. Define mathematical models for population growth, radioactive decay and other quantities that change in proportion to their size.

Key Ideas and Questions

1. How do you model populations that grow at a constant rate?
2. What are limitations of this model?
3. How does the logistic law make for a more realistic model for growth?
4. How are the half-life and decay rate of a radioactive substance related?

Vocabulary

Birth Rate	Radioactive Decay	Logistic Law
Death Rate	Decay Rate	Stable Population
Growth Rate	Half-Life	Population Dynamics
Mathusian	Logistic	
Population Growth	Population Model	
Ponzi Scheme	Carrying Capacity	

· ·

Overview

Growth and scaling are important considerations in any situation where there is a limit to resources. The ability to accurately predict future population size is important for the planning and allocation of resources.

Population Growth

One of the first models of population growth was suggested by Thomas Malthus. He assumed that over one unit of time, usually a year, the number of births and the number of deaths in a population would be proportional to the population. Thus, if the population is P, during one year

    number of births = $b \times P$ and number of deaths = $d \times P$, where
    b and d are constants called the **birth rate** and the **death rate**.

The net change in population over one year would be the number of births minus the number of deaths or

    net change = $b \times P - d \times P = (b\text{-}d) \times P = r \times P$ where $r = b\text{-}d$ is
    a new constant called the annual **growth rate**.

The new population is $P + r \times P$, or $(1 + r) \times P$.

The growth rate must be determined from data about the population, and it may change over time. With Malthus' model, for each year that passes, the effect on the population is to multiply by another factor of $(1 + r)$. When there is a constant growth rate, the result is *exponential growth* (see Section 1.7).

Malthusian Population Growth

If the population is initially P_0, then after m years the population will be

$$(1 + r)^m \times P_0$$

where r is the growth rate.

To apply the Malthusian model, it is necessary to know the growth rate. We present a formula for this growth rate below.

Formula for the Growth Rate

If the population changes from P to Q in m years and a Malthusian model is assumed, then the growth rate r is given by

$$r = \left(\frac{Q}{P} \right)^{1/m} - 1.$$

Radioactive Decay

Radioactive materials lose particles in proportion to the amount of material present. Thus, the behavior of radioactive materials is exactly modeled by the Malthusian model, except with a <u>negative</u> growth rate. The decay rate is a constant associated with the atomic structure of a particular substance.

Radioactive Decay

If a radioactive substance has an annual decay rate of d and there are initially A_0 units of the substance present, then after m years there will be

$$(1 - d)^m \times A_0$$

units of the radioactive substance present.

The decay rate is rarely mentioned in practice. Because the amount present decreases toward zero, there must be a time after which there is half of the original amount of the substance present. This is called the **half-life** of the substance. The decay rate can be found from the half-life and vice versa.

Decay Rate and Half-Life

If d is the annual decay rate of a substance and h is the half-life of the substance (in years), then

$$d = 1 - \left(\frac{1}{2} \right)^{1/h}.$$

The passing of each half-life has the effect of multiplying the amount of substance by $\frac{1}{2}$
This leads to the following rule.

Radioactive Decay and Half-Life

If initially there are A_0 units of a radioactive substance present, and
if the half-life of the substance is h, then after time m (measured
in the same time units as h) there will be

$$\left(\tfrac{1}{2} \right)^{(m/h)} \times A_0$$

units of the radioactive substance present.

Logistic Population Models

For a given environment with limited resources, there may be a maximum population
size that can be sustained. This population size is called the **carrying capacity** of the
environment. One model for population growth that considers the carrying capacity of
the environment is called the **logistic model.**

Suppose the initial population is P_0, the population after 1 breeding season is P_1, the
population after 2 breeding seasons is P_2 and so on. If there were no resource
pressure, then after $m + 1$ seasons the population would be

$P_{m+1} = (1 + r) \times P_m.$

In order to include the effect of the carrying capacity, c, a new rule must be devised.
One such rule is

$$P_{m+1} = (1+r) \times P_m - \frac{1+r}{c} \times P_m{}^2,$$

which is called the **logistic law** or **logistic equation**.

There is often some point at which the population stays the same size. This is a **stable
population**. A population governed by the logistic law typically behaves in this
manner.

Stable Population under the Logistic Law

If a population has a natural growth rate of r and the environment has a
carrying capacity of c, then the stable population size is

$$\frac{r \times c}{1 + r}$$

The study of the behavior of sizes of populations is called **population dynamics**. This
study is no mere academic exercise. Our survival may depend on human population
dynamics. Most people now appreciate the need to harvest natural resources on a
sustainable basis. The problem for our society is to understand the parameters that
enter into the dynamics of a population so the best possible decisions can be made.

Suggestions and Comments for Odd-numbered problems

9. The first method would be to use the formula for growth rate to find the annual rate and then use the Malthusian formula with a time period of 20 years. The second method uses a time period of 10 years rather than 1 year. How many of the 10-year periods are involved?

11. Since 15 is not a multiple of 8, you must find the yearly growth rate. What value is used as current population in making the projection?

13. When does $(1 + r)^n = 2$? Use trial and error with different values of n to estimate the answer.

19. and 21.
First, use the formula for the decay rate and half-life to find the annual decay rate. In the second part of the problem, either of the other two formulas for radioactive decay can be used. If the time period is more than the half-life, the amount of material remaining must be less than half of the original. If the time period is more than twice the half-life, the amount remaining must be less than one-fourth, and so on.

23. through 29.
You can save a significant amount of time making a sequence of calculations for population levels if you have a calculator that lets you use the result of your last calculation in your next one. For example, with the Malthusian model, the population in the next year is obtained by multiplying the last population calculated by a factor of 1 + r. That is, if the growth rate is 3%, you can repeat the steps "× 1.03" as many times as needed. You don't have to evaluate the formula $P_0 \times (1.03)^n$ for each value of n. The calculations for the logistic law are more complicated, but some planning and organization will reduce the time substantially. Graphing calculators typically have a feature that enables you to display the entire set-up (numbers and operations) from the last calculation. Another feature lets you recall the previous answer (the population P_m) and even substitute it in place of the last population value. That is, you only have to set the problem up once; then you can recall it and make substitutions of the values that change from step to step.

31. Make the following assignments:
$$P_m = 1500, P_{m+1} = 2500, \text{ and } P_{m+2} = 3500$$
Use the formula to find the growth rate and then make the calculations for the next six breeding seasons. Since the population increases from 1500 to 2500 in the first breeding season, the growth rate must be more than 67%.

33. Using the Malthusian model, the problem becomes:
$$\text{When does } N = 10 \times (1.4)^n = 260,000,000?$$
Trial and error can be used to estimate the answer.

Solutions to Odd-numbered Problems

Note: Population figures are almost always a recent estimate, so exact values don't have much merit. All answers will be calculated and then rounded to a convenient or appropriate value.

1. The Malthusian model for population growth is $P = (1 + r)^m \times P_0$, where r is the growth rate and m is the number of time periods involved. $r = 3\% = 0.03$.
   After 15 years, the population will be $P = (1 + 0.03)^{15} \times 50,000 = 77,898$
   At the current rate of growth, the population will be about 78,000 in 15 years.

3. For the first city, $r = 4\% = 0.04$ and $m = 20$; $1 + r = 1.04$.
   For the second city, $r = 3\% = 0.03$ and $m = 20$; $1 + r = 1.03$.
   Populations:
   $$P1 = (1.04)^{20} \times 20,000 = 43,822$$
   $$P2 = (1.03)^{20} \times 30,000 = 54,183$$
   The second city is 1.24 times the size of the first.

5. We assume a constant growth rate of 0.7%, or $r = 0.007$; $1 + r = 1.007$.
   $P_0 = 255,000,000$, and $m = 2020 - 1990 = 30$.

   $P = (1.007)^{30} \times 255,000,000 = 314,357841$
   The estimated population in 2020 will be about 314 million.

7. At a constant 3% growth rate, and $m = 1996 - 1626 = 370$, the value of the goods would be about $V = (1.03)^{370} \times \$24 = \$1,348,914$. At a 4% growth rate the value would be about $(1.04)^{370} \times \$24 = \$48,144,511$.

9. Method 1:
   If we consider the yearly growth rate, then $r = (\frac{33000}{20000})^{1/10} - 1 = 0.05135$.

   In 20 more years, the population will be $(1.05135)^{20} \times 33,0000 = 89,838$.

   Method 2:
   If we look at growth periods of 10 years, then $r = \frac{33000}{20000} - 1 = 0.65$.

   In 2 more decades, the population will be $(1.65)^2 \times 33,000 = 89,843$.

   Note: The differences are due to rounding in method 1.
   The greater the rounding, the greater the differences will be.
   The population in 20 years will be about 90,000.

11. First, we find the yearly growth rate. $r = (\frac{35500}{28000})^{1/8} - 1 = 0.03011$.

    In 15 years, the population is calculated as
    $$P = (1.03011)^{15} \times 35,500 = 55,397$$
    The population will be about 55,500.

13. We can re-phrase the question as: When does $(1 + r)^n = 2$?

    If $r = 0.7\%$ or 0.007, this becomes: When does $(1.007)^n = 2$?

    We can get an estimate through trial and error:

    $$(1.007)^{50} = 1.417$$
    $$(1.007)^{80} = 1.747$$
    $$(1.007)^{90} = 1.873$$
    $$(1.007)^{100} = 2.009$$

    At the current rate of growth, the population of the United States will double in about 100 years.

15. The formula for the annual decay rate is $d = 1 - (\frac{1}{2})^{1/h}$, where h is the half-life. For

    Strontium 90, this means $d = 1 - (\frac{1}{2})^{1/28} = 1 - 0.97555 = 0.02445$

    The annual decay rate for Strontium 90 is about 2.4%.

17. We use the formula $A_m = A_0 \times (\frac{1}{2})^{m/h}$, where h is the half-life.

    For the Plutonium 241, we have $m = 1994 - 1950 = 44$, and $h = 13$.

    $$A_{50} = 100 \times (\frac{1}{2})^{44/13} = 9.57 \text{ grams}$$

19. (a) Half-life to Decay Rate formula: $d = 1 - (\frac{1}{2})^{1/h}$.

    $$d = 1 - (\frac{1}{2})^{1/2.6} = 0.234 \text{ or } 23.4\%$$

    (b) We use the half-life formula $A = A_0 \times (\frac{1}{2})^{m/h}$, with $m = 55$.

    $$A = 200 \times (\frac{1}{2})^{55/2.6} = 8.572 \times 10^{-5}, \text{ or about } 0.000086 \text{ grams}$$

21. (a) Half-life to Decay Rate formula: $d = 1 - (\frac{1}{2})^{1/h}$.

    $$d = 1 - (\frac{1}{2})^{1/8.5} = 0.0783 \text{ or } 7.83\%$$

    (b) We use the half-life formula $A = A_0 \times (\frac{1}{2})^{m/h}$, with $m = 30$.

    $$A = 100 \times (\frac{1}{2})^{30/8.5} = 8.6605, \text{ or about } 8.7 \text{ grams}$$

23. Population governed by logistic law with $r = 0.5$, $c = 2000$, and $P_0 = 1000$

    (a)   The population for the next 10 breeding seasons:

Season	Population	Season	Population
0	1000		
1	750	6	669
2	703	7	668
3	684	8	667
4	675	9	667
5	671	10	667

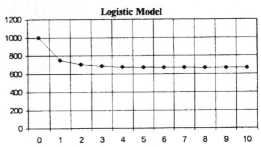

    (b) The population stabilizes at 667.

25. Population governed by logistic law with r = 2.2, c = 2000, and P_0 = 1000

    (a) The population for the next 10 breeding seasons:

Season	Population	Season	Population
0	1000		
1	1600	6	1026
2	1024	7	1599
3	1599	8	1026
4	1026	9	1599
5	1599	10	1026

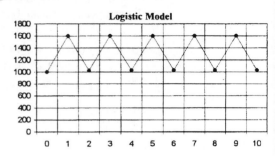

    (b) The population cycles between the values of 1026 and 1599.

27. **(a)** Logistic law model with r = 0.6, c = 2000, and P_0 = 1000.

    (b) Malthusian model with r = 0.6 and P_0 = 1000.

Season	Population (Logistic)	Population (Malthusian)
0	1000	1000
1	800	1600
2	768	2560
3	757	4096
4	753	6554
5	751	10486
6	750	16777
7	750	26844
8	750	42950
9	750	68719
10	750	109951

    (c)

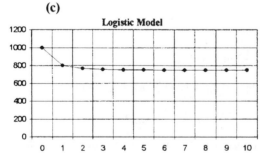
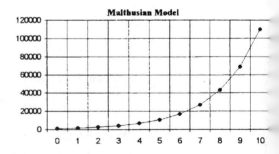

29. (a) Logistic law model with $r = 0.35$, $c = 2000$, and $P_0 = 1000$.

(b) Malthusian model with $r = 0.35$ and $P_0 = 1000$.

Season	Population (Logistic)	Population (Malthusian)
0	1000	1000
1	675	1350
2	604	1823
3	569	2460
4	550	3322
5	538	4484
6	531	6053
7	527	8127
8	524	11032
9	522	14894
10	521	20107

(c)

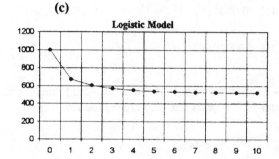

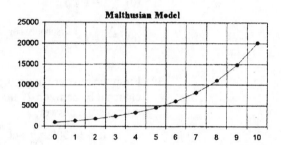

31. We make the following assignments:

$$P_m = 1500, \ P_{m+1} = 2500, \text{ and } P_{m+2} = 3500$$

(a) Carrying capacity: $c = \dfrac{(1500)^2 \times 3500 - (2500)^3}{1500 \times 3500 - (2500)^2} = 7750$

Growth Rate: $r = \dfrac{(1500)^2 \times 3500 - (2500)^3}{1500 \times 2500 \times (1500 - 2500)} - 1 = 1.06667$

(b) $P_{m+3} = (2.06667) \times 3500 - \dfrac{2.06667}{7750} \times (3500)^2 = 3967$

(c) We call $P_m = P_0$; the population for the first nine seasons

Season Population Season Population

Season	Population	Season	Population
0	1500		
1	2500	6	4000
2	3500	7	4000
3	3967	8	4000
4	4000	9	4000
5	4000		

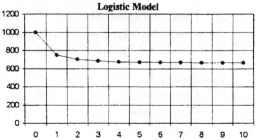

The population changes smoothly and stabilizes at 4000.

As a check, we compute the following: Stable population $= \dfrac{r \times c}{1 + r} = \dfrac{1.06667 \times 7750}{2.06667} = 4000$

33. Given $r = 0.4$, $1 + r = 1.4$. The question may be phrased as:

   When does $N = 10 \times (1.4)^n = 260,000,000$?

   We will estimate with trial and error:

   For $n = 30$, $N \approx 240,000$
          $n = 40$, $N \approx 7,000,000$
          $n = 50$, $N \approx 200,000,000$
          $n = 51$, $N \approx 280,000,000$

   It will take 51 iterations to reach the population size. Since each iteration is 90 days, this is more than 12.5 years.

37. We use the formula $\dfrac{A}{A_0} = (\tfrac{1}{2})^{m/h}$, where $h = 5600$

   (a) $\dfrac{A}{A_0} = (\tfrac{1}{2})^{(10000/5600)} \approx 29\%$ (0.29003)

   (b) $\dfrac{A}{A_0} = (\tfrac{1}{2})^{(20000/5600)} \approx 8.4\%$ (0.08412)

   (c) $\dfrac{A}{A_0} = (\tfrac{1}{2})^{(50000/5600)} \approx 0.2\%$ (0.00205)

39. From the answers in problem 37, we know that 2% of the C^{14} should remain when the charcoal be between 20,000 and 50,000 years old. We use "Guess and Test" to estimate the answer,

   $n = 30,000$, $\dfrac{A}{A_0} = (\tfrac{1}{2})^{(30000/5600)} \approx 2.44\%$

   $n = 35,000$, $\dfrac{A}{A_0} = (\tfrac{1}{2})^{(35000/5600)} \approx 1.3\%$

   To the nearest 5,000 years, we would estimate the charcoal to be approximately 30,000 years old.

Section 13.4   Scaling Physical Objects

Goals
1. Model physical objects and compute some quantities related to size, area and volume.

Key Ideas and Questions
1. What is the relationship between density, weight, and volume?
2. Why does the pressure exerted by similar objects scale like the scaling factor?

Vocabulary

Density	Pressure Rule	Tensile Strength
Weight Rule	PSI (Pounds per Square Inch)	Fastener Rule
Pressure		

. . . .

Overview

We have learned that if the shape of an object is kept the same while the linear dimensions are increased by a scaling factor of r, then areas are increased by r^2 and volumes are increased by r^3. These are important relationships for physical objects, including humans.

Weight

It is sometimes not practical to weigh an object by putting it on a scale. In such cases we estimate the volume of the object and then multiply by the density of the material from which it is made. The **density** of a material is the ratio of the material's weight to its volume. In other words, the density is the weight of a unit volume of the material. Density of materials can be used to solve many applied problems. Densities of common materials are readily available in tables and charts.

If the size of an object is increased by a scaling factor or r, then the volume is increased by a factor of r^3. Thus, if the object is composed of the same material, the weight of the object is also increased by a factor of r^3.

The Weight Rule

If the linear dimensions of a physical object are scaled by a factor of r
and a new object is composed of the same material as the old, then
[weight of the new] $= r^3 \times$ [weight of the old]

Pressure

The **pressure** on a surface is the force on the surface divided by the area of the surface. The common English unit for pressure is pounds per square inch, abbreviated **psi**.

The Pressure Rule

If the linear dimensions of a physical object are increased by a scaling factor of r and the new object is composed of the same material as the old, then

[pressure on the bottom of the new] $= r \times$ [pressure on the bottom of the old]

Bolts

A bolt is a wire rod with a head at one end and threads at the other. Two plates can be fastened together using a bolt with a nut on the threaded end. If the plates are pulled apart, the bolt is stretched. The ability of the bolt to resist this stretching is called its **tensile strength**. The tensile strength of a bolt is measured as the maximum force the bolt can survive divided by the cross-sectional area of the bolt.

Fastener Rule

If the linear dimensions of a structure are scaled by a factor of r,
then the diameters of the fasteners holding the structure together should be
scaled by $r^{3/2}$.

Modeling and Large Creatures

Scaling of objects correctly is of great importance in design work. Engineers often do experiments on a small scale. Sometimes a scale model is the only sensible procedure.

Large animals typically have relatively thicker bones than small animals because of the greater than proportional stress their weight puts on their bones. In this way, bones and joints are like fasteners holding a structure together.

. . . .

Suggestions and Comments for Odd-numbered problems

5. Find the weight of the water and assume none is "lost" when it freezes.

9. Although the internal structure (bones, organs, etc.) will not all scale by the same value (bones of larger people need to be proportionately thicker), the average density in large people is still basically the same as in smaller people. That is why the weight rule can be applied to people.

11. There are two ways to approach the problem:
If Earth and the moon are made of the same materials, then the masses are proportional to the volumes. However, if Earth and the moon are made of the same material, the weight rule can be applied.

13. If the average densities were the same, the ratio of the volumes would be the same as the ratio of the masses How does the volume of Venus compare to the volume of Earth?

19. Use w for the weight per cubic foot of the material and calculate the pressure on the base of the object. The answers will contain w in the expressions, but they can be easily compared (as a ratio or otherwise).

Solutions to Odd-numbered Problems

1. (a) A 1-foot cube has a volume of 12^3 in^3 = 1728 in^3.

   The weight of a 1-inch cube is $\frac{500}{17280}$ = 0.289 lb or 4.63 oz.

   (b) A 1-yard cube has a volume of 3^3 ft^3 = 27 ft^3.
   The weight of a 1-yard cube is 27×500 lb = 13,500 lbs.

3. All dimensions are in feet.
   The volume of the block is V = $3 \times 1 \times 1.5$ = 4.5 ft^3.
   The weight of the block is 4.5×57.4 lbs = 258.3 lbs.

5. The weight of the 10 gallons of water is W = 10×8.3 lbs = 83 lbs.
   From problem 3, we know a cubic foot of ice weighs 57.4 lbs.

   The volume of the ice is V = $\frac{83}{57.4}$ = 1.446 ft^3.

7. From Example 13.12, it was estimated that a typical human being has a volume of 2.4 ft^3. Most fall in a range from 1.6 ft^3. to 3.2 ft^3.

   (a) At 500 lbs per ft^3, a life-size statue of the typical human made of steel would weigh 2.4×500 = 1200 lb

   (b) Marble weighs 170 lbs per ft^3. A life-size statue made of marble would weigh 2.4×170 = 408 lb

9. First, we change the heights to inches: Smith, 76; Olajuwon, 84.

   The scaling factor for height is r = $\frac{84}{76}$ = 1.1053.

   That means the scaling factor for volume and weight is $r^3 = (1.1053)^3 \approx 1.35$.

   The weight rule applies to similar objects of the same composition or density. If Smith had been scaled up to Olajuwon's height, he would weigh 1.35×190 = 256.5 lb, or about the same as Olajuwon.

11. For the Earth, V = $\frac{4}{3} \times \pi \times (3964)^3 \approx 2.61 \times 10^{11}$ mi^3.

   For the moon, V = $\frac{4}{3} \times \pi \times (1080)^3 \approx 5.28 \times 10^9$ mi^3.

   If we assume that mass is proportional to volume, then the ratio of the masses is the same as the ratio of the volumes.

   The mass of Earth is $\dfrac{2.61 \times 10^{11}}{5.28 \times 10^9} \approx 49$ times the mass of the moon.

13. If the two planets had the same average density, then the ratio of the volumes would be the same as the ratio of the masses. Since the mass of Venus is compared to that of Earth, we will compare the volumes.

    For lengths, the scaling factor for Venus with respect to Earth is $\frac{3760}{3964}$.

    This means the scaling factor for volume is $(\frac{3760}{3964})^3 = 0.8534$.

    With equal densities, the ratio of the masses would also have this value. Since the ratio of Venus's mass to that of Earth is less than this value, Earth must have the higher density.

15. The 1-foot cube of steel weighs 500 lbs and is resting on an area of 1 ft^2, or 144 in^2.

    The pressure is $\frac{500}{144} = 3.47$ lb per in^2 or 3.47 psi.

17. The base of the cone has an area of $A = \pi \times r^2 = \pi \times (5)^2 = 78.54$ ft^2.

    The cone has a volume of $V = \frac{1}{3} \times \pi \times (5)^2 \times 8 = 209.44$ ft^3.

    The weight of the cone is $170 \times 209.44 = 35{,}605$ lbs.

    The pressure is $\frac{35605}{78.54} = 453.3$ pounds per square foot (or 3.15 psi).

19. Surprisingly, the pressure on the base of each is the same.
    Let w represent the weight per cubic foot of the material. Then, the weight of each object is w times its volume in cubic feet.

    For the cone, Area of Base $= \pi \times r^2 = \pi \times (4)^2 = 50.265$

    $$\text{Volume} = \frac{1}{3} \times \pi \times (4)^2 \times 6 = 100.53; \text{weight} = w \times 100.53$$

    $$\text{Pressure} = \frac{w \times 100.53}{50.265} = w \times 2.$$

    For the pyramid, Area of Base $= s^2 = 8^2 = 64$

    $$\text{Volume} = \frac{1}{3} \times (8)^2 \times 6 = 128; \text{weight} = w \times 128$$

    $$\text{Pressure} = \frac{w \times 128}{64} = w \times 2.$$

21. The volume of the wooden pole is $V = \pi r^2 h = \pi \times 1^2 \times 40 = 40 \times \pi$ ft^3.
    The weight of the pole is $60 \times (40 \times \pi)$ lbs $= 2400 \times \pi$ lbs.
    The area of the base of the pole is $\pi \times (1)^2 = \pi$ ft^2.

    The pressure on the base is $\frac{\text{weight}}{\text{area}} = \frac{2400 \times p}{p} = 2400$ lbs per ft^2.

    2400 lbs per ft^2 = 16.7 psi

23. The scale for the model refers to the scaling factor for length.
    The scaling factor for capacity (or volume) is the cube of the scaling factor for length.

$$\text{Length of model} = \frac{1}{48} \times 198 \text{ in.} = 4.125 \text{ in.}$$

$$\text{Volume of fuel tank} = (\frac{1}{48})^3 \times 18 \text{ gal} = 0.000163 \text{ gallons}$$

25. **(a)** The real locomotive will be 87 times as long as the HO-gauge model.

    (b) The volume of the real locomotive will be $(87)^3 = 658,503$ times as large as the volume of the HO-gauge model.

    (c) If the real locomotive and the model have the same materials, we will assume that the weights are proportional to the volumes. The weight of the real locomotive will be 658,503 times that of the model.

27. A $\frac{1}{4}$ - *inch scale* model has a scaling factor for length of $\frac{1}{48}$.

    Area has a scaling factor of $(\frac{1}{48})^2$, so the area of the living space in the architect's model is $(\frac{1}{48})^2 \times 3580 \text{ ft}^2. = 1.554 \text{ ft}^2$.

    The length of the swimming pool in the model is $\frac{1}{48} \times 25 = 0.521$ m. or 1.709 ft.

Chapter 13 Review Problems

Solutions to Odd-numbered Problems

1. $\overline{AB} \leftrightarrow \overline{DE}$, so the scaling factor of $\triangle ABC$ with respect to $\triangle DEF$ is $\frac{5}{15} = \frac{1}{3}$

   Side lengths for $\triangle ABC$:

AB = 5	Given
AC = 3	$\overline{AC} \leftrightarrow \overline{DF}$, so AC $= \frac{1}{3} \times 9$
BC = 7	$\overline{BC} \leftrightarrow \overline{EF}$, so BC $= \frac{1}{3} \times 21$

   Perimeter of $\triangle ABC = 5 + 3 + 7 = 15$
   Perimeter of $\triangle DEF = 15 + 9 + 21 = 45$

   Check: perimeter($\triangle ABC$) $= \frac{1}{3} \times$ perimeter($\triangle DEF$)

3. The ratio of the shadows' lengths should be the same as the ratio of the objects' heights if both are vertical. If the objects are vertical then we have two right triangles. Two right triangles are similar if they have one other pair of equal angles. The light rays are parallel, and strike the ground at the same angle. The height of the tree can be found by:

$$\frac{\text{height of tree}}{\text{height of man}} = \frac{\text{length of tree's shadow}}{\text{length of man's shadow}}$$

$$\frac{\text{height of tree}}{6} = \frac{20}{2}$$

The height of the tree is 60 ft.

5.

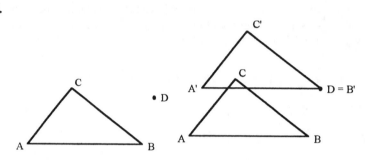

7.

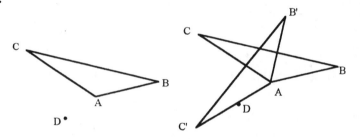

9. One possibility:
   (i)   reflect across AC
   (ii) rotate about 90° clockwise about the image of B (or C or A)
   (iii) translate the image of C to F
   Note: The order of the steps can vary. Also, line of reflection can differ, and the points used for rotation may differ.

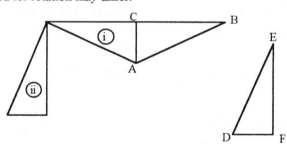

9. Continued

The "simplest" alternative:

(i) translate B to E

(ii) reflect across line L which bisects $\angle DEP$.

Note: Finding the line L may seem difficult, but it can be done with a compass and straight edge.

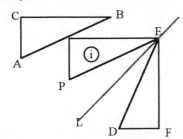

11. The scaling factor from the small triangle to the large triangle is $\frac{4.6}{2} = 2.3$.

The scaling factor for area is $(2.3)^2 = 5.29$

The area of the larger triangle is $5.29 \times 1.73 = 9.15$

13. The growth rate is r = 0.005.

The Malthusian model for growth is $160,000 \times (1.005)^m$.

The future predicted populations are:

10 years: 168,182

50 years: 205,316

100 years: 263,466

15. If the growth rate is 8% (0.08), in 10 years the value will be

$V = 20,000 \times (1.08)^{10} = \$43,178.50$ or about \$43,000

17. Using the half-life formula;

$d = 1 - (\frac{1}{2})^{1/h}$ or $d = 1 - (\frac{1}{2})^{1/2.5} = 0.242$ or 24.2%

19. $P = P_0 \times (1 + r) = 20,000 \times 1.04 = 20,800$

21. The current population is 140,000, the growth rate is 4%, and the carrying capacity is 400,000. We want the population in the following year.

The logistic model is $P_{m+1} = (1 + r) \times P_m - \frac{1 + r}{c} \times (P_m)^2$

The population will be $P = (1.04) \times 140,000 - \frac{1.04}{40000} \times (140,000)^2 = 96,640$

23. From problem 22, the weight of a cubic water is given as 62.4 lbs.

If a person weighs 140 lbs, we estimate the volume as $\frac{140}{62.4} = 2.244$ ft^3.

25. The weight of the block is $170 \times$ Volume $= 170 \times (20 \times 20 \times 2) = 136,000$ lbs.

If the block is put on a pedestal that is 1 foot square, then the area of the base is 144 in^2. The pressure is $\frac{136000}{144} = 944$ psi.